CELL STRUCTURE, FUNCTION AND METABOLISM

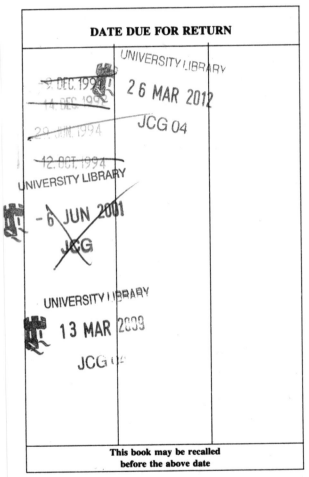

CELL STRUCTURE, FUNCTION AND METABOLISM

Edited by Norman Cohen

BIOLOGY: FORM AND FUNCTION

Hodder & Stoughton The Open University

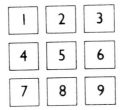

Cover illustrations

1 Schematic view of Singer and Nicholson's model of membrane structure.

2 Pseudocolour electron micrograph of chloroplast. ($\times$ 25 000)

3 Schematic representation of part of haemoglobin molecule

4 Section of large intestine of a mammal stained with alcian blue and Van Gieson. ($\times$ 20)

5 Dark-background autoradiogram showing sites of incorporation of ^{3}H-phenylalanine in gill tissue of the bivalve mollusc *Mya arenaria*. ($\times$ 400)

6 Pseudocolour electron micrograph of rat liver showing mitochondria. ($\times$ 75 000)

7 Immunocytochemical staining for GABA (γ-aminobutyric acid) in the central nervous system of the chick. ($\times$ 1 000)

8 Section of thyroid gland showing secretory follicles containing magenta-stained colloid (thyroglobulin). ($\times$ 125)

9 Columnar epithelium on a villus from the small intestine of a rat. A histochemical reaction has stained mucus-producing cells a greenish colour ($\times$ 600)

Back cover: Side view of DNA, with the individual atoms represented as dot-covered spheres.

British Library Cataloguing in Publication Data
Cell structure, function and metabolism.
1. Organisms. Cells. Physiology
I. Cohen, Norman   II. Series
574.876

ISBN 0–340–53188–6

First published 1991

Published by Hodder and Stoughton Educational, a division of Hodder and Stoughton Ltd, Mill Road, Dunton Green, Sevenoaks, Kent TN13 2YA, in association with the Open University, Walton Hall, Milton Keynes, MK7 6AA.

Designed by the Graphic Design Group of the Open University.

Typeset by Wearside Tradespools, Fulwell, Sunderland, printed in Great Britain by Thomson Litho Ltd, East Kilbride

The text forms part of an Open University course. Further information on Open University courses may be obtained from the Admissions Office, The Open University, P.O. Box 48, Walton Hall, Milton Keynes, MK7 6AB.

1.1

CONTENTS

PREFACE

Cell Structure, Function and Metabolism is the second in a series of five volumes that provide a general introduction to biology. It is designed so that it can be read on its own (like any other textbook) or studied as part of *S203, Biology: Form and Function*, a second level course for Open University students. As well as the five books, the course consists of five associated study texts, 30 television programmes, several audiocassettes and a series of home experiments. As is the case with other Open University courses, students of S203 are required both to complete written assignments during the year and to sit an examination at the end of the course.

In this book, each subject is introduced in a way that makes it readily accessible to readers without any specific knowledge of that area. The major learning objectives are listed at the end of each chapter, and there are questions (with answers given at the end of the book) which allow readers to assess how well they have achieved these objectives. Key words are identified in **bold** type both in the text where they are explained and also in the index, for ease of reference. A 'further reading' list is included for those who wish to pursue certain topics beyond the limits of this book.

INTRODUCTION

Perhaps the single most striking feature of the living world is the great diversity of organisms. Yet all organisms, whether minute unicellular ones such as bacteria and protista or large multicellular ones such as whales or trees, have one thing in common—*the living cell*. That is not to say that all cells are the same. Even in a single organism, such as, say, a cat or a human, there are many types of cell—red blood cells, hair follicle cells, nerve cells, muscle cells and so on—each with particular functions. But whatever the source—cat, human, shark or sparrow—granted minor differences, nerve cells are nerve cells, muscle cells are muscle cells and so on. And all cells, irrespective of their particular functions or sources—be they plant, animal, bacteria or whatever—have more major features in common than they have unique ones. It is this sharing of features that allows us to speak of *the typical cell*.

There are arguably three features that characterize the typical cell (i.e. unite nearly all cells):

☐ It interacts with its immediate surrounding environment (be that the outside world as, say, for a bacterium, or other cells as perhaps for a single cell in a multicellular organism) from which it obtains nutrients.

☐ It can convert nutrients into a wide variety of other chemical substances. By so doing, the cell can obtain both energy needed for numerous cellular processes and the chemical 'building blocks' from which it can ultimately synthesize large cellular components. This ability of the cell to synthesize its own components allows it to replace 'worn-out' parts—the cell is to some extent a 'self-repairing' machine.

☐ It can synthesize enough cellular components to grow and reproduce. The cell enlarges, divides to give two cells which can enlarge, divide and so on.

This book is about two of these features—how the typical cell obtains its nutrients from its environment and how it converts these into other chemicals thereby obtaining energy and vital cellular building blocks. The first of these concerns the cell's boundary with its surrounding environment—the cell membrane. The second depends on intermediary metabolism—the cellular network of chemical reactions whereby energy and building blocks are obtained from nutrients. Though not explicitly dealt with in this book, the third feature, the construction from building blocks of macromolecules and large cellular components—the stuff of self-repair, growth and reproduction—is also dependent on the cell's ability to obtain and convert nutrients.

But this book is not only concerned with how the typical cell carries out its vital functions. It also considers how we know what we do know about cells—the evidence, and the experimental approaches used to gather it. Underpinning all these approaches is one central philosophy: that *structure is related to function*. Like a city, where particular buildings and systems are designed for specific functions, a cell contains components and systems whose structures are intimately related to the cellular functions that they perform. However, for a cell, we do not speak of 'design', just 'trial and error' evolution by natural selection. With the tight relationship between structure and function as our guide, we can argue that by describing and understanding the detailed structure of something we can better understand its function and how it performs that function.

With this in mind, the book begins with a description of the relatively gross structures of major cellular components, notably the main cell organelles, as revealed by microscopy. Chapter 1 outlines the essential features of the two principle forms of microscopy—light microscopy and electron microscopy. Chapter 2 discusses what these techniques, chiefly the latter, tell us about the structure of major cellular components. In Chapter 3 we descend one more level in scale, to one generally even beyond the reach of the electron microscope. Here we discuss the structures of the three classes of cellular macromolecules—proteins, nucleic acids and polysaccharides—and the structure of lipids. These molecules largely make up the major cellular components (e.g. organelles) dealt with in Chapter 2.

The rather static view of the cell, given by the largely structural picture of the first three chapters, gives way to a consideration of the more dynamic cell, that we know it to be, as typified by the chemical transformations of intermediary metabolism. But here too the relationship between structure and function provides the key to understanding, indeed most powerfully. For in Chapter 4 we show how detailed knowledge of the structure of enzymes, the catalytic workhorses of the cell, can provide insight into how intermediary metabolism operates—how the cell can carry out a vast array of chemical interconversions at rapid rates and under, what are for chemical reactions, mild conditions. This theme continues in Chapter 5, where enzyme structure is shown to provide the cell with a way of regulating and controlling the vast network of intermediary metabolism, thus avoiding chemical chaos and making the cell appropriately responsive to its surrounding environment without being overwhelmed by it. These features of enzyme structure, developed in Chapters 4 and 5, are then put to work in Chapter 6 to help explain some of the central pathways of intermediary metabolism—those concerned with utilizing carbon compounds, notably glucose, as sources of cellular energy.

Finally, having shown that the structure–function approach can help elucidate how intermediary metabolism works, we apply it to the other of our two features of the typical cell: how the cell obtains nutrients from, and interacts with, its environment. Chapter 7 deals with the boundary between the cell and its surroundings—the cell membrane—and with the internal membranes of the cell (e.g. those around organelles), and demonstrates once again that detailed structure is finely tuned to specific function. This further reveals that not only do cell membranes function in dynamic processes, such as the uptake of nutrients from the environment, but in so doing the structural components themselves, the membranes, may undergo dynamic changes—membranes can pinch off, flow and fuse together.

Thus by the end of the book two things should be evident. The typical cell is a highly active entity—performing a variety of dynamic functions and itself constantly changing, yet always essentially the same. For, like a city where specific buildings or roads may come and go, or an ocean with its cycle of tides, evaporation and rain, a living cell is in a constant state of flux yet always recognizably that cell. And, though, like the 'average English family with 1.8 children', no such thing as a 'typical cell' exists, all cells, whatever their particular source, type and hence function—cat, dog, plant or bacterium, nerve or muscle—are variations on the, thematic, typical cell. It is by the approach developed in this book, an unveiling of the relationship between structure and function, that we can, in principle, hope ultimately to mentally 'reconstruct' the operations of this typical cell.

CELL MICROSCOPY ◆ CHAPTER 1 ◆

1.1 INTRODUCTION TO THE STUDY OF CELLS

There is a very wide diversity amongst living organisms, in size, shape, colour, behaviour and habitat. Despite this evident diversity there are also similarities. The fundamental similarity is known historically as the cell theory. This states that all living organisms—plants, animals and protistans—are composed of one or more cells and the products of cells. A cell was, and still is, thought of as a single unit of living material. The living material of which cells are made came to be called protoplasm. Each cell was discovered to be surrounded by a membrane and to contain a nucleus. The material inside the nucleus was thought of as being distinct from the rest of the cell and was called karyoplasm—now termed nucleoplasm—while the material outside the nucleus, but inside the cell, was termed cytoplasm.

As investigations continued into the chemistry and structure of living organisms it became obvious that there are great similarities between cells—of whatever origin—in their chemical composition, chemical and biochemical behaviour and in their detailed structure.

This chapter and the next are concerned with the latter aspect, the structure of cells, and in particular the structure of cells which have a nucleus, eukaryotic cells.

Living cells are generally small, delicate and transparent so that finding out what they are made of, what is inside them and how they work is far from easy. It is possible to examine and experiment on living cells directly, keeping them alive and healthy, using cell culture (*in vitro*) techniques, but this usually requires expensive apparatus and great care must be taken to control such variables as temperature, pH and the supply of nutrients. *In vitro* techniques are important because some aspects of cell biology can only be investigated with living cells, cell movement for example. However, an examination of the detailed structure of cells usually involves preparing the cells in ways which kill them. This is of course very unfortunate, and means that great care has to be taken in applying knowledge gained from the study of cells that are dead, to those that are living.

The small size of cells and the lack of contrast between their structural components are two particular problems that have to be overcome. Microscopes of a variety of different kinds can be used to produce a magnified image of cells. Some make use of differences in the way structures in cells affect light passing through them; phase contrast and interference microscopes are of this type. Dark-field microscopes make use of differences in the way light is scattered by structures inside cells. Polarizing microscopy is based on the way in which different parts of cells affect polarized light. By far the most common type of light microscope is called the compound or, simply, light microscope, and it is with the images obtained using this, and a quite different type of microscope, the electron microscope, that the rest of this chapter is concerned.

Lack of contrast between the various structures inside cells make them more or less transparent to light. For light microscopy this problem can be overcome by using dyes which impart colour to subcellular structures. Some of these dyes can be used on living cells without damaging them, but the majority are used on dead cells.

For electron microscopy the colour-imparting dyes are replaced by chemicals which interfere with the passage of electrons with which the specimen is 'illuminated'. The end results are similar to those obtained by using dyes: contrast is enhanced allowing previously invisible structures to be seen or photographed.

Preparing cells for microscopic examination may cause them to change so that the image obtained may not in fact be an accurate or reliable guide to the structure of untreated, living cells. Even if methods are devised that prevent or slow down these changes, the images that can be seen may still be difficult to relate to intact cells.

Before going on to learn about the detailed structure of living cells (the subject of Chapter 2), it is necessary to consider the ways in which they are prepared for study and examined, because our understanding of cell structure is mainly based on inferences from the study of prepared, dead material.

1.2 MICROSCOPES

To understand both the limitation of light microscopy and the major advantage of electron microscopy requires an understanding of a property called the resolution or *resolving power* of the instruments.

You may be familiar with one type of light microscope, and you have probably seen electron micrographs produced from two types of electron microscope, the transmission electron microscope (TEM) and the scanning electron microscope (SEM).

The main advantage that electron microscopes have over light microscopes is their greater **resolving power**. A microscope with a high resolving power enables you to see two small objects close together as two distinct objects, whereas with a microscope of low resolving power, the image you see will appear to be that of a single object. During the years in which light microscopes have been in use, there have been considerable improvements in both the magnification they can provide and their resolving power. However, resolution is limited by both practical and theoretical considerations. Resolving power is inversely proportional to the wavelength of the light used to form the image. By using the best quality microscopes and using light in the visible spectrum with a wavelength of $0.5 \ \mu m \ (0.5 \times 10^{-6} \ m)$, it is possible to resolve two points $0.2 \ \mu m \ (0.2 \times 10^{-6} \ m)$ apart. To improve resolution beyond this, it is clear that 'light' of shorter wavelength still must be used.

The electron microscope provides this increased resolving power. Basically, the idea behind the electron microscope is that a beam of electrons is directed at the specimen and 'illuminates' it, much as ordinary light does in the light microscope. Light propagates as waves, but interacts with particulate matter, and the same is true of electrons.

The wavelength of an electron beam depends on the momentum of the electrons and this in turn depends on their velocity. Because the electron beam can be greatly accelerated to very high velocities the wavelength can be made extremely short. For instance, an electron wavelength of $5 \times 10^{-12} \ m$ can be produced, which is about 10^5 times smaller than the wavelength of visible light. By making use of these properties, the resolving power of an electron microscope can be thousands of times greater than that of a light microscope, and a modern high quality high transmission electron microscope (TEM) can easily resolve points down to about 0.1 nm $(0.1 \times 10^{-9}\text{m})$ apart. The image produced of the object being studied is enlarged just as the image produced in a light microscope is, but in the electron microscope, magnifications of up to 1 000 000 are technically possible, compared with magnifications of up to 1 500 with light microscopes. Figure 1.1 compares a light and a transmission electron microscope.

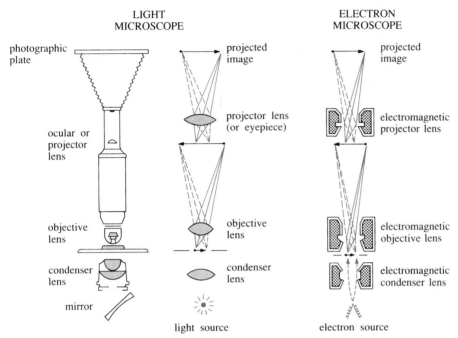

Figure 1.1 Comparison of image formation in the light microscope and the transmission electron microscope. (In practice, the TEM has the electron source at the top and the image is formed at the bottom.)

The visible light (either sunlight or artificial light) used with the light microscope is replaced in the TEM by a beam of electrons emitted, under vacuum, from a filament. The electron microscope must operate at a vacuum because the electrons would be deflected if any other particles were present in the column. The high-energy electron beam is focused by electromagnetic lenses instead of optical lenses, and the image is viewed on a fluorescent screen.

The scanning electron microscope (SEM) has a lower resolving power than the TEM. The main use of the SEM is in examination of the *surfaces* of specimens. In an SEM, a beam of electrons scans the surface of a thick (unsectioned) specimen, which is normally first coated with a thin film of metallic gold or palladium. An image of the object is built up line by line by the electrons deflected from the surface of the specimen, which are converted into light flashes and amplified to give an image on a fluorescent screen.

1.3 PREPARATIVE TECHNIQUES AND PROCEDURES

1.3.1 Fixation

Fixation is the rapid killing and preservation of biological material. Correct fixation is essential. The material must be fixed to preserve the three-dimensional arrangement of constituents of the tissue and of the contents of the cells. It will also prevent **autolysis** (digestion of the cell by its own enzymes) and bacterial or fungal attack, and make the tissue resistant to any damage that might be caused by later procedures.

Fixation may be by either physical or chemical means. Physical methods involve immersing the specimen in liquid nitrogen, thereby freezing it so rapidly that ice crystals, which would disrupt and distort the tissue, do not form. This method of fixation is often essential if it is necessary to preserve the tissue structure and to prevent any damage occurring to the enzymic components of cells. Frozen tissue is of course only fixed while frozen, and if brought to room temperature, would rapidly undergo autolysis. Thus, if it is necessary to keep permanent preparations of frozen sections, they must be chemically fixed (after thawing).

Chemical methods of fixation involve the use of reagents that will stabilize both proteins and lipids, which are the major structural components of cells. If the specimen is to be examined with the light microscope, chemicals (fixatives) that will precipitate proteins are used. If the specimen is to be examined with the electron microscope, it is essential to use fixatives whose reactions lead to the formation of precipitates that do not obscure the structure of the cell organelles. The most widely used fixatives for light microscopy include acetic acid, alcohol, acetone–formaldehyde and mercuric chloride, while formaldehyde and glutaraldehyde (which fix proteins) and osmium tetroxide (which fixes lipids) are used for electron microscopy.

The fixatives used in microscopy are usually dissolved in a solution similar in ionic concentration and pH to the fluids in the tissue to be fixed. Unless this is done, the specimen might shrink or swell during fixation and so be distorted. It might even rupture.

It is essential to fix the cells of the tissue as rapidly as possible. The smaller the tissue volume, the more rapidly the fixative will reach the centre of the specimen so that fixation of the whole tissue will be uniform. The optimum sample volume for electron microscopy is about $1\,\text{mm}^3$. Thus for very small organisms (e.g. protistans) fixation may be achieved by direct immersion in fixative. For higher organisms (animals and plants) a small sample of tissue must be cut from the donor, or in the case of animals the tissue may be perfused with fixative.

If fixation is carried out at a low temperature (e.g. $4\,^\circ\text{C}$), post-mortem changes and the loss of tissue components are kept to a minimum. However, the rate of penetration of the fixative into the tissue will be lower, so that fixation time must be increased.

1.3.2 Embedding

You may know from your own experience with the light microscope that even quite small specimens may still be too thick to allow enough light through them for details of structure to be seen. To obtain thin enough specimens from them, cells, tissues, organs or whole organisms have to be sliced into sections each about 5–$10\,\mu\text{m}$ thick. Such thin sections require mechanical support during their preparation: this is achieved by infiltrating the specimen in some suitable medium, which subsequently hardens, and cutting the **embedded** sections of the specimen in the supporting medium. The supporting medium used depends on the thickness of the section required. For relatively thick sections of plant material, specimens may simply be supported by sandwiching them between easily cut material such as raw carrot or the pith from the centre of elder tree twigs. Thinner sections can be cut if the specimen is embedded in a medium such as paraffin wax, and for the ultra-thin sections required for electron microscopy, a more rigid support such as the epoxy resin Araldite—which can withstand the effects of electrons—must be used.

Before a specimen can be embedded in wax, all the water present must be removed, because water and wax do not mix. This procedure is called dehydration and is very often done by passing the specimen first through a series of baths of increasing concentrations of ethanol, which replaces water in the specimen. The ethanol is then replaced with a wax solvent which is finally replaced by molten paraffin wax. The wax-impregnated specimen can be hardened by cooling and then sectioned. A similar dehydration procedure is used to embed material in epoxy resin, but instead of a wax solvent, a resin solvent such as propylene oxide is used. The resin-impregnated specimen is then warmed in an oven to speed up polymerization of the resin.

1.3.3 Sectioning

To produce thin sections (1–20 μm for light microscopy, but 50–100 nm for electron microscopy), an instrument called a **microtome** is used. All microtomes consist of a specimen holder, a sharp cutting edge and a means of regulating the thickness of the section being cut. The sharp cutting edge may be that of a steel razor for wax-embedded specimens, or of a glass or diamond knife for resin-embedded specimens. Sections of frozen specimens are cut with a microtome mounted in a freezer maintained at −20 °C. When the sections have been cut, they are supported by being mounted either on glass microscope slides for examination with the light microscope or on a grid of fine copper strands for electron microscopy.

1.3.4 Staining

Thin sections of cells or tissues are usually transparent or nearly so. To overcome this lack of contrast it is usually necessary to stain (dye) the sections before they are examined. Many of the stains in common use are dissolved in water, so for these, sections prepared from wax-embedded material must first be treated with a wax solvent and then rehydrated (by passing them through decreasing concentrations of ethanol) before they can be stained—the reverse of the procedures used to embed the material.

For light microscopy, most of the commonly used stains are organic aromatic dyes originally produced for use in the textile industry. Some, like toluidine blue, colour all tissues; others colour only particular tissues, parts of tissues or components of cells. Of these more specific stains, there are essentially two groups: basic and acidic. The specificity of these stains depends on the difference in charge of different cell components. A commonly used basic stain is haematoxylin—the colour-imparting (chromogenic) group is cationic (positively charged) and reacts primarily with negatively charged molecules, such as the nucleic acids in the nucleus, to produce a blue colour. Haematoxylin is most frequently used in combination with a second dye, eosin. In an acidic solution, the chromogenic group of eosin is anionic (negatively charged) and it will react with basic groups in the cell, which are largely in the cytoplasm, staining them red. Haematoxylin and eosin are used together as routine stains for most animal tissues and you often find this staining technique referred to as 'H and E'.

A wide variety of specific chemicals in cells and tissue sections can be identified by staining the specimen with substances that react specifically with one or other of the chemical components in the specimen, forming a coloured compound. These histochemical techniques, as they are called, are very numerous: one of them is discussed in more detail in the following chapter (Section 2.6).

In electron microscopy different stains are used to stain different cell components, as they are in light microscopy. This is because in the electron microscope the formation of an image depends on a beam of electrons, which passes through the section and hits a fluorescent screen. The image is formed by the removal of electrons from this beam by the specimen. The electrons pass easily through the resin and unstained tissue because the major constituents of cells (and resins)—carbon, hydrogen and oxygen—are all small atoms and thus offer little impediment to the electrons so the image is faint and with low contrast. By staining particular cell components with elements of high atomic number like lead and uranium, their contrast is increased because their electron density is increased.

Uranium binds preferentially to nucleic acids and proteins, whereas lead binds to lipids. Hence when stained with lead and uranium, cell components rich in lipids and nucleic acids respectively appear dark or electron-dense, and areas of cytoplasm or the resin surrounding the tissue appear light. Osmium (although used primarily as a fixative—Section 1.3.1) imparts electron density to lipids and nucleic acids, and to membranes as well.

The double-staining technique of uranium and lead used in electron microscopy is, to some extent, what H and E is to light microscopy.

1.4 THE INTERPRETATION OF IMAGES

There is no simple way to be sure that the image seen down a microscope, or in a photograph taken of such an image, corresponds closely to any reality in the living cell. To appreciate this, remember what has been done in the earlier stages of preparation. The specimen will have been immersed in a fixative, dehydrated, impregnated with wax, sectioned, rehydrated, stained, washed and then impregnated with a mounting medium—all before it is ready for final examination with a light microscope. If an electron microscope is used, the specimen will have been fixed, dehydrated, embedded in resin, sectioned, stained, then dehydrated in a vacuum before being bombarded with a beam of electrons. These procedures can cause shrinkage, expansion, or other distortions of the specimen or parts of it. Prolonged immersion in ethanol can extract some tissue and cell components more than others. The specimen may be compressed or torn during sectioning. Such alterations in cell structure are called **artefacts**. If, however, similar images are seen after using a variety of different techniques of fixation, dehydration, embedding and staining, it is reasonable to conclude that we are looking at a real structure. But the problems of interpreting the image produced by any microscope do not end even if artefacts are eliminated.

◇ Suppose a section through a cell, a two-dimensional image, contained a series of outlines as shown in Figure 1.2. What could the three-dimensional shape(s) of the sectioned structure(s) be?

◈ There are many possibilities: for instance a number of tubes of different diameters and cross-sectional shapes; tubes of different diameters but similar cross-sectional shape; spheres or pear-shaped structures (Figure 1.3). You may have thought of other solutions. The important point is that from a single two-dimensional image, it is impossible to decide.

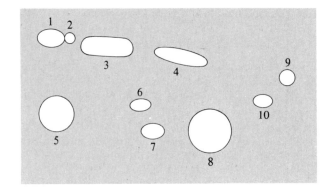

Figure 1.2

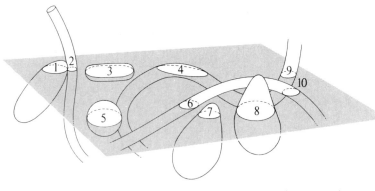

Figure 1.3

It is clear then that a two-dimensional image may not be a good representation of the complete structure from which the section has been taken. The outline of the components in a two-dimensional section will obviously depend on the location at which the section is taken (this is true of structures in cells whether of regular or irregular shape). To take a real example, examine the electron micrographs taken of a series of sections through a structure in part of the brain of a chicken (Plates 1.1a–e).

The structure of interest is a synapse (synapses are the points of contact between nerve cells). A synapse consists of a swelling at the end of a nerve fibre, called a pre-synaptic bouton (indicated by the dashed blue outline in Plate 1.1a), and a post-synaptic region (which in Plate 1.1a is a dendritic spine). In between the pre- and post-synaptic regions there is a synaptic cleft. This consists of a synaptic junction, shown between the black solid arrows in Plate 1.1a, and in part of this there is a prominent zone where the membranes are darkened and thickened (between the blue arrows); this is the *active zone* of the synapse. The pre-synaptic bouton consists of a number of components including a mitochondrion and synaptic vesicles. These vesicles contain the chemical(s) responsible for transmission of information from one nerve cell to another, which occurs across the active zone of the synapse.

However, the image of the synapse, and in particular the active zone, is deceptive for it depends very much on where the section is cut through the synapse. Plates 1b–e are electron micrographs taken from serial sections of 65 nm thickness cut immediately after the 65 nm section shown in Plate 1.1a.

Examine Plate 1.1b and compare it with Plate 1.1a.

◇ What are the major differences between Plates 1.1a and 1.1b?

◈ In Plate 1.1b there are more synaptic vesicles and the active zone of the synapse (between the blue arrows) is smaller.

Examine Plate 1.1c and compare it with Plate 1.1b.

◇ What are major differences between Plates 1.1b and 1.1c?

In Plate 1.1c the appearance of the active zone of the synapse is different from that in Plate 1.1b (and 1.1a). Also, the mitochondrion has disappeared and synaptic vesicles now almost entirely fill the synaptic bouton.

In Plate 1.1d the appearance of the active zone of the synapse has changed again: it is longer than in Plate 1.1c and a possible indentation (large white arrow) becomes visible in the middle. In Plate 1.1e it is clear that the indentation is indeed a perforation in the active zone. This is not an artefact. Each of Plates 1.1a–e are actually side-on views of the active zone, which is a slightly curved disc. A possible representation of the active zone of the synapse, as would be seen if one could stand inside the dendritic spine (Plate 1.1a), is shown in Plate 1.1f which also indicates where each of the sections shown in Plates 1.1a–e have been cut. Of course, there is clearly considerable variation seen in the shape of the synapse, depending on the level at which the section is cut. To obtain a true picture of the three-dimensional shape one would need to cut serial sections through the entire structure and 'reconstruct' it to see its shape. This has frequently been done in the past using for example polystyrene shapes cut to the precise shape of the synapse in each section. These cut-outs can be stuck together to show the true three-dimensional shape of the object.

It is now possible to view cells, or their components, in the light or electron microscope and use a computer-aided device to draw around the shape of the structures of interest in serial sections. The computer can then use a reconstruction programme to construct the shape of the structures in three dimensions.

Even if one can establish with confidence the precise shape and form of structures in animal or plant material viewed in two-dimensional sections, there are several major questions that remain to be answered. For instance, what is the three-dimensional arrangement of the components, how many of the structures are there and what is their total volume in the intact living tissue? As an illustration of why such information may be important and the pitfalls in extrapolating from two-dimensional photographs, we take the example of the effect of ageing on the number of nerve cells in the human brain.

There is some evidence to suggest that there is a decline in the number of nerve cells in the human cerebral cortex (the 'thinking' part of the brain) in old age. This is often cited in the popular press in discussions of ageing. If one could find a way to prevent cell death, perhaps some of the defects in mental ability which are associated with ageing might be prevented. But is there really a decrease in cell number in the brain in old age?

Much of the information on cell number in human brains comes from studies carried out on analyses of tissue from the brains of old people who have died. There are many reasons why estimates of cell number obtained in such circumstances may not be valid. Most importantly, people often appear malnourished at death, perhaps due to the debilitating effects of diseases such as cancer which may have prevented them from eating in the terminal stages of their illness. Few elderly people who die look well nourished. Studies in rats have shown that undernourishment, particularly in young animals, has severe effects on both brain cell number and the contact zones between nerve cells—the synapses. So the finding that there is a decline in both the number of brain cells and the number of synapses in ageing humans based on post-mortem data may not reflect a purely age-related phenomenon.

More importantly, estimates of the number of brain cells may be affected by the overall volume of the brain. If this shrinks, as may happen in old age without a commensurate decrease in the volume of individual cells, then there will *appear* to be more brain cells. Alternatively if the overall volume of the brain increases due to the brain tissue swelling but cell volume does not change, then any estimate of the number of cells in a given volume of brain tissue will be smaller than before. Thus measurements consisting solely of cell number per unit volume in the brains of ageing animals or humans will be misleading without knowing factors such as the overall brain volume and individual cell diameter.

The name sometimes given to studies in which precise estimates of numbers of structures, their position and dimensions are made is **stereology**. This is a body of mathematical procedures relating two-dimensional measurements obtainable on sections of the structure to three-dimensional parameters defining the structure.

We do not deal here with the numerous parameters which can be estimated on sections by stereological means, but we will consider three that are commonly measured i.e. number, volume and diameter.

In stereological terms, the *number* of cells or components is usually expressed with reference to the volume in which they are contained, i.e. number per unit volume—the numerical density (abbreviated as N_V). *Volume* is normally expressed as volume density i.e. volume of a feature per unit sample volume (abbreviated as V_V). *Diameter* is expressed as the mean diameter of the feature (abbreviated $\bar{D}$). How can these stereological measurements be made in practice? Consider Figure 1.4 which shows a theoretical section cut through a number of profiles of uniform diameter (in living tissues the situation is rarely quite so simple).

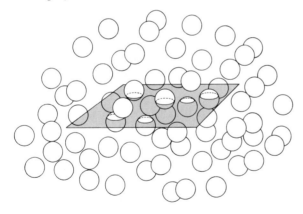

Figure 1.4 Population of spheres intersected by a horizontal plane.

Clearly the section will not cut all of these in their centre, but at various planes ranging in diameter from a maximum value for sections passing through the centre of the sphere, to a value close to zero for a section passing through the surface of the sphere. If the diameters of the profiles visible (d) are measured then a mean value ($\bar{d}$) can be obtained and the mean *structure* diameter ($\bar{D}$) can be calculated from the formula:

$$\bar{D} = \bar{d} \times 4/\pi$$

To calculate the number of uniform structures per unit volume in Figure 1.4 we make use of the mean value $\bar{D}$ calculated above. The procedure is simply to count the numbers of structures cut by the section shown in Figure 1.4 and divide by the total area of the section. This gives the number of structures per unit area, abbreviated as N_A.

The number per unit volume of tissue (the numerical density) is the mean number per unit area divided by the mean structure diameter ($\bar{D}$):

i.e. $N_V = N_A / \bar{D}$

This equation assumes that the structure which is being measured and counted has a diameter greater than the section thickness. If it does not then a correction factor will also have to be introduced to take account of section thickness, but we will not go into this here.

To calculate the volume density (V_V) of the structure which we are examining, i.e. the volume per unit sample volume, several different methods can be used. Perhaps one of the simplest, which has been used for so long that it may be described as an almost classical technique, is to superimpose a test grid consisting of a number of cross lines which give a number of intersections or points. Examine Figure 1.5 which shows such a test grid superimposed over an electron micrograph of rat liver in which the components of interest are mitochondria.

The volume density of the mitochondria may be found by counting the number of points (intersections) over the mitochondria and dividing this value by the total number of points on the grid.

i.e. V_V = points over mitochondria/total points over micrograph.

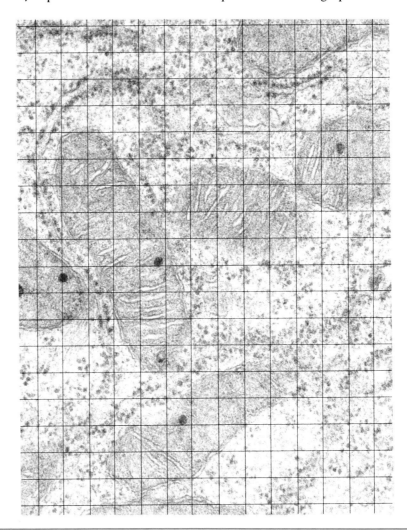

Figure 1.5 The measurement of the volume density, V_V, of cytoplasmic components (here, mitochondria in rat liver) by point counting. The intersections of the grid are taken as points. (magnification ×57 000)

◇ Calculate this for yourself.

◆ Number of points over mitochondria = 88
Total number of points over electron micrograph = 266

Therefore $V_V = \dfrac{88}{266} \approx 0.33$.

In modern stereological studies much of the tedium of making measurements and collecting data can be removed by the use of computers, though this does not guarantee more accuracy.

There are many examples that show how easy it is to misinterpret microscope images. Unfortunately, there seems to be no sure way of avoiding mistakes: vigilance and practice in the art of interpretation are the best recipes for success and avoidance of error. The camera, whether used to photograph a specimen with a light microscope or with an electron microscope or in the hands of a photographer of a beauty contest, does not lie, but the microscopist's preparative procedures, like cosmetics, can alter 'reality'.

SUMMARY OF CHAPTER I

1 Living organisms are made of one or more cells and the products of cells.

2 Because cells are small, techniques for their study commonly involve some means of producing a magnified image of the cell or cells; however, living cells are generally more or less transparent so they must first be fixed (killed) and stained to impart contrast to the structures inside them.

3 A wide variety of preparative treatments are available but all, to a greater or lesser extent, incur risks that the magnified image which is studied may reflect the results of the treatments used as well as showing real details of the cell's structure.

Question 1 (*Objectives 1.1, 1.2 and 1.5*) In which of the following statments is the resolving power of a microscope correctly defined or used?

(a) Resolving power means the ability to make out fine detail.

(b) A microscope with which two images of a single object are obtained is said to have a high resolving power.

(c) The higher the magnification of a microscope, the greater is its resolving power, because at higher magnification, images can be obtained of smaller objects.

(d) The smaller the distance between two objects, the greater the resolving power of a microscope must be in order to continue to 'see' them as two separate objects.

Question 2 (*Objectives 1.1, 1.3 and 1.4*) (a) In what order are the following processes carried out when preparing a specimen for microscopic examination? Embedding; dehydration; rehydration; mounting; fixation; staining; sectioning.

(b) What are the main differences in technique that can be related directly to the use of light or of electron microscopy?

Question 3 (*Objective 1.1 and 1.4*) In which of the following circumstances would it be more advantageous to use a transmission electron microscope (TEM) than to use a light microscope? Briefly explain why.

(a) Examining an aphid from a pot plant in order to identify it.

(b) Examining a kidney cell to look at its cell membrane.

(c) Examining a cell to estimate the number of mitochondria it contains.

(d) Examining sections of skin to look for the pattern of capillaries.

Question 4 (*Objective 1.1 and 1.5*) In a section of a cell, the shape shown at the top left of Figure 1.6 was obtained. From which of the objects shown as A–E in Figure 1.6 could it have been obtained?

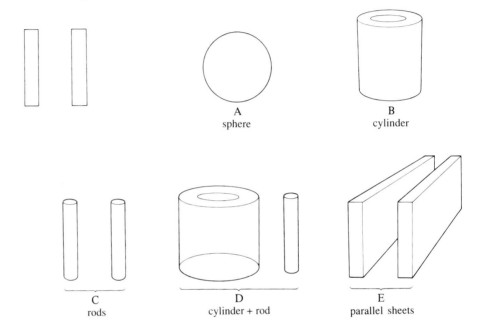

Figure 1.6 For use with Question 4.

OBJECTIVES FOR CHAPTER 1

Now that you have completed this chapter, you should be able to:

1.1 Define and use, or recognize definitions and applications of, each of the terms printed in **bold** in the text.

1.2 Define resolving power of a microscope. (*Question 1*)

1.3 Outline the important steps in the preparation of specimens for examination by means of the light and electron microscopes and state the main principles involved in each stage. (*Question 2*)

1.4 Explain the similarity between the light microscope and the transmission electron microscope, and outline possible advantages of the electron microscope over the light microscope. (*Questions 2 and 3*)

1.5 Outline some of the difficulties in interpreting microscope images. (*Questions 1 and 4*)

CELL STRUCTURE AND FUNCTION ◆ CHAPTER 2 ◆

2.1 THE DETAILED STRUCTURE OF EUKARYOTIC CELLS

The previous chapter dealt with the techniques used to prepare cells for microscopic examination and the problems of interpretation of the images obtained. Such microscope studies have enabled considerable information to be obtained about the detailed structure of cells and the functions of the various cell components: it is the structure and function of cells that is the subject of the present chapter.

Every living cell is surrounded by a *cell membrane*. This cannot be seen clearly in Plate 2.1 but in Plate 2.2 it is the fine black line at the edge of the cell. Inside the cell membrane is the **protoplasm**. This is a general term, now perhaps somewhat old-fashioned, used to describe the living matter itself. Protoplasm contains a variety of components of various kinds. Those that are surrounded by a membrane or membranes are called **organelles** or subcellular organelles. The material surrounding the nucleus and inside the cell membrane is called **cytoplasm**. This in turn can be thought of as an aqueous medium, the **cytosol**, in which there are a variety of structures, some bounded by membranes.

Examine Plates 2.1 and 2.3 so that you may become familiar with the general appearance of cells.

The large, roughly circular structure near the middle of Plate 2.1 is the *nucleus* of the cell. Note that it is distinct from its surrounding material. The nucleus is enclosed by *nuclear membranes*; these can be seen more clearly in Plate 2.4. The nucleus in Plate 2.1 appears to have only one definite structure inside it, the prominent, roughly circular *nucleolus*.

In the material surrounding the nucleus in Plate 2.1 there are a number of sausage-shaped structures. These are *mitochondria* (sing. mitochondrion). In Plate 2.5 you can see that they are bounded by a double membrane and that the interior is crossed by fine membrane-bounded, parallel-sided, leaf-like structures.

Around the mitochondria in Plate 2.4 there is a complicated branching network of parallel-sided, membrane-bounded spaces, studded with black granules. All this is part of the *endoplasmic reticulum*, an extensive elaborate series of flattened interconnecting sacs whose outer surface is covered with *ribosomes*, the small black granules. Notice that some ribosomes are apparently not connected to the endoplasmic reticulum. The endoplasmic reticulum can be seen in some parts of Plate 2.1 but the ribosomes are too small to be seen easily.

Near the lower left-hand edge of the nucleus in Plate 2.3 is the *Golgi apparatus*. This consists of a stack of flattened membranous sacs surrounded by numerous vesicles of various sizes and can be seen very clearly in Plate 2.6.

◇ The cell in Plate 2.1 does not appear to have a Golgi apparatus. How do you account for this?

◆ You might suggest that this cell lacks a Golgi apparatus, or more likely, that although the cell has a Golgi apparatus the section shown in the plate does not include it. You might even suggest that the cell has a Golgi apparatus but that it has not been preserved or made visible by the techniques used to prepare the cell for examination. However, the reality is that no definite answer can be given to the question on the basis of the appearance of a single specimen.

An organelle, found only in plant cells which can photosynthesize, is the *chloroplast* (Plate 2.7). This organelle, like the mitochondrion, is bounded by a double membrane. Inside chloroplasts is a complicated array of membranous sacs. These are not attached to the inner chloroplast membrane.

At various places inside the cell in Plate 2.1 you can see clear circular areas. These are *vacuoles* or *vesicles*, generally small in most animal cells (as here), but mature plant cells have a very large fluid-filled vacuole occupying most of the cell volume.

The remainder of the cell is filled with a granular material which, even at the higher magnification used in Plate 2.4, appears to be structureless; this is the cytosol.

Surrounding each plant cell, outside the cell membrane, there is a *cell wall*; this can be seen in Plate 2.8. The cell wall, composed of cellulose, provides mechanical strength to plant tissues.

There is no one type of cell that can really serve to illustrate the typical cell, its ultrastructure and its organelles. Many cells contain most of the organelles described in this chapter, but not in the same proportions, and both the organelles and the gross outline of the cell vary in shape depending on the function of the cell and its location within the organism. However, we can draw an idealized animal and plant cell (Figure 2.1)—idealized in that it contains many of the organelles described in this chapter. Of the organelles shown, only limited treatment is given in this chapter to chloroplasts, mitochondria and membranes. However the structure and function of mitochondria and membranes are explored in more detail in Chapters 6 and 7 respectively.

As discussed in Chapter 1, it is usually necessary to fix (kill) cells before preparing them for microscopic examination; but you should remember that *living cells are not static*. Many cells and their component parts have short lifespans: for example, each human red blood cell lasts on average 120 days. For reasons we do not go into here, it is more usual to refer to **half-life** of cells, subcellular organelles and other cell components; that is, the time taken to remove and replace half the population: the half-life of human red blood cells is 60 days. Mitochondria in rat liver cells have a half-life of 8 days, and some proteins in rat brain cells have a half-life of 15–20 days while others have a half-life of hours or even minutes.

Far from being static and fixed, organisms, their tissues, cells, cellular organelles and individual chemical components constantly change: they are highly dynamic. Cells synthesize new materials for maintenance, growth and export, and this requires raw materials, which must be brought into the cell across membranes. Inside cells, molecules, macromolecules, aggregations of macromolecules and cell organelles are constantly being broken down and recycled, and waste materials are produced that must be exported.

The image 'seen' through a light microscope or an electron microscope is a picture frozen in time, rather like a photograph of motorway traffic. The cars and lorries may appear static, but another photograph taken a few minutes later will be different.

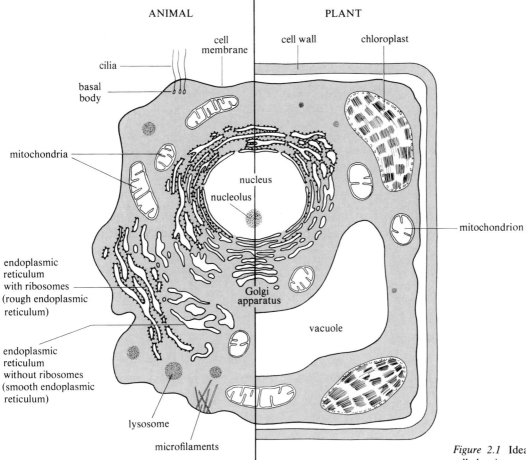

ANIMAL   PLANT

Figure 2.1 Idealized plant and animal cell showing many of the organelles described in this chapter.

In general, the lifespan of multicellular animals is greater than that of the cells they are made of. Consequently there is a constant turnover of cells; most of those that die are replaced but in some tissues, cells may not always be replaced (e.g. nerve cells in the nervous system).

The turnover of cells in a tissue can be studied by the technique of **autoradiography**. This is a technique in which particular components of cells or tissues are made radioactive, enabling them to be detected (see Box 2.1).

Plates 2.9a and 2.9b show the results of an autoradiographic study of the incorporation of ^{3}H-thymidine into the epithelial cells that line the intestine of a mouse. ^{3}H-thymidine, like non-radioactive thymidine, becomes incorporated into DNA in dividing cells, and as you can see this is located in the nuclei of the cells. Once inside the cell ^{3}H-thymidine stays there because DNA molecules do not leave the nucleus.

Examine Plates 2.9a and 2.9b. Both are of part of the wall of the intestine of mice. In this study the mice were injected with a small amount of ^{3}H-thymidine and then killed, one after 8 hours (Plate 2.9a) and one after 36 hours (Plate 2.9b). A similar section (not shown) prepared 48 hours after injection with ^{3}H-thymidine did not contain any radioactively labelled nuclei.

◇ What are the major differences in the distribution of nuclei containing radioactive thymidine in Plates 2.9a and 2.9b?

Eight hours after injection, nuclei containing ³H-thymidine are present principally at the bottom of the folds in the intestine but 36 hours after injection, labelled nuclei are present at the top of the folds, not at the bottom.

◇ What conclusions can you draw from the changes in distribution of incorporated ³H-thymidine in the 48 hours after injection?

You might have concluded that the labelled DNA has been passed from cells at the bottom of the folds to cells at the top of the folds. However, DNA does not leave the nucleus. Alternatively, you might have concluded that the ³H-thymidine moved from nucleus to nucleus, because you remembered that chemical compounds in cells are constantly being broken down and new molecules synthesized. DNA is, however, exceptional in that it is not broken down once synthesized, unless the cell itself dies. Finally, you might have concluded (correctly) that the cells containing the labelled nuclei moved from the bottom to the top of the folds and were then thrown off into the intestine—a migration completed in less than 48 hours. Notice that the marked cells have been replaced by a new population of cells at the bottom of the folds.

Box 2.1 Autoradiography

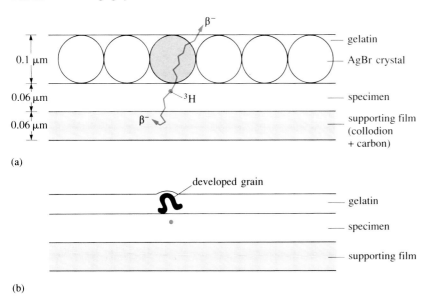

(a)

(b)

Autoradiography (Figure 2.2), a technique which can be used in both light and electron microscope studies, uses substances labelled with radioisotopes, most commonly isotopes—such as ¹⁴C or ³H—that emit β-particles (a weakly penetrating form of radiation which is stopped by metal foil); however, other isotopes, including those which emit the more penetrative γ-particles, may be used in certain circumstances.

These substances are introduced into animals or plants and become incorporated into cells. A section is prepared of tissues whose cells contain radioisotopes, and this is coated with a photographic emulsion (silver bromide in gelatin). When the radioactive particles emitted from the cells in the section hit the emulsion, a grain of metallic silver is produced, which appears as a black dot when the emulsion is developed. These dots, which are present in the emulsion, are seen to be in a different focal plane from the specimen, when viewed at high magnification.

Interpretation of electron microscope autoradiographs is complicated by the fact that tracks of particles may be distant from the atom emitting them, making pinpointing of the labelled molecule difficult. Autoradiography allows the fate of labelled compounds in cells and tissues to be determined, and enables particular components to be seen.

Figure 2.2 Diagrammatic representation of an electron microscope autoradiography preparation. (a) *During exposure* An emulsion of silver bromide crystals in a gelatin matrix covers the section. A β-particle, from a tritium (³H) point source in the specimen, has hit one of the crystals (light blue). (b) *During examination and after processing* The exposed crystal has been developed into a filament of silver and the non-exposed crystals have been dissolved. The total thickness has decreased because the silver bromide occupied approximately half the volume of the emulsion.

2.2 THE CELL MEMBRANE

The existence of a **cell membrane** has been known for long time. It is known to be of considerable importance in maintaining the integrity of the cell, and to be selectively permeable in allowing molecules into and out of the cell. Plate 2.10 is a high-power electron micrograph of part of the outer membrane of an intestinal cell. The macromolecular composition of the cell membrane will be discussed in Chapter 3, but here attention is concentrated on its appearance.

Cell membranes are between 7.5 and 10 nm thick. In a typical 9 nm membrane, there is an inner electron-dense layer of 3.5 nm, a middle clear layer of 3.0 nm and an outer dense layer of 2.5 nm. Where two animal cells are in close contact, the outer surfaces of their membranes are usually separated by a space 11–15 nm wide, of low electron density.

Conventional electron microscopy of thin sections shows the cell membrane to have a uniform three-layered appearance, but an electron microscope technique called **freeze-fracture** (or **freeze-etching**), has shown that it is far from uniform. In this technique, the specimen is frozen, then fractured, the exposed surface allowed to sublime and the structure of this surface subsequently examined using the electron microscope (see Box 2.2).

The cell membrane frequently splits into two parts, down the centre of the middle layer. This suggests there is a plane of weakness here, which would occur if the membrane was made up of a bilayer of lipids (fatty molecules) with proteins in and on the bilayer (as it is thought to be—see Chapter 3). The

---

Box 2.2 Freeze-fracture (freeze-etching)

Freeze-fracture (freeze-etching) enables a three-dimensional image to be formed, and its most significant application is in the study of membrane function (see Chapter 7). In freeze-fracture, the specimen is fixed by being rapidly frozen in liquid nitrogen (at about −200°C); it is then placed in a special sectioning apparatus in a vacuum and fractured with a sharp knife (Figure 2.3a), the fracture line passing through the plane of the cell membranes.

The exposed surface of the fractured specimen is allowed to etch slightly (Figure 2.3b), by partial sublimation at the surface, hence the name freeze-etching. Carbon, or carbon with a heavy metal (e.g. platinum), is then deposited ('shadowed') onto one of the exposed surfaces thereby making a replica (Figure 2.3c) of the tissue. The specimen itself is then digested away leaving the replica, which is so thin it can be viewed in the transmission electron microscope (Figure 2.3d).

Figure 2.3 The stages involved in freeze–fracture.

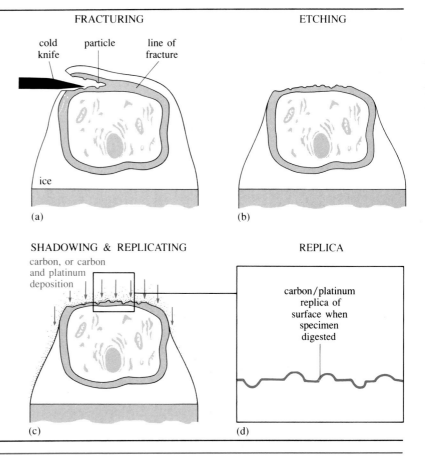

FRACTURING

cold knife   particle   line of fracture

ice

(a)

ETCHING

(b)

SHADOWING & REPLICATING

carbon, or carbon and platinum deposition

(c)

REPLICA

carbon/platinum replica of surface when specimen digested

(d)

inner faces of cell membranes do not appear to be smooth, but are covered with particles, 6–12 nm in diameter, which are thought to be the proteins extending through the lipid bilayer (Plate 2.11).

The membranes surrounding the nucleus and the membranes of the endoplasmic reticulum and the Golgi apparatus are all very similar to the cell membrane in appearance. However, these have a thickness of only 5–7 nm, and there are differences in chemical composition from that of the cell membrane (see Chapter 7). Such differences in membrane structure, you will see later, are reflected in important functional differences.

2.2.1 Variations of the cell membrane

Cells organized into tissues interact with their neighbours and this interaction is often reflected in variations in membrane structure in the region of the interaction. The cells lining the small intestine (epithelial cells) provide a good illustration of this.

The cell surface next to the cavity, the **lumen**, of the small intestine is the **apical cell membrane**, which has undergone extensive outfolding into finger-like or tubular structures known as **microvilli**, and this increases the surface area available for absorption. The membranes on the side of the cell, the **lateral cell membranes**, are held together by a number of specialized junctions; in epithelial cells each specialized junction is called a **junctional complex** (Figure 2.4, Plate 2.12).

On the upper part of the lateral cell membrane there is a **tight junction**. Tight junctions are regions where the membranes of adjacent cells fuse together and occur just below the apical surfaces of the cells. Specialized proteins in the two adjacent membranes form structures called sealing strands which run in a belt-like fashion around the cells. The presence of tight junctions seals the apical region of epithelial cells from other regions, preventing the movement of fluids from the lumen into the intercellular space. The numbers of tight junctions and sealing strands vary from tissue to tissue and may even vary from time to time in the same tissue giving rise to varying degrees of 'tightness'.

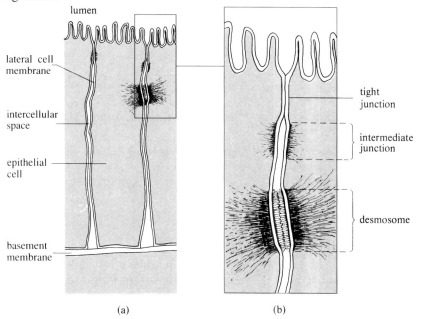

Figure 2.4 The components of a junctional complex between cells in the small intestine.

(a)   (b)

In some epithelial tissue, for example frog skin, the tight junction is indeed tight and impermeable, but in epithelial tissues in the intestine and kidney tubule cells, the tight junctions are selectively permeable, allowing water and certain ions through.

The innermost of the junctions is the **desmosome** of which there are two types, *belt desmosomes* and *spot desmosomes*. Belt desmosomes are areas where belt-like bands of microfilaments and intermediate filaments (see Section 2.10) run around the inner surface of the cell membranes of adjacent cells. They are connected by a further series of filaments running *between* adjacent cells. Belt desmosomes seem to be important in relation to changes in cell shape. Spot desmosomes occur particularly in tissues subject to mechanical stress, in skin cells for instance and in cardiac muscle cells. Each spot desmosome is circular in outline. They are the anchoring points for a system of intermediate filaments which form the internal skeleton of each cell and points at which further filaments run between cells anchoring neighbouring cells to each other.

Another type of junction, the **gap junction**, is found in epithelial and other tissues (Plate 2.13). Gap junctions are believed to be important in communication between cells. They appear to be areas that allow the movement of ions and small molecules (relative molecular mass, M_r, 600 or less) between cells.

In plants, membranes of adjacent cells are normally separated from each other by the cell wall. However intercellular contact between adjacent cells is made via channels running through the cell walls (Plate 2.14). These channels are 500–100 nm in diameter, are lined by the cell membrane, and often contain endoplasmic reticulum. They are called **plasmodesmata**, and provide a pathway for movement of materials from cell to cell.

The rigid **cell wall** outside the cell membrane is one of the main features which distinguishes plant cells from animal cells. Bacterial cells also have an extracellular wall, which is responsible for maintaining the shape of the cell.

2.2.2 Bulk transport: endocytosis

Transport of ions and soluble molecules across cell membranes occurs by passive and active processes (as discussed in Chapter 7). However, there are two other important means of transport of materials into and out of cells; both are dealt with in this chapter. **Endocytosis** (often referred to as *bulk transport*) is the intake of extracellular fluid and particulate material, ranging in size from macromolecules, like the iron-containing protein ferritin (M_r 500 000), to whole cells, like the bacteria engulfed by **macrophages** (a type of white blood cell) in vertebrates. **Exocytosis**, the bulk transport of material *out* of cells, is dealt with in Section 2.5.

Broadly speaking we can distinguish between three types of endocytosis. **Phagocytosis** involves the digestion of large particles, even whole microbial cells (Plate 2.2). **Pinocytosis** involves the ingestion of small particles and fluid. *Receptor-mediated endocytosis* (dealt with in Chapter 7) involves large particles, notably proteins, but also has the important feature of being highly selective; for example, only specific proteins can gain entry to the cell in this way, while others are excluded.

Endocytosis involves part of the cell membrane being drawn into the cell along with the particles or fluid to be ingested. This membrane is pinched off to form a membrane-bound vesicle within the cell, while at the same time the cell membrane, as a whole, reseals. Inside the cell, the fate of this, endocytic, vesicle depends on the type of endocytosis involved and the material it contains. In some cases the endocytic vesicle ultimately fuses with an organelle called a lysosome, after which processing of the material ingested into the cell can occur (see Section 2.6).

Endocytosis is also the means by which many simple organisms obtain much of their food.

2.3 THE NUCLEUS

Prokaryotic cells do not have a nucleus, but most protistan, plant and animal cells do (Figure 2.1). Most cells have a single nucleus but some have more than one, for example, rat liver cells may have two or even three nuclei per cell. Others have none when mature; for example, the mature human red blood cell and an important type of transport cell in vascular plants called a sieve tube cell.

A single membrane around the nucleus can be seen in cells examined in the light microscope. However on examination in the electron microscope (Plate 2.4), this turns out to be composed of *two* membranes, each 5–7 nm thick, and separated by a space of 11–40 nm. This has led to the **nuclear membranes** often being referred to as the **nuclear envelope**. The inner membrane is in contact with the nuclear contents, while the outer membrane appears to be continuous with the endoplasmic reticulum (described in Section 2.4) and often has ribosomes on its outer surface. The nucleus and the cytoplasm are connected through numerous **nuclear pores**, which can be seen particularly clearly in freeze-fracture preparations (Plate 2.15).

In the centre of the nucleus is a large dense region, the **nucleolus**, which is rich in proteins and RNA, and seems to be chiefly concerned with the synthesis of ribosomal RNA (rRNA) and the production of ribosomes.

2.4 ENDOPLASMIC RETICULUM

This consists of interconnected flattened sacs (called cisternae) or of interconnected tubules or vesicles. When ribosomes are present on the surface of the **endoplasmic reticulum** (ER), it is termed **rough endoplasmic reticulum** (RER) (Plate 2.4), and when absent, the ER is termed **smooth endoplasmic reticulum** (SER) (Plate 2.16). **Ribosomes** are small electron-dense bodies composed of rRNA and protein, that are involved in protein synthesis. ER appears to be connected to the nuclear membrane; indeed, it is believed to be formed from it. The membranes of the ER and nucleus form an extensive membranous system in the cell. RER is particularly well developed in cells that actively synthesize and export proteins. The path of the newly synthesized protein in such cells can be followed by autoradiography. A short time after radioactive protein precursors like ^{14}C-leucine (an amino acid) are supplied to the cell, radioactivity is located first on the surface of the RER, then within it, then in the Golgi apparatus, and then in the secretory granules containing the enzymes—we return to the link between ER and Golgi in the next section (and again in Chapter 7).

However, ER is not just a series of channels concerned with transport of materials. It contains a number of enzymes of importance in cell metabolism, and alteration or additional processing of proteins destined for export can occur within the ER cisternae before they are secreted.

In SER, the cisternae are more tubular than sac-like. SER appears to form from RER that has lost its ribosomal covering. It is extensive in steroid hormone secreting cells, such as the cells of the adrenal cortex or the testes. It is also present in liver cells, and here it is thought to have a role in drug detoxification (inactivation).

2.5 GOLGI APPARATUS

The **Golgi apparatus** was named after Camillo Golgi, who in 1898 used a metal (silver) impregnation technique to stain nerve cells and observed a structure that became known as the Golgi apparatus.* Electron microscopy reveals the Golgi apparatus as a number of flattened cisternae stacked together (Plate 2.6). The number of cisternae per stack and the number of stacks per cell is very varied.

There appear to be differences between the cisternal sacs. Parts of the ER next to the Golgi appear to break off, fuse together and contribute to the bottom cisternae of the Golgi. This led to the idea that there is a forming or *cis* face of the Golgi next to the ER, and a releasing or *trans* face at the top of the cisternal stack. Figure 2.5 illustrates this. It is interesting that while all Golgi membranes have the layered appearance, those next to the ER are 5–7 nm thick (like ER and nuclear membranes), but those on the releasing face are 7–9 nm, thick, the same as the cell membrane. This transformation from an intracellular type of membrane to one similar to the cell membrane appears to be an essential step in the preparation of vesicles that can fuse with the cell membrane—in other words, in enabling materials synthesized within the cell to be exported. Thus the Golgi apparatus is concerned with packing the materials processed in the ER for export. Secretory cells have many Golgi stacks, while non-secretory cells have few Golgi stacks per cell.

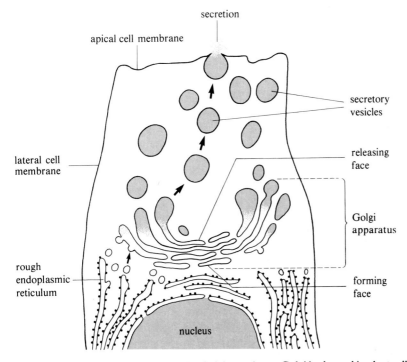

Figure 2.5 A route into and out of the Golgi area.

*In animal cells, this is often referred to as the Golgi complex or Golgi body, and in plant cells, it may also be called the dictyosome.

Autoradiographic studies demonstrate the route of newly made exportable proteins in the cell, from synthesis on the ribosomes through the ER and Golgi cisternae to Golgi vesicles. However, autoradiography can show that the Golgi is itself involved in the addition of carbohydrate molecules to certain proteins synthesized on ribosomes, resulting in the formation of glycoproteins (see Chapter 3). The synthetic functions of the Golgi in joining particular sequences of sugars together is more marked in plants than in animals, probably because polysaccharide is exported by plant cells to their cell walls. Autoradiography of cells supplied with labelled glucose shows the progress of polysaccharide synthesis in Golgi, followed by export in its vesicles very clearly. Vesicles leaving the Golgi fuse with the cell membrane by a process that is, in essence, the reverse of endocytosis—exocytosis. The path of movement of proteins through the cell following synthesis on the ribosomes and ending in fusion with the cell membrane is summarized in Figure 2.5.

Material cannot be continually added to the cell membrane without causing it to change shape. In some cells, after exocytosis occurs, the cell membrane is effectively taken back into the cell by endocytosis.

2.6 LYSOSOMES

Lysosomes are membrane-bound organelles that contain a variety of enzymes originally produced on ribosomes within the cell. Lysosomes appear as electron-dense bodies of up to 400 nm in diameter (see Plate 2.17). They have few characteristic features; consequently their identification in electron micrographs is difficult. The presence of a single outer membrane and a positive reaction for the enzyme acid phosphatase are the two criteria usually used. Acid phosphatase is detected by **enzyme histochemistry**, a technique which allows the location of an enzyme within a cell or tissue to be visualized (Box 2.3).

Lysosomes have a number of functions:

□ They are responsible for the intracellular digestion of material taken up by endocytosis, for example, food and pathogenic organisms.

□ They also break down cell components; as for example in the change of a tadpole into a juvenile frog, where the tadpole tail is broken down and absorbed.

□ In humans, the uterus, which weighs around 2 kg at full term, is invaded by phagocytic cells rich in lysosomes after the baby's birth; these reduce the uterus to its non-pregnant weight of about 50 g in nine days or so.

□ In normal animal cells, some of the proteins synthesized contain 'errors' i.e. amino acid sequences that do not correspond to mRNA sequences; lysosomes are responsible for the removal of these faulty proteins.

In all these activities macromolecules are broken down by digestive enzymes inside the lysosome to low M_r products which then pass across the lysosome membrane into the cytosol.

It is important that lysosomes do not rupture and release their contents inside living cells.

◇ What would happen if they did?

◆ The lysosomal enzymes would start to digest the cell.

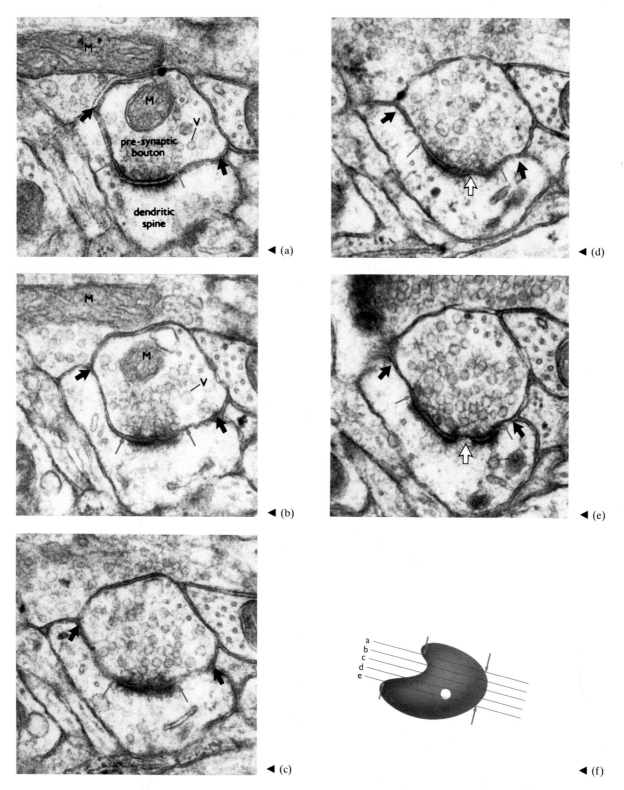

Plate 1.1(a)–(e) Five serial sections cut through a synapse from the brain of a chicken (all at magnification ×34 000). The synaptic cleft is the region between the solid black arrows. The active zone of the synapse is shown between the blue arrows. M = mitochondrion; V = synaptic vesicle. A perforation in the active zone can be seen in (d) and (e) (large white arrow). (f) is a diagrammatic representation of the entire active zone as viewed from inside the dendritic spine. Again, the blue arrows indicate the extent of the active zone in each of the five sections (a)–(e).

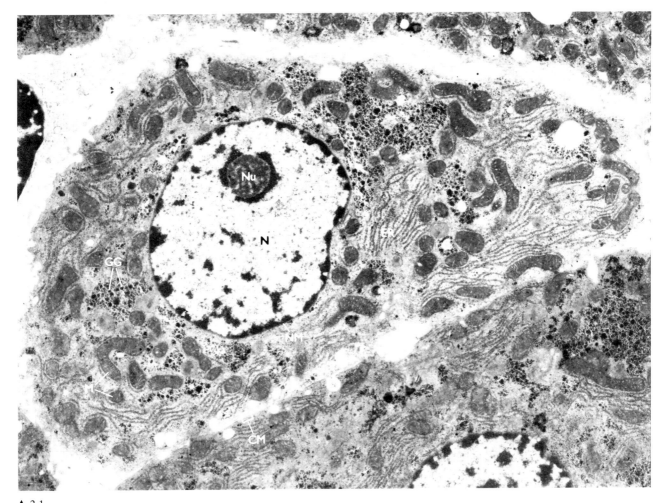

▲ 2.1

▼ 2.2

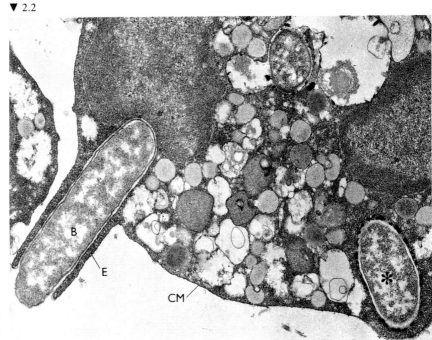

Plate 2.1 Electron micrograph of a whole rat liver cell and parts of two other cells, showing nucleus (N), nucleolus (Nu), nuclear membranes (NM), mitochondria (M), endoplasmic reticulum (ER), glycogen granules (GG) and cell membrane (CM). (×7000) Because this micrograph is shown at a relatively low magnification, the cell membrane is not distinct, so here the label CM refers to the location of the cell membrane.

Plate 2.2 Electron micrograph showing phagocytosis of a bacterium by a macrophage (a type of white blood cell). Note the evagination (E) of the cell membrane (CM) around the bacterium (B) and another bacterium enclosed within the cell (✳). (×20 000)

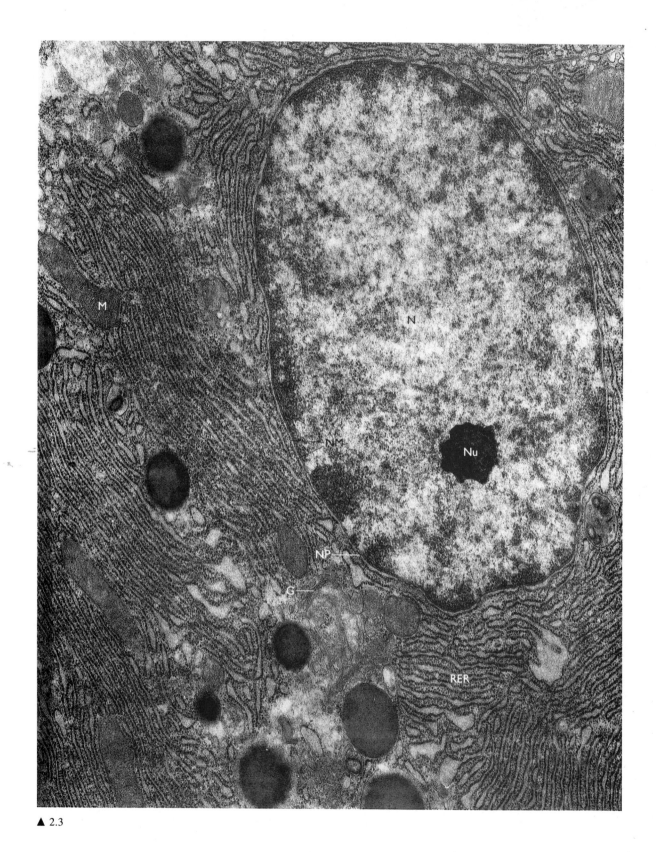

▲ 2.3

Plate 2.3 Electron micrograph of a liver cell showing nucleus (N), nucleolus (Nu), nuclear membranes (NM), nuclear pore (NP), mitochondria (M), rough endoplasmic with ribosomes (RER) and Golgi apparatus (G). (×20 000)

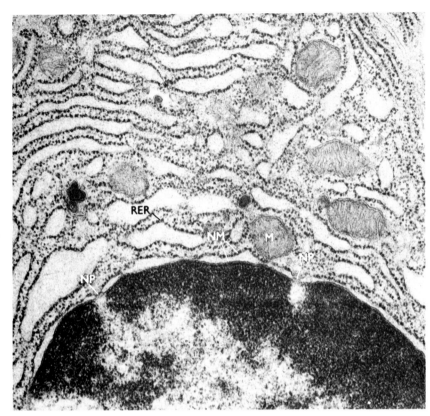

▲ 2.4

Plate 2.4 Electron micrograph of a liver cell showing nuclear membranes (NM) with ribosomes on the cytoplasmic face, nuclear pores (NP), mitochondria (M) and extensive rough endoplasmic (RER). (×30 500)

Plate 2.5 Electron micrograph of mitochondria in a rat liver cell. (×68 500)

2.5 ▼

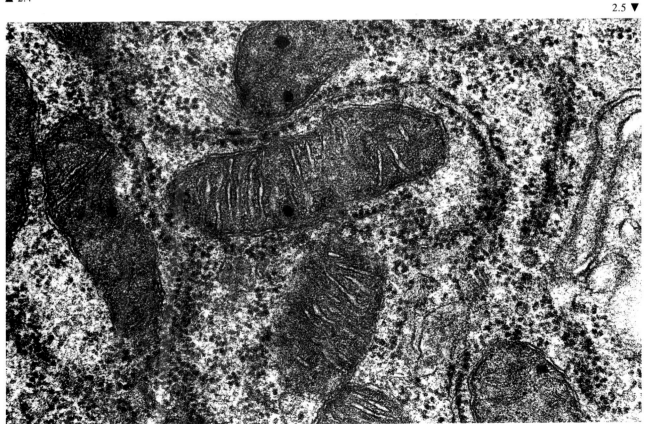

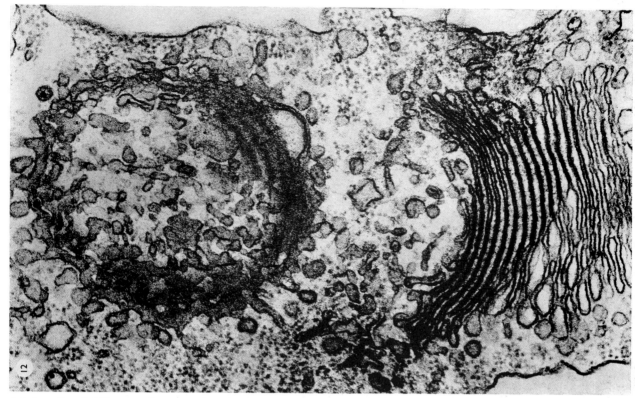

▲ 2.6   *Plate 2.6* Electron micrograph of the Golgi apparatus (right). (×75 000)

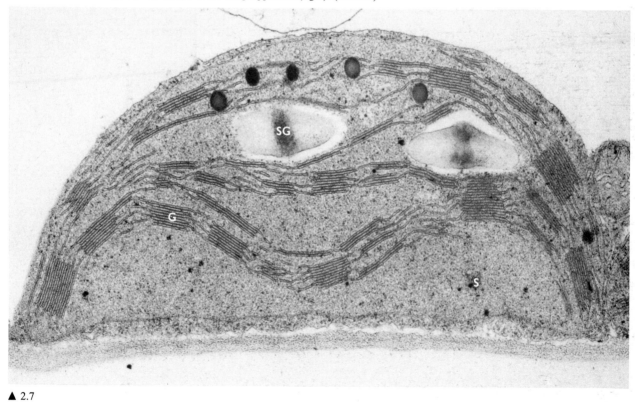

▲ 2.7

Plate 2.7 Electron micrograph of a chloroplast showing the grana (G), i.e. stacks of thylakoids, within the stroma (S) and two starch granules (SG). (×25 000)

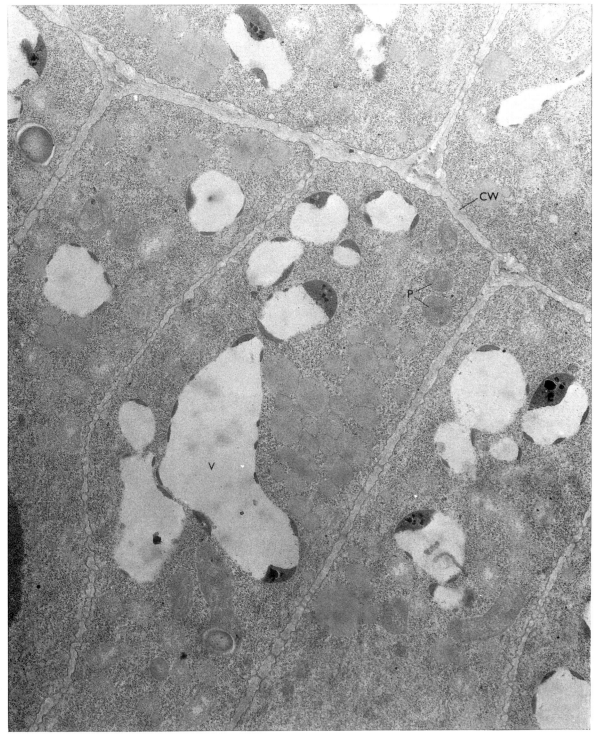

▲ 2.8

Plate 2.8 Electron micrograph of an immature plant cell showing proplastids (P), cell wall (CW) and vacuoles (V). (×2000)

Plate 2.9 Light micrographs of sections of the intestines of mice injected with thymidine: (a) 8 hours after injection; (b) 36 hours after injection. TN, radioactively tagged nuclei; V, villus. (×70)

Plate 2.10 Electron micrograph of cell membranes of intestinal epithelial cells showing the three-layered membrane: a dense outer layer (DOL), a clear inner layer (CIL) and a dense inner layer (DIL). (× 150 000)

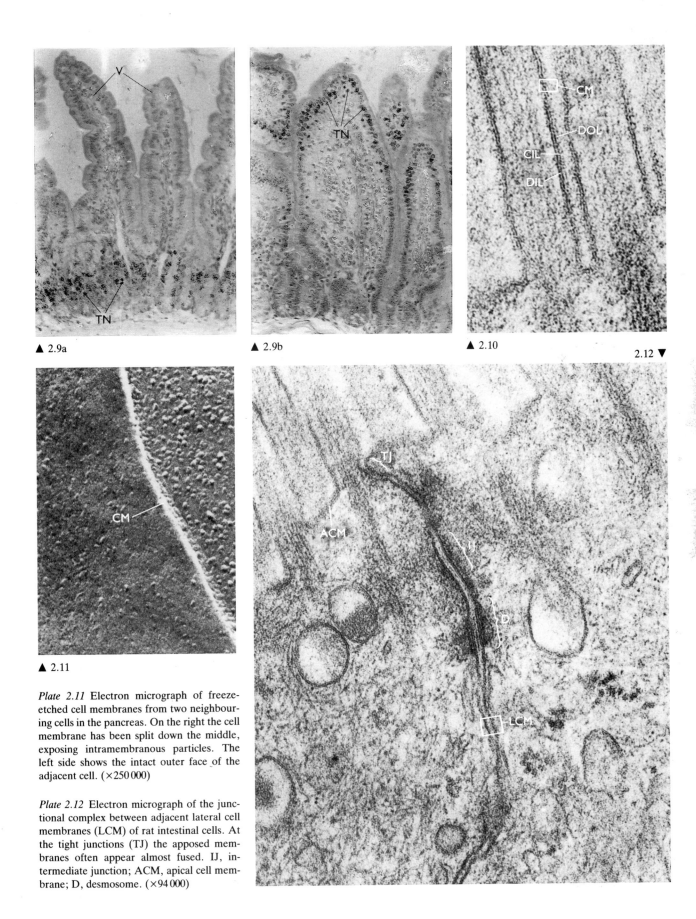

▲ 2.9a

▲ 2.9b

▲ 2.10

2.12 ▼

▲ 2.11

Plate 2.11 Electron micrograph of freeze-etched cell membranes from two neighbouring cells in the pancreas. On the right the cell membrane has been split down the middle, exposing intramembranous particles. The left side shows the intact outer face of the adjacent cell. (×250 000)

Plate 2.12 Electron micrograph of the junctional complex between adjacent lateral cell membranes (LCM) of rat intestinal cells. At the tight junctions (TJ) the apposed membranes often appear almost fused. IJ, intermediate junction; ACM, apical cell membrane; D, desmosome. (×94 000)

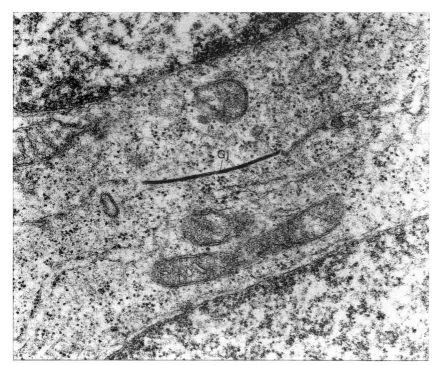

▲ 2.13

Plate 2.13 Electron micrograph of a gap junction (GJ) between ovarian granulosa cells. (×26 000)

Plate 2.14 Electron micrograph of the boundary between adjacent cells of corn (*Zea*) showing plasmodesmata (*arrowed*) and mitochondria (M). (×18 500)

2.14 ▼

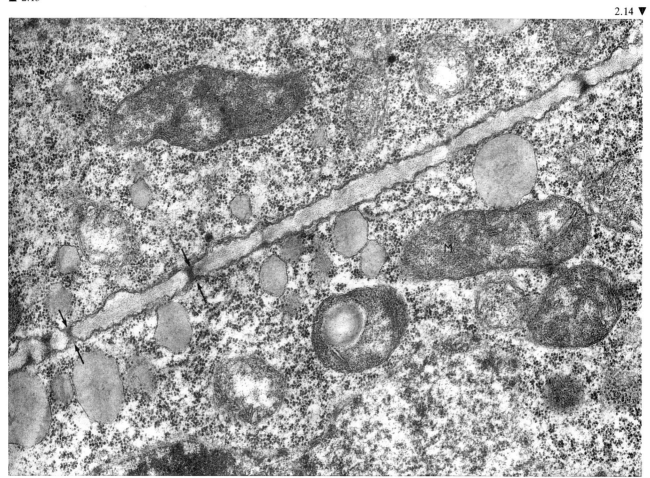

▲ 2.15

Plate 2.15 Electron micrograph of a freeze-etched preparation from onion. Note the nuclear pores (*arrowed*).
N, nucleus; C, cytoplasm; G, Golgi apparatus; M, mitochondria. (×28 000)

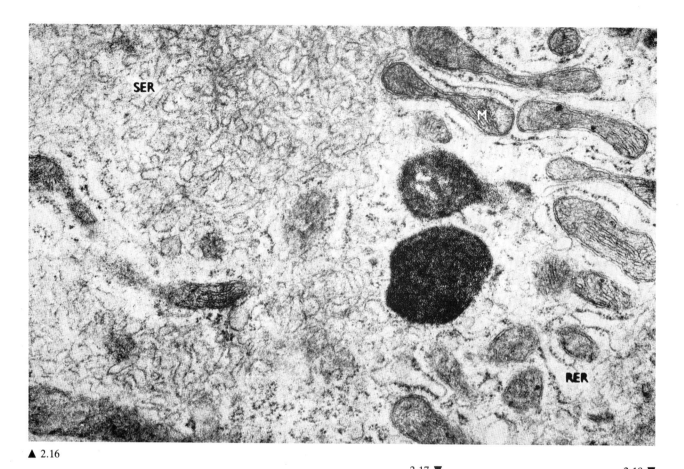

▲ 2.16

2.17 ▼

2.18 ▼

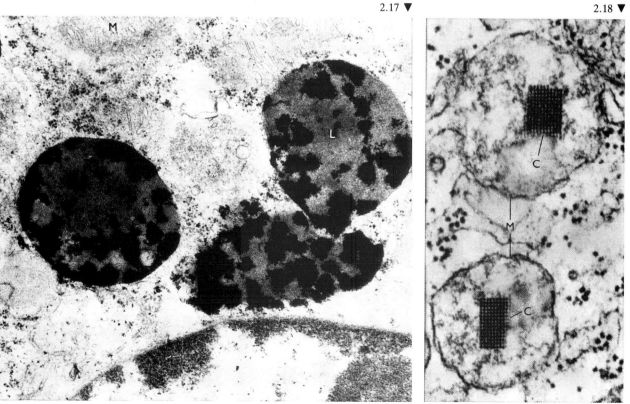

Plate 2.16 Electron micrograph of a cell from human intestine. The left-hand side of the field of view is packed with twisted cisternae of the smooth endoplasmic (SER); at the bottom and right is some rough endoplasmic reticulum (RER); also on the right are mitochondria (M). (×29 000)

Plate 2.17 Electron micrograph of a liver cell showing lysosomes (L) containing the densely staining lead sulphate reaction product, indicating the site of acid phosphatase activity, and a mitochondrion (M). (× 25 000)

Plate 2.18 Electron micrograph of two peroxisomes from a cell of the grass plant, *Avena*, showing the single external membrane (M) and the core of densely staining, crystal-like material (C). (×104 000)

Plate 2.19 (a) Electron micrograph of isolated cristae from mitochondria swollen by placing in a dilute solution and negatively stained with phosphotungstate. (×150 000)

(b) An enlargement of (a) showing the F_1 particles (*arrowed*) (×650 000)

Plate 2.20 (a) Cultured rat cell stained with anti-actin antibody. (b) Cultured rat cells stained with anti-myosin antibody. (×60)

Plate 2.21 Electron micrograph of a pair of centrioles in a Chinese hamster cell, at right angles to each other. TSC, transversely sectioned centriole showing nine triplets of subunits skewed towards the centre; LSC, longitudinally sectioned centriole. (×31 000)

Plate 2.22 Electron micrograph of cilia in transverse section showing the '9+2' arrangement of tubules (T). Note the short arms on the A tubules of each doublet, and the membrane (M) surrounding the cilium. (×210 000)

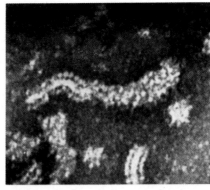

▲ 2.19a

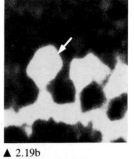

▲ 2.19b

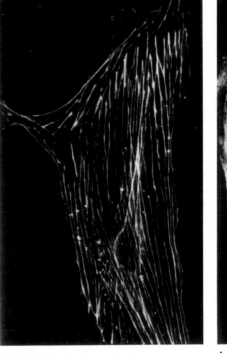

▲ 2.20a

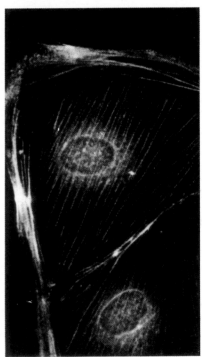

▲ 2.20b

▼ 2.21

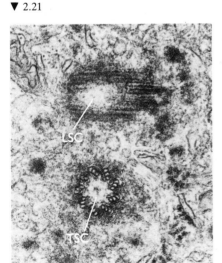

2.22 ▼

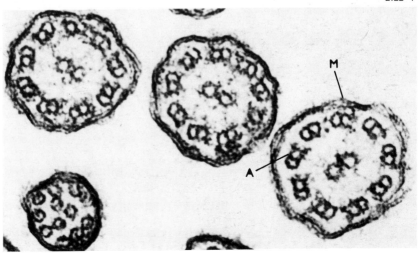

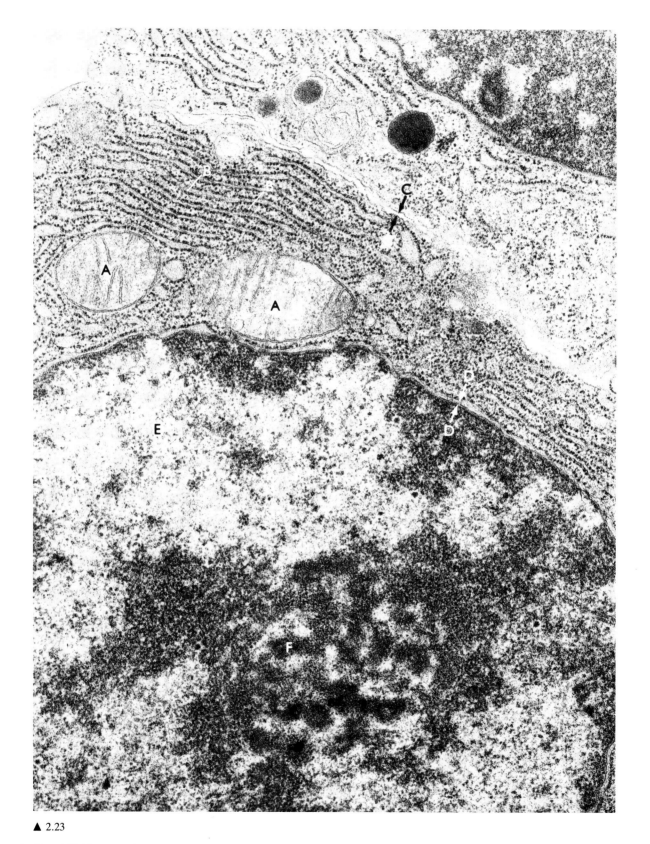

▲ 2.23

Plate 2.23 Electron micrograph from a piece of animal tissue fixed in glutaraldehyde and osmium tetroxide and stained with lead and uranium salts. (×42 000) (For Question 1 of Chapter 2)

Box 2.3 Enzyme histochemistry

This technique identifies particular enzymes in a tissue, the specificity of an enzyme forming the basis of the technique. Some principles of enzyme histochemistry can be illustrated by the techniques used to demonstrate the presence of acid phosphatase in lysosomes. Because phosphatases are enzymes that liberate phosphate ions from organic phosphate esters, acid phosphatases can be demonstrated in a frozen tissue section by thawing the section, then incubating it in an acid medium at pH 5.5 in a solution containing a substrate that is a phosphoric acid ester e.g. sodium β-glycerophosphate. The demonstration of the enzyme depends on the capture of the liberated phosphate ions. If lead acetate is present in the original incubating solution, the lead ions will combine with liberated phosphate ions to produce insoluble lead phosphate, $Pb_3(PO_4)_2$, at the site of acid phosphatase activity. $Pb_3(PO_4)_2$ is relatively colourless but by treating the sections with ammonium sulphide, a black precipitate of lead sulphide is formed instead, and this is visible when the slide is viewed in the light microscope.

$$Pb_3(PO_4)_2 + 3(NH_4)_2S \rightarrow 3PbS + 2(NH_4)_3PO_4$$

Enzyme histochemical techniques can also be applied in electron microscopy. The reaction product, $Pb_3(PO_4)_2$, produced as a result of acid phosphatase activity, is electron-dense and can be viewed in the electron microscope without further treatment, where it appears as a dense deposit in lysosomes (see Plate 2.17). For electron microscopy, the histochemical reaction is normally carried out on a thin slice of tissue (100 μm thick), which is then embedded in resin and processed for electron microscope examination (see Chapter 1).

Histochemical techniques can be made quantitative by measuring the optical density of the end-product in the stained slide and relating this to the optical density of the reaction product of a known concentration of the material under investigation.

In certain degenerative diseases of humans, for example rheumatoid arthritis, enzymes released by the breakdown of lysosomes from macrophages may be a significant factor—in attacking living cells and tissues.

Lysosomes, though in some senses 'destructive' organelles, have a highly constructive role to play. Lysosomes can contribute to hormone production, an example of which can be seen by looking at the formation of thyroxin, a hormone affecting a wide range of physiological activities including metabolic rate (Figure 2.6).

The functional units of the thyroid gland are spherical follicles consisting of a single-layered wall of epithelial cells surrounding a lumen. The lumen

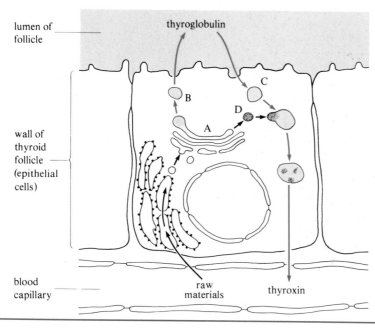

Figure 2.6 Participation of lysosomes in the processing of thyroglobulin and the release of thyroxin.

contains a store of a globular protein, thyroglobulin, which is secreted by the epithelial cells. When the thyroid gland is stimulated by thyroid stimulating hormone from the pituitary gland, the thyroglobulin is taken up into the thyroid follicle cells by endocytosis. The endocytic vesicle subsequently fuses with a lysosome, and thyroglobulin is degraded by lysosomal proteases (enzymes that hydrolyse proteins—Chapter 4) to thyroxin, which can then be transported to the nearest blood capillary.

◇ What are structures A–D in Figure 2.6?

◈ A, Golgi apparatus; B, secretory vesicle (containing thyroglobulin); C, endocytic vesicle; D, lysosome about to fuse with an endocytic vesicle.

2.7 PEROXISOMES

Peroxisomes are also called microbodies or glyoxysomes. They are structures of 0.5–1.0 μm diameter containing a dense core and bounded by a single membrane 6.5–8 nm thick. They are believed to arise from outgrowths of the endoplasmic reticulum, and are present in most animal or plant cells. Plate 2.18 shows peroxisomes in a plant tissue.

The name peroxisome comes from the presence in these bodies of enzymes that catalyse the breakdown of hydrogen peroxide (H_2O_2). The enzymes purified from peroxisomes include D-amino acid oxidase, which breaks down D-amino acids producing H_2O_2 among other products, and catalase, which breaks down H_2O_2. The role of peroxisomes in animal cells seems to be primarily in detoxification. In plant cells that photosynthesize, they are important in the process of photorespiration in which the two-carbon compound glycolate, produced in photosynthesis, is converted to the amino acid glycine, which can then enter mitochondria and be metabolized further. And in plant cells that store fats, they are involved in the conversion of fatty acids to carbohydrates via a metabolic pathway called the glyoxylate cycle— hence the alternative name of glyoxysomes for these organelles.

2.8 MITOCHONDRIA

Mitochondria vary in size, but typically they are up to 0.5 μm wide and 7 μm long. Studies of living mammalian cells by specialized techniques, such as time-lapse cinephotography, show that mitochondria are capable of considerable movement, changing in both shape and position within the cell.

Plate 2.5 shows some mitochondria in a liver cell; they are surrounded by two membranes, an outer membrane 5–7 nm thick (the same thickness as other internal membranes), separated by a space of 6–8 nm from an inner membrane, which is thrown into complex infoldings called **cristae**.

The mitochondrial *matrix*, the space surrounded by the inner membrane, contains enzymes of the tricarboxylic acid (TCA) cycle and those involved in fatty acid oxidation. The inner membrane is responsible for oxidative phosphorylation.

The inner membrane is the same thickness as the outer membrane in intact mitochondria. When the inner membrane is isolated and fragmented, 9 nm diameter spheres on the end of stalks 4 nm high (Plates 2.19a and 2.19b) spaced at regular intervals of 10 nm along the membrane can be seen. These are known as elementary or F_1 particles, and there may be 10^4–10^5 of them in each mitochondrion. The F_1 particle contains an enzyme that participates in the synthesis of adenosine triphosphate (ATP) (see Chapter 6). The reason the F_1 particles cannot be seen in the intact mitochondrion is that they are apparently an integral part of the inner membrane and only appear as stalked particles when the membrane structure is disturbed.

The number of cristae per mitochondrion tends to be greater in cells in which intense metabolic activity occurs.

◇ Why is this so?

◆ The inner membrane of the mitochondrion is the site of ATP production in the cell, so the greater the number of cristae, the greater the surface area on which ATP production can occur.

The mitochondria themselves are often found concentrated in regions of the cell associated with intense metabolic activity. In the cells of the intestine, for example, energy is required for the transport of materials from the lumen of the intestine into the cells. Accordingly, in these cells, mitochondria are concentrated just beneath the apical cell membrane. The orientation of the cristae inside the mitochondria seems to vary between cells from different parts of the body. For instance, the cristae run transversally in the mitochondria of the human liver cells, but lengthways in those of nerve cells and they are tubular in cells of the adrenal cortex. Whether these differences have any functional significance is not known.

Apart from their important functional role in cells, mitochondria are interesting because they are one of the two types of subcellular organelle containing DNA, the other being *plastids*, found in plant cells. These organelles are in fact self-replicating and possess considerable autonomy.

2.9 PLASTIDS

Plant cells, such as those found in the shoot or root tip, contain organelles surrounded by a double membrane that at first sight could be mistaken for large mitochondria. These are the proplastids—the simplest form of a class of organelles found in almost all plant cells (Plate 2.8). According to the position of a differentiating cell in the plant, the proplastids will develop into one of a variety of mature **plastids**, such as **chloroplasts** in photosynthetic tissue, amyloplasts (where starch is stored) in root or other storage tissues, and chromoplasts (containing colour pigments) in petals and fruits.

In a cell from a leaf, the chloroplasts (see Plate 2.7), of which there may be several hundred, are each about $6\,\mu m$ long by $3\,\mu m$ wide. The internal structure of a chloroplast is a complex of membranes running through a matrix. The matrix, or **stroma**, consists of a concentrated solution of protein including all the enzymes needed to convert carbon dioxide into sugar. Some chloroplasts contain granules of starch (Plate 2.7), the product of photosynthesis, which is stored prior to export. Ribosomes are also present in the stroma, and sometimes a DNA-containing structure can also be seen.

The chloroplast membranes form a system of sacs or **thylakoids**, whose contents are separated from the stroma. Some of the thylakoids run for quite long distances through the stroma, and from the micrographs, could be either paired sheets or tubes of membrane. These are the first to be formed during the differentiation of the proplastid into a chloroplast. Some chloroplasts do not develop beyond this point, although most also form **grana** (sing. granum), structures in which several thylakoids are stacked on top of each other (Plate 2.7). The end result is a large area of interconnected membranes in the chloroplast. These membranes contain pigments (mainly the green photosynthetic pigment **chlorophyll**) whose function is to trap the light energy required to make molecules (ATP and NADPH) used up by the reactions taking place in the stroma.

Other types of plastid do not have such well developed membrane systems as the chloroplasts. The predominant feature of amyloplasts is the presence of many starch granules. These granules may simply be storage products, as in potato tubers, but they may be involved in the detection of gravity in certain specialized cells (amyloplasts are the densest organelles).

Plastids in many cells are responsible for colour. Where these cells are non-photosynthetic, as in the carrot root or tomato fruit, the plastids are called chromoplasts.

2.10 THE CYTOSKELETON

Some living cells are capable of altering their shape. The **cytoskeleton**, a lattice-like array of fibres and fine tubes that occur in the cytoplasm, is involved in this, and in the maintainence of cell shape.

There are three components of the cytoskeleton, *microtubules*, *microfilaments* and *intermediate filaments*.

Microtubules are straight, unbranched tubes of indefinite length and about 25 nm outside diameter. Their principal component is the protein tubulin. Microtubules are not fixed and permanent structures in the cell but can change in number and length depending on the internal state of the cell. A variety of functions has been suggested for microtubules; the shape of cells seems to depend on them, they are involved in cell movement, in the movement of cilia and flagella (see Section 2.11), in transport within the cell of granules and vesicles and with the movement of chromosomes during cell division.

Microfilaments are polymers of the protein actin and are thinner than microtubules, being about 6 nm in diameter. Like microtubules, microfilaments can grow or shrink depending on the internal state of the cell. They are connected to the cell membrane, to microtubules, and to one another. Along the length of microfilaments are regions at which their structure is more complicated; here the protein myosin is involved. The result is a complex called actomyosin. The extent of the microfilament network can be seen when cultured mammalian cells are stained with anti-actin, or anti-myosin antibodies (Plate 2.20) by the technique of **immunocytochemistry**. This technique is based on the ability of some cellular components to react specifically with antibodies to which a 'marker' substance has been attached (see Box 2.4).

Actomyosin is the contractile unit that forms the basis of muscle contraction, although in skeletal and cardiac muscle cells actomyosin is much more highly organized than in the cytoskeleton. So the distribution of this complex through the cytoplasm gives it the properties of a diffuse muscle; in fact, it is sometimes referred to as the cytomusculature.

Box 2.4 Immunocytochemistry

Immunocytochemistry is based on the detection of antigens by antibodies (Chapter 3, Section 3.2). Antigens are usually large cellular molecules such as proteins, polysaccharides and nucleic acids but smaller cell constituents may also be antigenic. When antigens are introduced (or injected) into an animal they activate lymphocytes (a type of white blood cell) to produce antibodies, which are specific to the antigen. Antibodies are often produced by injecting a particular antigen into a large mammal, such as a rabbit. After a time, a blood sample is taken from the rabbit and the antibody can be separated from the blood cells.

In immunocytochemistry this antibody (which is referred to as a primary antibody) can be added to cells in culture, or to a tissue section which contains the antigen. However, the antibody at this stage cannot be seen. Therefore, before adding it to the tissue containing the antigen the antibody is first conjugated to a fluorescent dye such as fluorescein. When the fluorescein-labelled antibody has bound to the antigenic sites in the tissue it can be visualized in a fluorescence microscope (e.g. as shown in Plate 2.20). Alternative methods of antibody visualization include the use of the enzyme peroxidase to directly detect the antibody. Peroxidase is chemically linked to the antibody and then reacted (in the presence of hydrogen peroxide) with a substance called diaminobenzidine, the DAB reaction, which produces a dark end-product that is easily seen in a light microscope.

Indirect methods of antibody visualization can now be used including the peroxidase–antiperoxidase (PAP) method which involves intermediate, non-covalent links between the primary antibody and the dark end-product of the DAB reaction. The purpose of the indirect methods is, essentially, to enable a more specific and denser localization of the primary antibody.

Microfilaments are connected to the cell membrane and the mechanical forces generated in the actomyosin complexes can result in the whole shape of the cell changing or in more local changes involved in phagocytosis. Movement of cytoplasm within cells, called **cyclosis**, and streaming movement of whole cells, called amoeboid movement, involve mechanical forces generated within the microfilament network.

Intermediate filaments are about 10 nm in diameter, which is in between the thickness of the microfilaments and the microtubules. There seem to be several different kinds of intermediate filaments, differing biochemically from each other and from microfilaments and microtubules. Like the other elements of the cytoskeleton, intermediate filaments help to determine cell shape. One type, in which the filaments are made from the protein keratin, are involved in holding sheets of epithelial cells together; and in skin cells these form a tough resistant covering to the surface of the body of animals.

2.11 CENTRIOLES, CILIA AND FLAGELLA

Centrioles are found in most animal and some plant cells. These are cylindrical structures, 0.15 μm in diameter and 0.5 μm long. The wall of the centriole consists of nine sets of tubular structures, each composed of a triplet of subunits skewed towards the centre (Plate 2.21). Usually there is a pair of centrioles, at right angles to each other and frequently located close to the Golgi apparatus.

Cilia and **flagella** are structures that extend from the surface of some cells that can bend, thus causing movement (Figure 2.7, Plate 2.22). They resemble centrioles in having nine sets (this time, pairs) of tubules arranged in a cylinder of 0.5 μm diameter, and they form in association with *basal bodies* (see Figure 2.1), which are believed to be homologous with centrioles. From the basal body, fibres which extend deep into the cytoplasm anchor the cilium

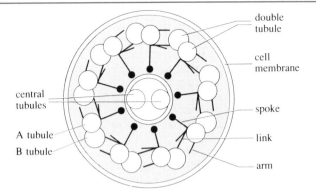

Figure 2.7 Structure of a cilium or flagellum seen in cross-section. Each doublet consists of an A and B tubule; two arms are attached to the A tubule. The A tubule is a complete circle in cross-section, whereas part of the wall of the B tubule is shared with the A. Spokes seem to occur at intervals along the length of each doublet; perhaps they connect the doublets to the sheath surrounding the central tubules.

or flagellum in the cytoplasm. In cilia and flagella, there are two additional tubules in a sheath in the centre of the cylinder (Plate 2.22). Cilia and flagella are said to have a '9 + 2' structure as opposed to the '9 + 0' seen in centrioles and basal bodies. The tubules of each pair (doublet), called A and B, have two short arms attached to the A tubule. Spoke-like connections also occur between the outer tubules and between the outer tubules and the central tubules. Mature cilia or flagella are covered by an extension of the cell membrane.

How cilia or flagella beat to exert a force is not entirely certain. However, it is known that the arms of the tubules contain a protein called dynein that can break down ATP, and cilia or flagella isolated from the cell will continue to beat if ATP is added. It is likely that the ATP breakdown releases energy, and this causes sliding of the tubules in the wall of the cilia—the doublets move longitudinally with respect to one another and the central doublet, and it is this movement that causes bending of the cilium (Figure 2.8).

Figure 2.8 The sliding of doublets during the motion of a cilium on the gills of a mussel. One of the two tubules of each doublet protrudes further into the tip of each cilium than does the other. Thus cross-sections at an appropriate constant distance from the tip show a change in tubule pattern during bending from all doublets to doublets and single tubules. The entire doublets slide with respect to each other; but the two members of each doublet retain their positions relative to each other.

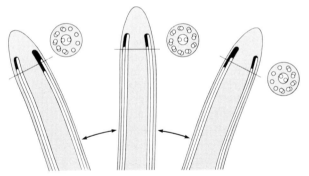

In multicellular animals, cilia generally have the function of moving fluid or particles over the surface of cells, whereas flagella and cilia function as locomotory organelles in many protistans and unicellular algae. A cell with cilia has many of them, whereas a cell or unicellular organism with a flagellum usually has a few or just one. A flagellum is usually a much larger structure than a cilium.

2.12 COMPARTMENTATION

As you have learned in this chapter, eukaryotic cells consists of organelles surrounded by cytosol and are bounded by a cell membrane. One of the striking features about the way in which cells function is the extent to which biochemical reactions are localized within the cell. For example the powerful digestive enzymes inside lysosomes do not normally leak into the surrounding cytosol. The products of digestion inside lysosomes do however pass through the lysosome membrane into the cytosol. The digestive reactions are localized and contained. Localization is a consequence of differences between the membranes separating the various organelles from their surroundings and the membranes within the organelles. For example the outer membrane of mitochondria is permeable to some but not all of the molecules occurring in the surrounding cytosol. The inner membrane like the outer membrane is selectively permeable so that the composition of the mitochondrial matrix is different from that between the inner and outer membranes and different from the surrounding cytosol. In effect the membranes within eukaryotic cells divide the cell into a series of *compartments*.

◇ What is the advantage of this?

◆ The reactants required in a particular process are concentrated within a small region, bounded by a membrane and in a different environment from the rest of the cell contents.

However, the activities of different organelles must be integrated, for example, during the production and export of macromolecules from cells.

◇ What subcellular structures are involved in these processes?

◆ Ribosomes, endoplasmic reticulum and the Golgi apparatus.

Compartmentation is thus essential for efficient metabolism within the cell. You will come across compartmentation again in later chapters.

2.13 PROKARYOTIC AND EUKARYOTIC CELLS COMPARED

The cells which you have been studying so far in the chapter, eukaryotic cells, are not the smallest nor, in terms of their structure, the simplest cells. The type of cell characteristic of the bacteria and cyanobacteria—the prokaryotic cell—is in many ways less complex and may be smaller than the type of cell characteristic of protistans, plants and animals—the eukaryotic cell. In spite of this, two points should be emphasized: first that prokaryotic cells may be as complex in terms of their biochemistry as eukaryotic cells, and second that prokaryotic cells have a much longer evolutionary history than eukaryotic cells—three billion years compared with less than one billion years. So it is quite wrong to think of prokaryotic cells as 'simpler' in the sense of being primitive or as being inefficient. They are neither, but in a number of ways they are different.

□ Prokaryotic cells are generally surrounded by a cell wall; in this they resemble plant cells, but chemically the two differ—plant cell walls contain the carbohydrate cellulose but the cell walls of prokaryotes do not.

□ Both types of cell are bounded by a cell membrane.

□ Eukaryotic cells contain a nucleus, mitochondria, endoplasmic reticulum, Golgi apparatus and may contain chloroplasts, all of which are membranous structures. In prokaryotic cells intracellular membranes are usually absent; there are no subcellular organelles in the cell and the processes of endocytosis and exocytosis which involve membranes do not occur.

□ Like eukaryotes, prokaryotes contain DNA, but in prokaryotes this is not separated from the cytoplasm of the cell as it is in eukaryotes by nuclear membranes.

☐ Both types of cell contain ribosomes, but there are differences between the ribosomes of the two types of cell.

☐ Some bacterial cells are flagellated, but the internal structure of the prokaryotic flagellum differs from that of the eukaryotic flagellum.

☐ A cytoskeleton is universally present in eukaryotic cells but absent from prokaryotic cells.

The major differences between the two types of cell are summarized in Table 2.1.

Table 2.1 A comparison between prokaryotic and eukaryotic cells

Feature	Prokaryotic cells (bacteria, cyanobacteria)	Eukaryotic cells (protista, plants, animals)
Cell wall	present—does not contain cellulose	present—contains cellulose (plant cells)
Cell membrane	present	present
Intracellular membranes	usually absent	present
Subcellular organelles	absent	present
Flagellum (when present)	contains a single fibril	contains more than one fibril
Cytoskeleton	absent	present

SUMMARY OF CHAPTER 2

1 Within eukaryotic cells, there is a remarkable uniformity of plan: normally one central nucleus connected to a membranous system of endoplasmic reticulum and Golgi apparatus, and a variety of structures—lysosomes, peroxisomes, mitochondria, ribosomes, centrioles, and cilia or flagella, and in plant cells, plastids.

2 The endoplasmic reticulum, the Golgi apparatus and the lysosomes play an important functional role in the production and 'packaging' of materials for export out of the cell. Mitochondria and chloroplasts are both involved in making energy available in the cell.

3 The cell is dynamic—it is constantly changing, and its components and the cell itself are continually being renewed.

4 There is a network of fine tubules and filaments in the cell, the cyto-skeleton, which is responsible for maintaining cell shape, and which allows many cells to change their shape.

Question 1 (*Objectives 2.1 and 2.2*) Plate 2.23 is a high magnification electron micrograph from a section of animal tissue. Identify each of the items (A–F) indicated on the plate, giving reasons for the conclusions you reach.

Question 2 (*Objectives 2.1 and 2.3*) The ionic (salt) composition of the fixation medium can have a crucial effect on the appearance of structures in electron micrographs. If the ionic concentration is too high, tissues will shrink, but if it is too low, they will swell and eventually burst. Mitochondria are critical indicators of this.

Examine Figure 2.9 and say which of the mitochondria has been fixed in: (a) distilled water (+fixatives); (b) salt solution with a concentration ten times that of the body fluids of the animal (+fixatives); and (c) salt solution with a concentration equivalent to that of the body fluids of the animal (+fixatives).

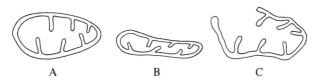

A   B   C

Figure 2.9 Mitochondria fixed in media of different ionic composition. (For Question 2)

Question 3 (*Objectives 2.1 and 2.2*) What processes would probably be occurring to a major extent in each of the following cells?

(a) A cell in which there was a great deal of rough endoplasmic reticulum.

(b) A cell in which there were many Golgi stacks.

(c) A cell in which the apical cell membrane contained microvilli.

(d) A cell in which there was a great deal of smooth endoplasmic reticulum.

OBJECTIVES FOR CHAPTER 2

Now that you have completed this chapter, you should be able to:

2.1 Define and use, or recognize definitions and applications of, each of the terms printed in **bold** in the text.

2.2 Identify the main organelles of eukaryotic cells in photomicrographs of stained preparations and state their functions. (*Questions 1 and 3*)

2.3 Outline the problems, including those caused by artefacts, in interpreting photomicrographs in terms of structure. (*Question 2*)

MACROMOLECULES

◆ CHAPTER 3 ◆

3.1 MACROMOLECULES AND THEIR ASSEMBLY INTO CELL STRUCTURES

Microscopy, as you have seen, is a very useful way of looking at cell structures, but it gives us little idea of the functions of the different components or how they work. For instance, what dictates the permeability of membranes? How do the properties of their outside surfaces enable cells to recognize each other? To answer these and other questions, we need to know the chemical make-up of the structures we saw in the electron micrographs of the last chapter.

Cells contain many small molecules. These include water, which makes up about 80% of living matter, inorganic ions, such as those of sodium and magnesium, which are vital for cell function and whose concentrations are therefore carefully controlled, and the myriads of organic molecules that are the substrates and products of metabolism. However, we concentrate initially in this chapter on the large molecules formed by joining together small organic building blocks. The three major classes of these macromolecules are the proteins, nucleic acids and polysaccharides. Just as important are the lipids, which strictly speaking are not macromolecules, because their average relative molecular mass (M_r) is less than a thousand, which contrasts with the true macromolecules whose M_r may be up to several millions. Lipids tend to form aggregates and for this reason are usually considered along with the macromolecules.

The cell membrane consists largely of lipids and proteins, some of them carrying short sugar chains that are responsible for intercellular recognition. The contents of the cell are subdivided by membranes into various compartments, each with its characteristic macromolecular components. The nucleus, apparently homogeneous under the light microscope apart from the nucleolus, contains not only DNA, but also RNA and large amounts of protein. DNA, together with protein, forms a complex known as chromatin. The cytosol too contains recognizable aggregates of macromolecules, such as the ribosomes, each of which consists of about 60 precisely interlocked protein and RNA molecules. Lipoprotein membranes, ribosomes, DNA and granules of the polysaccharide starch are often apparent in the chloroplast, and just visible in very high power micrographs are molecules of a large protein—the major enzyme of this organelle.

From this brief description, you should appreciate firstly that each class of macromolecule can be found in a number of different locations within the cell, and secondly that functional units (e.g. organelles) are very often assemblies of several separate macromolecules. Structures like the ribosome and the chloroplast are specific assemblies large enough to be seen under the electron microscope.

We begin this chapter by breaking down the organelles you met in the preceding chapter into their chemical constituents, for in order to understand how these are made up and function, we need to determine how the individual components contribute to the function. The main bulk of the chapter is a review of the different classes of macromolecules, their composition and structure. The chapter finishes by asking how the separated macromolecules interact to make the assemblies we started with. What causes the molecules of living organisms to aggregate into the structures we see under the microscope? How do such aggregates interact to give the properties characteristic of our well-known organelles? We tackle these questions briefly for two such assemblies.

Although this chapter concentrates on macromolecular structure, we will emphasize the relationship between chemical structure and biological func-

Table 3.1 Specificity, support and storage roles of macromolecules

Role	Class	Examples
Specificity	protein	enzymes: catalytic power enables biochemical reactions to occur rapidly and be controlled
		antibodies: enable the body to recognize and remove foreign matter
		receptors: enable particular tissues to bind and respond to hormones (which may themselves be proteins)
		carrier molecules: e.g. haemoglobin, the carrier of O_2 and CO_2 in the blood
	nucleic acid	DNA: base-pairing ensures faithful replication
		RNA: mRNA carries information from nucleus to cytoplasm
	polysaccharide	blood groups: A, B and O differ only in the sugar molecules carried on the cell surface
Support	protein	fibroin: an inextensible structural protein found in silk
		collagen and keratin: more elastic proteins found in tendons and skin or hair
	nucleic acid	rRNA: needed to maintain ribosome structure
	polysaccharide	chitin: forms exoskeleton of insects and cell walls of most fungi
		cellulose: the strengthening material in most plants
		hyaluronic acid: the lubricant in skeletal joints
		mucilage: slimy secretion that protects seaweeds from drying out at low tide
	lipid	the major components of membranes
Storage	protein	seed proteins: food reserves of many plants
		precursor proteins: are often made and stored in cells in an inactive form until needed
	polysaccharide	glycogen: an easily mobilized energy store in animals
		starch: a major storage material in plants
	lipid	adipose tissue: an energy store in animals
		oil-seeded plants: e.g. sunflower and oil-seed rape

tion wherever possible; such links will, however, become clearer as you read the rest of this book. Table 3.1 summarizes much of what is to come, grouping functions into those concerned with specificity, support or storage. The table is principally for reference at present—do not expect to understand it all at this stage.

3.2 ISOLATING MACROMOLECULES FROM CELLS

The first step in studying any biological molecule is to extract it from cells and to separate the one substance of interest from all the other cellular components. This is by no means as simple as it may seem! Consider for example cells of the bacterium *Escherichia coli* which may contain somewhere in the region of 2 000 different proteins. To study just one of these, you have to prepare a homogeneous solution of this which is free of the other 1 999 proteins, not to mention all the other cell components. Another problem is that macromolecules are rather susceptible to destruction and damage during extraction procedures.

◇ Consider a suspension of animal cells from which you wish to isolate a particular macromolecule. During preparation, a number of lysosomes are disrupted. What might be the consequence of this?

◆ Damage to lysosomes would release enzymes which could digest the macromolecule you are attempting to isolate.

The structure of most macromolecules is also disrupted by heat, or by acidic or alkaline conditions, and long DNA molecules are broken into shorter lengths if an extract is stirred too vigorously. The key to success is therefore to use mild conditions during the extraction of macromolecules from cell preparations, but even then it is not easy to be sure that the molecule remains exactly as it was in the intact cell.

Whatever method is used for purification, there has to be some way of following the progress of the separation. It is fairly easy to measure how much of a particular macromolecular class is present, since there are rapid methods for measuring, for example, the protein concentration of a solution. If we are interested in a single macromolecule we also need to measure how much of it is present.

◇ How might the presence of a specific macromolecule in a mixture of many others be detected and measured?

◆ We could measure a unique property of the molecule in question.

If, for example, an enzyme which has a precise catalytic activity was being purified, then this could be measured or assayed relatively easily. But what if the macromolecule has some other function? Many macromolecules have specific binding sites for other molecules, and this property can be used in an assay. A molecule that binds to a macromolecule at a specific site is known as a **ligand**. The more of a ligand which binds in a particular preparation, then the more of the specific macromolecule is present.

A particularly sensitive technique that uses this principle is termed an immunoassay. This makes use of the natural defence mechanisms of higher animals, whereby foreign material entering the body is neutralized by interaction with antibodies (immunoglobulins, discussed further in Section 3.4.4). Antibodies are proteins secreted into blood, with binding sites specific for the foreign material (known as the antigen) that induced their production (see Chapter 2, Box 2.4). Most macromolecules are antigens, and if they are injected into an animal (experimentally, usually a rabbit) they induce its white blood cells to secrete specific antibodies. This antibody can be isolated and used in an assay for the detection of the specific macromolecule (antigen). The beauty of this assay is that it is highly specific for the antigen and it is highly sensitive, so that very small amounts can be measured.

Having established what we mean by an assay, we can now move on to look at the various techniques used for separating macromolecules. There are a large number of these, but here we will consider just a small sample, to give you some of the basic principles. Macromolecules differ in terms of their solubility, the charge they carry, their size and/or their specificity—most methods used for separation exploit these differences.

3.2.1 Separation according to solubility

The first stage in an isolation procedure is to rupture a suspension of cells. The medium used to make this initial cell extract will depend on the solubility properties of the required macromolecule. As a rule of thumb, *like dissolves like*, so that charged (polar) compounds will tend to dissolve in polar solvents but not in non-polar ones. Generally speaking, proteins, nucleic acids and polysaccharides are polar molecules, whereas lipid molecules contain large regions which are non-polar.

◇ Suppose you wish to extract membrane lipids from a cell extract. Would an aqueous solution be suitable or not?

◈ It would be unsuitable. Organic solvents (non-polar) should be used. For example, fats dissolve easily in dry-cleaning fluid (e.g. carbon tetrachloride, CCl_4). So lipids in general can be readily separated from the other more polar cell components by using a non-polar solvent to make the initial cell extract.

If proteins are the required macromolecules, an aqueous solution is used for the initial extraction. A popular early step in separating proteins from one another is ammonium sulphate fractionation, where increasing amounts of the salt $(NH_4)_2SO_4$ are added. Since protein solubility depends on salt concentration, different proteins will precipitate at different stages, as the salt concentration is gradually increased.

3.2.2 Separation according to charge

Most biological macromolecules are electrically charged and thus move in an electric field. For example, if the terminals of a battery are connected to the opposite ends of a tube containing a solution of positively charged protein molecules, the molecules will move from the positive end of the tube to the negative end. The direction of motion obviously depends on the sign of the charge, but the rate of movement depends on the magnitude of the charge and (to a lesser extent) on the shape of the molecule. Why is shape important?

◇ Imagine pulling a ball through water by means of a piece of string. Will it move more or less easily than a cube of the same mass?

◈ For the same pull, the ball moves faster than the cube, since it is more compact and offers less resistance to flow.

Electrophoresis is the most widely used technique for separating macro-molecules on the basis of charge. It separates small quantities and is therefore used for analysis and is not usually suitable for the production of large quantities of material. In this process, a mixture of molecules is applied to a support medium (which may be absorbent paper, or a gel made from starch, or a synthetic polymer called polyacrylamide) that is saturated with solvent. When an electric field is applied, charged molecules move towards the oppositely charged electrode at a speed proportional to their charge (Figure 3.1). After a suitable time, the paper or gel is removed from the apparatus, and once the components have been located, it can be cut up and the separated macromolecules dissolved out.

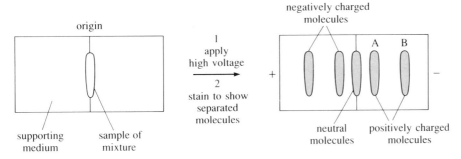

Figure 3.1 The technique of electrophoresis. A sample is placed at the origin, and the charged molecules separate under the influence of an electric field. Negatively charged molecules move towards the positive pole (anode); positively charged molecules move towards the negative pole (cathode). Afterwards staining can be used to locate the separated molecules.

◇ Study Figure 3.1. Which band, A or B, is likely to contain the more positively charged molecules?

◆ B, because it has moved further from the origin towards the cathode, i.e. negative pole. (This assumes no effect due to differences in shape.)

3.2.3 Separation according to size

Here we will consider just two of a wide variety of techniques which can be used to separate macromolecules of different sizes. The first is the most common type of electrophoresis used in molecular biology, gel electrophoresis, which is a modification of the procedure just described.

Gel electrophoresis can be performed so that the rate of movement depends only on the relative molecular mass (M_r), i.e. size, of the molecule. An experimental arrangement for gel electrophoresis of a sample of different sized DNA molecules is shown in Figure 3.2. A thin slab of agarose (a polysaccharide) gel is prepared containing small slots ('wells') into which samples are placed. An electric field is applied and all the DNA molecules,

Figure 3.2 Apparatus for (vertical) slab gel electrophoresis capable of running seven samples simultaneously. The liquid gel is allowed to harden while the plastic frame is still horizontal. An appropriately shaped mould is placed on top of the gel during hardening in order to make wells for the samples. After electrophoresis the slab is stained by removing the plastic frame and immersing the gel in a solution of the stain. The components of the sample appear, after staining, as bands (as shown in Figure 3.3). The region of the gel in which the components of one sample can move is called a lane; thus this gel has seven lanes.

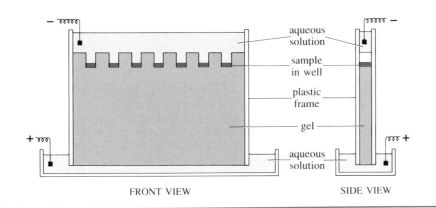

FRONT VIEW   SIDE VIEW

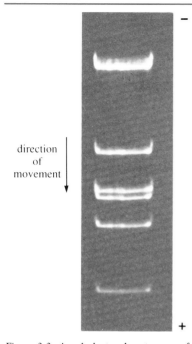

direction
of
movement

Figure 3.3 A gel electrophoretogram of six different molecules of DNA run in a single lane. The direction of movement is from top to bottom.

which are negatively charged, penetrate and move through the agarose. A gel is a complex network of molecules, and migrating macromolecules must squeeze through narrow, tortuous passages. The result is that smaller molecules pass through more easily and thus the migration rate increases as the M_r decreases. The distance moved, D, depends logarithmically on M_r and obeys Equation 3.1:

$$D = a - b \log M_r \qquad (3.1)$$

in which a and b are constants that depend on the solution, the gel concentration and the temperature. Figure 3.3 shows the result of electrophoresis of a mixture of DNA molecules.

◇ Examine Figure 3.3. Which band contains the smallest DNA molecules—the one at the top of the gel or the one at the bottom?

◈ Smaller molecules move through the gel more readily than larger ones, so the former therefore travel faster and further. The band at the bottom of the gel therefore contains the smallest of the DNA molecules visible on the gel, while the top band contains the largest molecules, since this has moved slowest.

Gel electrophoresis of proteins is in principle carried out in the same manner as that just described for DNA. The support material is usually polyacrylamide, hence the technique is commonly called PAGE (*polyacrylamide gel electrophoresis*). However, proteins can be *either* positively *or* negatively charged, and a sample containing several different proteins must be placed in a centrally located well in order that migration can occur in both directions. (This separates the molecules on the basis of charge, as shown in Figure 3.1.) However, proteins can be treated in such a way that they all carry the same charge.

◇ What feature of the molecules could then be used to separate them?

◈ The protein molecules could then be separated by electrophoresis on size difference alone.

In order to do this, the protein solution is treated with a detergent called sodium dodecyl sulphate (SDS). Most proteins bind the same mass of SDS per unit mass of protein, thereby giving each protein molecule the same charge per unit mass. Furthermore, when SDS is bound all proteins have nearly the same shape. This procedure is then termed SDS–PAGE. The net effect is that, as in the case of DNA separation, the migration rate increases as M_r decreases according to Equation 3.1 above. (A significant feature of SDS–PAGE is that, as it uses a detergent, it can be applied to membrane proteins, which are otherwise insoluble in water.)

As well as a method for separating macromolecules, gel electrophoresis is often used to check on the purity of preparations.

◇ How might this be achieved?

◈ The preparation is subjected to SDS–PAGE and the gel pattern examined. If there are several bands within a lane then the preparation is still a mixture of different molecules. If there is a single band, this suggests that the preparation is pure i.e. contains just one macromolecule.

◇ There is a problem with the latter interpretation. Can you suggest what this is? Remember that the proteins are treated with SDS to remove the effect of charge.

◆ SDS–PAGE can be used to separate on the basis of size. Two *different* molecules of the *same* size would appear as a single band.

◇ If, for example, a band is thought to contain two different proteins, how might these be separated?

◆ Electrophoresis could be used as shown in Figure 3.1. Here molecules are separated on the basis of charge. Although the two proteins have the same M_r, they *might* differ in terms of charge, and so could be separated.

The second technique for separating macromolecules on the basis of size that we shall consider is known as **gel filtration**. This involves the use of a 'column', a hollow glass or plastic cylinder packed with a very watery paste (tapioca-like) made from a powder and a liquid. This gel filtration column is packed with particles, each consisting of a three-dimensional water-filled network of polysaccharide fibres. This acts as a 'molecular sieve' since it is made by carefully controlled linking of polysaccharide chains—the more cross-linking, the smaller the pore size. When a mixture of macromolecules passes through a column, the largest molecules will be too big to penetrate between the closely packed fibres in the gel particles, but will pass through the larger spaces between the particles (Figure 3.4). They will be washed through quite quickly (washing through the column is termed *elution*). Addition of a volume of liquid equal to the volume between the particles will be sufficient to wash (elute) these molecules right down the column and they can be collected. Rather smaller molecules will be able to enter some of the larger pores and so will take longer to pass through, while molecules below a certain size can enter even the smallest pores and will take longer still to pass through. Different gels can be used with different pore sizes, so that passage down two or three different columns can improve separation.

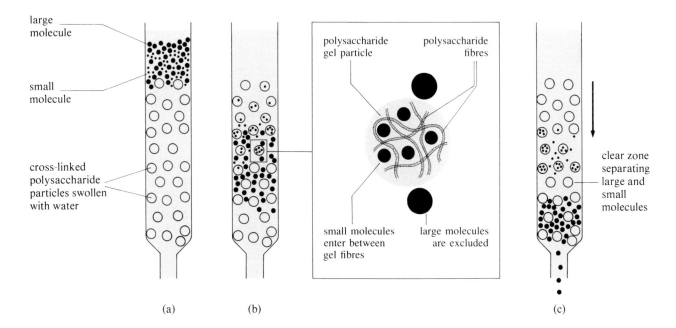

Figure 3.4a–c Gel filtration: (a) a solution containing large and small molecules is applied to the gel column; (b) during elution, small molecules take a longer route than large molecules, passing through gel particles (see the enlargement showing the interaction with gel particles); (c) large molecules emerge first.

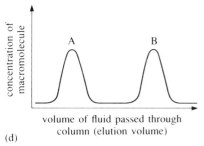

(d)

Figure 3.4d Elution profile after gel filtration.

◇ Figure 3.4d shows an elution profile, that is, the pattern of macromolecules in relation to the volume of fluid passed through the column. Which of the peaks represents the largest molecules?

◆ Peak A. These molecules were too large to enter the pores of the gel, and so were washed through the column with a relatively small volume of fluid. The smaller molecules of peak B entered the pores and so needed more fluid to wash them out of the column.

3.2.4 Separation according to specificity

The classical techniques for distinguishing between, and hence separating, macromolecules are based on crude broad differences in solubility, charge and size. Any one of these factors is unlikely to be unique to one particular macromolecule. Many enzymes, for example, are likely to be of roughly the same size. Many will share the same overall net charge, and so on. Purification of one particular type of molecule will require a combination of these techniques, as any particular molecule is quite likely to be unique in its combination of characteristics—solubility, charge and size—although not in any one of them.

A different approach to separation is based on the fact that has already been noted, namely that many macromolecules have specific binding sites for other molecules, known as a ligands. In the case of an enzyme, for example, the ligand is its substrate and the significant feature here is that the enzyme exhibits *specificity for its substrate*. In a mixture of enzymes extracted from a cell, it is quite likely that a given substrate will bind to only one (or at least to very few) enzymes. This property can be exploited in **affinity chromatography**—a technique used to separate a required macromolecule from a mixture that is based on the binding of a specific ligand to a macromolecule.

What is done, in essence, is that molecules of the ligand (substrate analogue in the case of an enzyme) of the macromolecule that one wishes to purify are chemically attached to some inert porous material (e.g. to a polysaccharide), which is then formed into a gel in a column as for gel filtration. In principle it only remains to pour the mixture containing the desired macromolecule down the column, whereupon all the components pass through unhindered save for the required macromolecule, which will bind to the ligand fixed to the gel (Figure 3.5). Later the macromolecule can be detached from the gel and isolated free of all other substances.

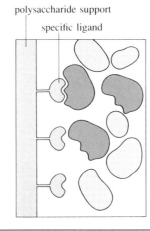
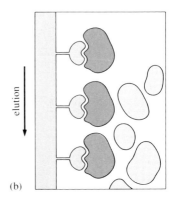

Figure 3.5 Affinity chromatography: (a) desired molecules (dark blue) bind to specific ligands on the gel; (b) un-adsorbed contaminants (pale blue) wash through the resin and are removed at the bottom of the column. Later the desired molecules are washed off the column and collected.

(a)   (b)

In principle such a simple and precise technique should eliminate the need for all the various possible combinations of the more hit-and-miss classical purification procedures outlined in Sections 3.2.1 to 3.2.3. One should merely, say, break open some cells that contain the desired macromolecule, remove the large cell debris (nuclei, cell membrane fragments, etc.) and pour the remaining solution down the appropriate affinity chromatography column. In practice, however, there are several drawbacks, only a selection of which will be considered here. Firstly, chemical attachment of the ligand to the gel may not be achieved easily. Secondly, although the ligand may be very specific for a given macromolecule, there may still be non-specific adsorption to the gel material of other components in the mixture. Thirdly, each macromolecule may be highly specific in terms of the ligands it will bind, but a given ligand may well bind to other macromolecules. To take one example: the enzyme hexokinase specifically has glucose as its substrate. However, a number of other enzymes can also act on glucose, glucose isomerase for one. Thus passing a crude mixture through a column to which a glucose analogue is attached might 'fish out' both hexokinase and glucose isomerase.

In practice, then, affinity chromatography is not quite so straightforward as it is in principle, but by choosing appropriate conditions it is possible to overcome the above-mentioned difficulties, often by using this technique in combination with the more classical ones. Given a very specific relationship between ligand and macromolecule, this can be a very powerful isolation procedure and is one that is gaining increasing use.

3.2.5 Characterizing the purified macromolecule

Having obtained a pure preparation of the desired macromolecule, we could now go on to determine its various features by using a whole host of techniques. Quite a lot of information on the physical and chemical properties of a macromolecule can be deduced from its behaviour during purification. The way a molecule behaves during electrophoresis, for example, will be strongly influenced by its properties. Size and shape are the most important factors here—for example, other factors being equal (such as charge after SDS treatment), you would expect a large molecule to move more slowly during electrophoresis than a small one. Size, or more precisely, relative molecular mass (M_r), may be deduced by comparing the behaviour of the macromolecule with the behaviour of standard molecules of known M_r.

Let us consider the determination of M_r for a recently isolated protein. The most convenient and a fairly accurate procedure for determining relative molecular mass of proteins is SDS–PAGE. It is performed by adding a series of proteins for which M_r is known to a lane adjacent to the lane containing the sample molecule of unknown M_r. The proteins are then subjected to SDS–PAGE and the distances travelled by the various proteins (relative to the solvent front) measured. This gives relative electrophoretic mobility, R_f which is plotted against M_r as shown in Figure 3.6.

◇ Study Figure 3.6 and determine the approximate M_r for the unknown protein X.

◆ M_r for X = 3×10^4 = 30 000.

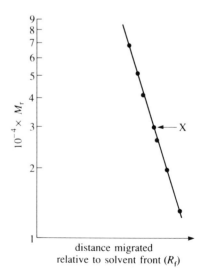

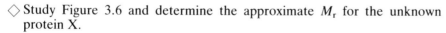

distance migrated
relative to solvent front (R_f)

Figure 3.6 Determination of M_r for an unknown protein X by comparing its mobility with that of six standard proteins of known M_r.

Summary of Sections 3.1 and 3.2

1 The contents of a cell are subdivided by membranes into various compartments, each with its characteristic macromolecular components. The main
classes of macromolecules are the proteins, nucleic acids and polysaccharides;
the lipids are usually grouped with them, although not technically macromolecules.

2 Macromolecules can be isolated and purified on the basis of differences in
solubility, charge, size and/or specificity. Electrophoresis is a widely used
technique for separation on the basis of charge or size. Gel filtration separates
on size differences. Separation using affinity chromatography is based on
differences in specificity.

Question 1 (*Objectives 3.1 and 3.2*)

(a) Suppose you have a mixture of a variety of proteins. What methods might
be used to isolate and purify one particular protein from the mixture?

(b) In addition to the purification techniques, what would you require to
check that your isolation has been successful?

Question 2 (*Objectives 3.1 and 3.2*) Figure 3.7 shows an elution profile from a
gel filtration experiment which attempted to isolate a particular protein,
avidin, from a mixture.

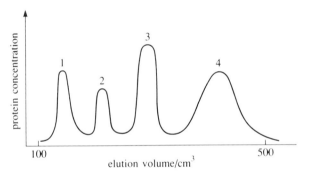

Figure 3.7 Elution profile from a gel
filtration experiment to isolate the protein avidin.

(a) How many proteins do you think the mixture contains?

(b) An antibody to avidin reacts with protein from peak 3 of the profile; what
do you conclude from this?

(c) The M_r of avidin is 68 000; is this larger or smaller than the protein of
peak 1?

(d) How might you check to see if peak 3 contains a single protein or a
mixture?

Question 3 (*Objectives 3.1 and 3.2*) You are trying to isolate a particular
DNA molecule from a suspension of bacterial cells. When your preparation is
subjected to gel electrophoresis you find that there are three bands. What
might you deduce from this?

3.3 GENERAL PRINCIPLES OF MACROMOLECULAR STRUCTURE

Once a macromolecule has been isolated and purified, its particular features can be studied. We now turn to consider the detailed description of macromolecules and will show how their biological properties and functions can be related to their precise structures. As you will see, the four groups of molecules to be considered—the proteins, nucleic acids, polysaccharides and lipids (the latter, though, are not strictly macromolecules as explained in Section 3.1)—present a wide diversity of structures, which in turn is reflected in a variety of biological roles. Fortunately there are some general principles which can be applied to macromolecular structure.

One of the most important is that a hierarchy can be recognized in the structure of most macromolecules. The four levels of hierarchy in a macromolecule are described as primary, secondary, tertiary and quaternary structures: the latter three are collectively referred to as higher-order structures. This structural hierarchy is demonstrated for proteins in Figure 3.8.

Figure 3.8 Structural hierarchy in a macromolecule. Here the example is a protein.

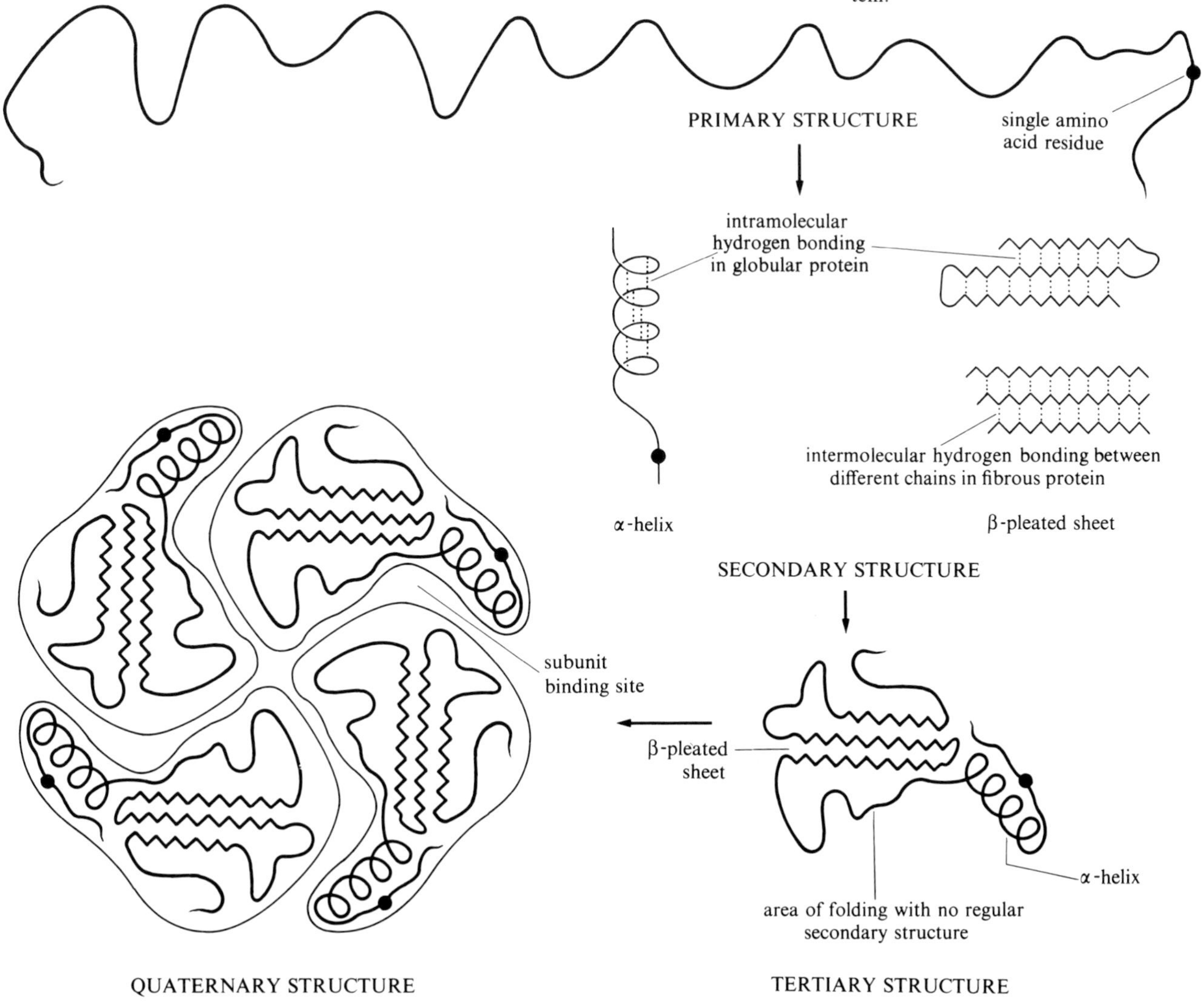

PRIMARY STRUCTURE

single amino acid residue

intramolecular hydrogen bonding in globular protein

intermolecular hydrogen bonding between different chains in fibrous protein

α-helix

β-pleated sheet

SECONDARY STRUCTURE

subunit binding site

β-pleated sheet

α-helix

area of folding with no regular secondary structure

QUATERNARY STRUCTURE

TERTIARY STRUCTURE

Each macromolecule consists of building blocks or **monomers**, different types of which go to form the major classes of macromolecule; so a protein has different monomers from a nucleic acid, and so on. The monomers are covalently joined together, so that the **primary structure** of a macromolecule describes the type and linear sequence of the monomers in the chain. For our protein example in Figure 3.8, the monomers are amino acids and so the primary structure is the sequence of amino acids.

The chain can now be folded in various ways. In some macromolecules portions of the chain fold in a very regular manner, described as the **secondary structure**. A helix is a very common form of secondary structure found in proteins and nucleic acids, although other folding patterns can also occur. The regular secondary structures, together with interspersed irregular regions, can fold to give **tertiary structure**, which is a very precise three-dimensional shape for each particular macromolecule. The highest order, **quaternary structure**, is formed when several chains come together in a specific fashion; the combining chains can be either similar or dissimilar. You should note that many macromolecules consist of a single chain, and therefore do not have quaternary structure.

◇ In our protein example of Figure 3.8, how many separate parts of tertiary structure have associated to form the quaternary structure?

◆ Four separate chains have combined to form a complex quaternary structure. Note that, in this example, all four are identical.

Our second general principle of macromolecular structure is that different types of bonds are involved in maintaining the hierarchy of structure just described. The monomers of the primary structure are held together by strong covalent bonds. In contrast, the higher-order structures are held together principally by weak bonds, which are non-covalent interactions. Since these are easily broken and reformed, they give a certain amount of flexibility, which is vital to molecular function. Three kinds of weak bonds are important in maintaining macromolecular structure: hydrogen bonds, hydrophobic interactions and ionic bonds. We shall consider each of these in turn.

The hydrogen bond

Hydrogen bonds may be found wherever there is an opportunity for a hydrogen atom to be shared between two other atoms (which are electronegative). In macromolecules of biological importance, these atoms are either oxygen or nitrogen; Figure 3.9 shows the ways in which hydrogen bonds can occur. The bonds are strongest (i.e. the bond energy is greatest) if the three atoms that participate in the bond lie in a straight line. Hydrogen bonds are particularly important in stabilizing the regular secondary structures of proteins and nucleic acids, notably the helices.

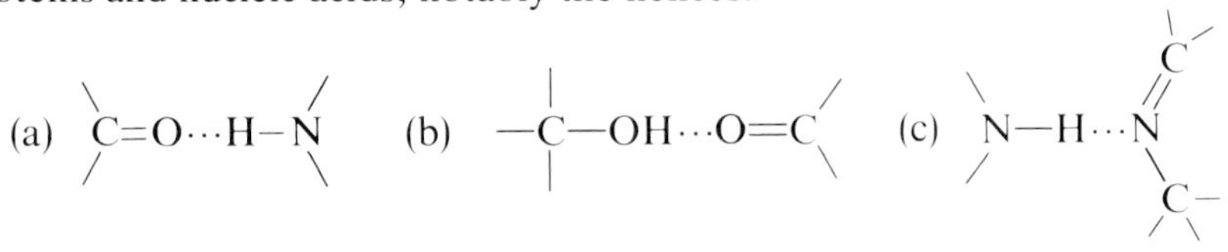

Figure 3.9 Hydrogen bonds (shown as rows of dots) found in biological systems. (a) A type found in proteins and nucleic acids. (b) A type found in proteins. (c) A type found in proteins and nucleic acids.

(a) $C=O \cdots H-N$ (b) $-C-OH \cdots O=C$ (c) $N-H \cdots N$

Hydrophobic interactions

A **hydrophobic interaction** occurs between two molecules that are somewhat insoluble in water (hydrophobic means 'water-hating'). The phenomenon is simple in that two molecules (which may be different) that are poorly soluble in water tend to associate. Such non-polar residues coalesce, squeezing out the water molecules trapped between them. This is shown, for example, by the lipid molecules in an oil droplet on the surface of an aqueous solution.

Many component chemical groups in nucleic acids and proteins have the hydrophobic property just described. For example, the bases of nucleic acids have flat (planar) organic rings carrying localized weak charges. The large, poorly soluble organic ring portions tend to cluster when placed in an aqueous environment. The most efficient kind of clustering is one in which the faces of the rings are close together, an arrangement known as base-stacking (Figure 3.10). Note that since the bases are adjacent in the chain, stacking gives some rigidity to a single strand. As you will see later, such stacking is important in maintaining the structure of DNA. Many amino acids, which are the building blocks of proteins, also have side chains (sometimes called R groups) with ring structures which are similarly relatively insoluble in water. Such side chains can form hydrophobic interactions; again these minimize contact with water. Since the amino acids are not necessarily adjacent in the primary structure, hydrophobic interactions tend to bring together distant hydrophobic parts of the protein chain. Similar hydrophobic interactions occur in polysaccharides, but it is in the lipids that hydrophobic interactions really come into their own.

The ionic bond

An ionic bond is the result of attraction between unlike charges. A few amino acid side chains have ionized, negatively charged, carboxyl groups, whilst some have positively charged groups. Ionic bonds can form between such groups, which tend to attract one another, viz:

$$-NH_3^+ \ldots\ldots\; ^-OOC-$$

This tends to bring together distant parts of the chain. (Ionic interactions can also be mutually repulsive, i.e. two like charges repel one another.) Ionic bonds are destroyed by extremes of pH, which can change the charge of the groups, and by high concentrations of salt, the ions of which shield the charged groups from one another.

The effect of the weak bonds (non-covalent interactions) described above is to constrain a linear chain to fold in such a way that different regions, which may be quite distant if the chain were straight, are brought together. Examples of the weak bonds are illustrated in a hypothetical molecule shown in Figure 3.11; as you can see they result in quite a high degree of folding and compaction of the linear chain.

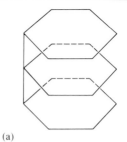

(a)

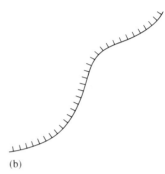

(b)

Figure 3.10 Hydrophobic interactions between ring structures result in stacking. (a) Three stacked rings. (b) Stacking in DNA results in an extended chain.

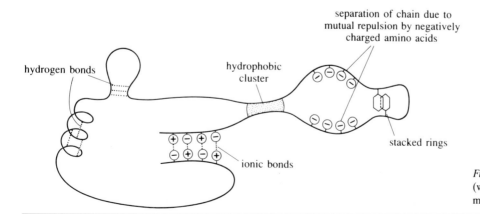

Figure 3.11 Non-covalent interactions (weak bonds) in a hypothetical macromolecule.

In the following sections we will discuss the precise structures of the different types of macromolecules that result as a consequence of these interactions; we start by looking at proteins.

3.4 PROTEINS

There are many thousands of different proteins, each of which has a unique three-dimensional structure. (As you will see later, this contrasts with DNA which has a universal three-dimensional structure, the double helix). This diversity of forms makes the study of proteins very complex, but it is this very diversity that enables these molecules to carry out the thousands of different processes required by a cell and makes proteins fascinating objects of study.

Here we shall see how the general principles of hierarchy of structure within a macromolecule, maintained by a mixture of covalent and weak bonds, apply to proteins.

3.4.1 Primary structure

A **protein** is a polymer of amino acids. Each amino acid can be thought of as a single carbon atom (the α-carbon) to which is attached a carboxyl group, an amino group and a side chain (R group) (Figure 3.12*). Each different amino acid has a different side chain: a short carbon chain or involving one or more rings. The simplest side chain is that of glycine: a hydrogen atom. The commonly occurring amino acids, i.e. those found in proteins, are shown in Table 3.2.

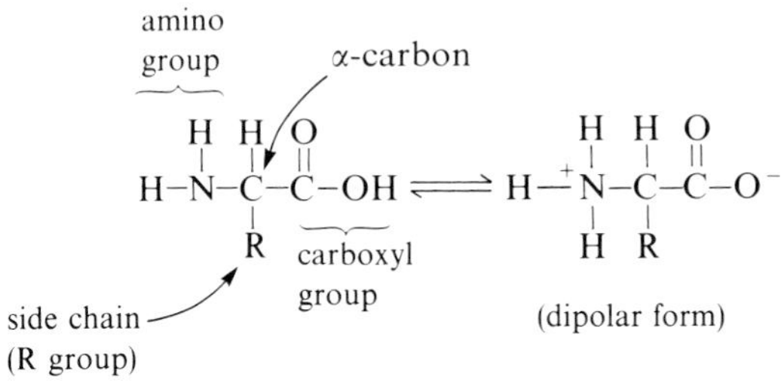

Figure 3.12 The general structure of an amino acid. The $-NH_2$ and $-COOH$ groups are involved in connecting amino acids one to another. The (predominant) dipolar structure is also shown.

The chemical properties of a protein can best be understood if we consider the nature of the amino acid side chains that are available for reaction with ions or compounds in their vicinity. As you can see from Table 3.2 there are a number of different side chains on amino acids.

Firstly, consider the carboxyl group ($-COOH$) of the side chains of aspartic and glutamic acids. This can dissociate (ionize) as follows:

$$-COOH \rightleftharpoons -COO^- + H^+$$

The dissociated form is therefore charged. The degree of ionization depends on pH; at pH 7 it will be almost fully ionized.

◇ How will the addition of H^+ (i.e. a decrease in pH) affect the degree of ionization and therefore the amount or number of charges?

---

*Note that amino acids exist predominantly in the dipolar form, $H_3\overset{+}{N} -CH(R)-COO^-$, but for simplicity are sometimes given as the neutral form, $NH_2-CH(R)-COOH$.

Table 3.2 Naturally occurring amino acids. Amino acids have the formula shown where R may be any of the 19 alternatives shown in this table. The charged side chain groups are shaded blue.

$$R$$
$$|$$
$$H_2N-CH-COOH$$

general formula for amino acids

Amino acid	Symbol	Side chain	Amino acid	Symbol	Side chain
alanine	Ala	CH_3	lysine	Lys	$\overset{+}{N}H_3$ — $(CH_2)_4$
arginine	Arg	H_2N $\overset{+}{N}H_2$ / C — NH — $(CH_2)_3$	methionine	Met	CH_3 — S — $(CH_2)_2$
asparagine	Asn	$CONH_2$ — CH_2	phenyl-alanine	Phe	(benzene ring) — CH_2
aspartic acid	Asp	COO^- — CH_2	proline*	Pro	*
cysteine	Cys	SH — CH_2	serine	Ser	OH — CH_2
glutamic acid	Glu	COO^- — $(CH_2)_2$	threonine	Thr	OH — $CH-CH_3$
glutamine	Gln	$CONH_2$ — $(CH_2)_2$	tryptophan	Trp	(indole ring) HN C=C H CH_2
glycine	Gly	H			
histidine	His	H — C / HN $\overset{+}{N}H$ HC=C CH_2	tyrosine	Tyr	OH (phenol ring) CH_2
isoleucine	Ile	CH_3 CH_2CH_3 CH	valine	Val	CH_3 CH_3 CH
leucine	Leu	CH_3 CH_3 CH CH_2			

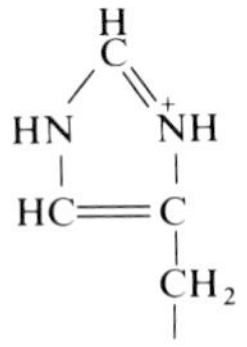

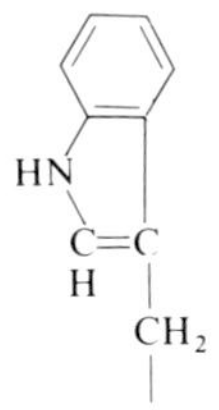

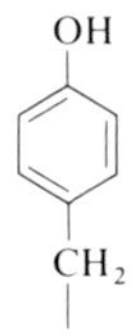

* Proline has no free —NH_2 group. Its formula is

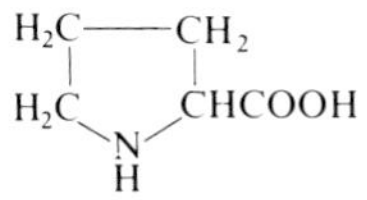

$$H_2C-CH_2$$
$$H_2C\quad CHCOOH$$
$$\underset{H}{N}$$

It is more properly called an *i*mino acid; it can still however form peptide links via its imino (>NH) group.

◆Lowering the pH reduces the degree of ionization (by shifting the equilibrium of the above reaction to the left, so that $-COOH$ predominates over $-COO^-$). If the amino acids are part of a protein, then this means that there will be fewer negative charges on the protein.

◇Conversely, what effect would adding alkali such as NaOH have?

◆This would tend to increase the overall negative charge on the protein.

In a similar manner, the amino group ($-NH_2$) of the side chain of lysine can be charged, but in this case it is the *un*dissociated group that is charged:

$$-NH_3^+ \rightleftharpoons -NH_2 + H^+$$

Again, the degree of ionization is pH-dependent.

Some amino acids, therefore, can carry a net negative charge, whilst others are positively charged. Another group of amino acids does not have charged side chains, but instead has non-polar hydrophobic groups; these can participate in hydrophobic interactions, as we shall see later.

◇Select an example of such an amino acid from Table 3.2.

◆Ile, Leu, Met, Phe and Val belong to this group.

A protein is therefore a charged molecule, and the overall charge depends on the relative number of the different sorts of amino acids, and also on the pH.

A protein is formed when amino acids are joined together. The amino group ($-NH_2$) of one amino acid reacts with the carboxyl group ($-COOH$) of another, to form a chemical bond known as a **peptide bond** (see Figure 3.13); a peptide group ($-NH-CO-$) is thereby produced. A sequence of amino acids joined together in this way forms a linear **polypeptide** chain. When the

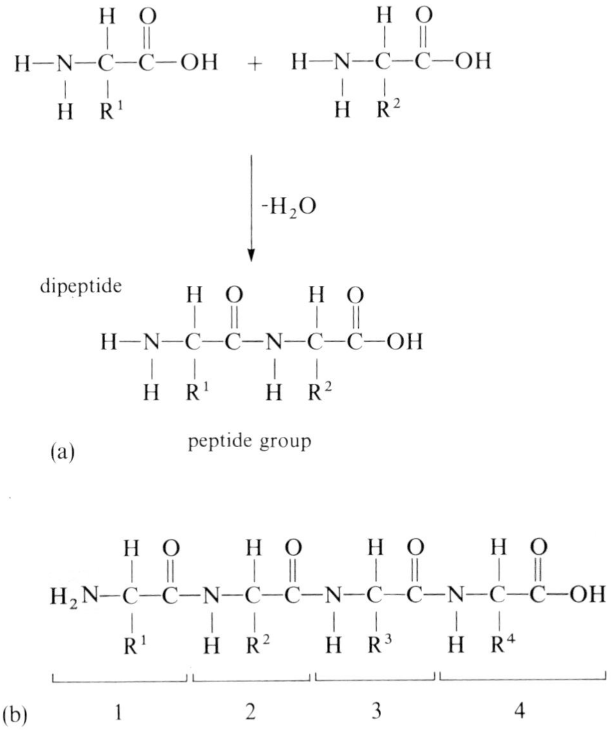

Figure 3.13 Formation of peptide bonds. (a) A dipeptide is formed when two amino acids are joined together via a peptide bond. (b) A tetrapeptide showing the alternation of α-carbon atoms (unshaded) and peptide groups (shaded blue). The four amino acids are numbered below the chain and the amino terminus and carboxyl terminus of the tetrapeptide are shaded grey.

number of amino acids in the chain joined by peptide bonds exceeds about 20, then the polypeptide is also known as a protein. A few amino acids joined together is termed a peptide. The distinction between peptide, polypeptide and protein is often arbitrary and the terms polypeptide and protein are used synonymously. A further complication is that a protein can consist of several *separate* polypeptide chains.

As you can see from Figure 3.13 the peptide molecule has two different ends: the one with the free $-NH_2$ group is called the **amino terminus**, while the other end with a free $-COOH$ group is known as the **carboxyl terminus**. The names of the ends are also abbreviated and called, respectively, the N-terminus and the C-terminus. By convention, the sequence of amino acids in a protein is given from the N-terminus to the C-terminus, reading from left to right.

As you saw from Table 3.2 there are 20 different commonly occurring amino acids that are found in proteins. Typical proteins have relative molecular masses (M_r) in the range 15 000 to 70 000. The average amino acid has an M_r of 110, so that typically proteins contain 135 to 635 amino acids. *Variations in the number and sequences of the amino acids in these chains give rise to a wide diversity of proteins.* The linear sequence of amino acids in a given protein is its primary structure.

3.4.2 Secondary structure

The length of a typical protein when fully extended, is between 100 and 500 nm (1 nm = 10^{-9} m). A few long fibrous proteins are in this range, such as myosin (a key protein involved in the contraction of muscle), with a length of 160 nm, and collagen, 300 nm long.

◇ Most proteins in their natural state have their longest dimensions in the range 4 to 8 nm and yet consist of 20 or more amino acids; what can you conclude from this?

◆ The primary structure of the protein must be folded in some way.

So a protein is much more than a mere linear sequence of amino acids, since this chain can be folded to give higher-order structures, but an important point to note is that the primary structure, i.e. the amino acid sequence, determines the folding. We shall return to this topic a little later. For the moment let us consider what sorts of folding patterns can be achieved. Most proteins contain at least some regions where the polypeptide chain is folded in a regular manner, and these are stabilized by interactions involving the backbone atoms rather than the side chains mentioned earlier. The weak bonds holding them together are mainly hydrogen bonds between non-adjacent $>C=O$ and $>NH$ groups of the peptide-bonded backbone.

Proteins adopt various secondary structures, divided into those in which the chain is coiled into a helix—the α-helix—and those in which the chain remains fairly extended but with parts of it aligned side by side to form sheets—β-structures.

The α-helix

In an **α-helix** the polypeptide chain follows a helical path that is stabilized by hydrogen bonding between peptide groups. Each peptide group is hydrogen-bonded to two other peptide groups, one three units (amino acid residues) ahead and one three units behind in the chain. As you can see from Figure

3.14, the α-helix is a relatively regular structure, a cylinder if you like, which is quite stable. It has a diameter of 0.23 nm and contains 3.6 amino acids per turn.

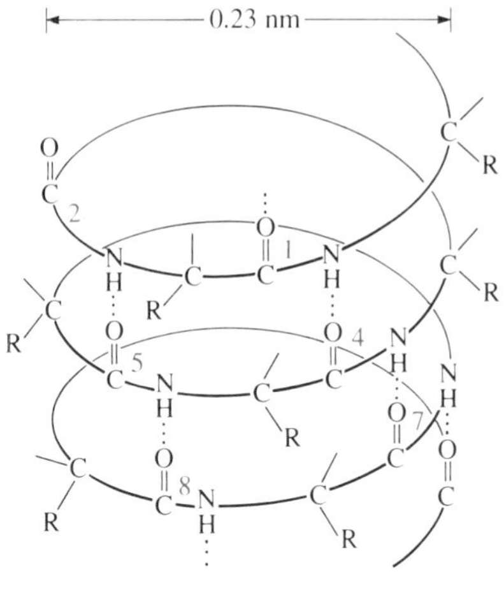

Figure 3.14 Some features of an α-helix, showing how the structure is stabilized by hydrogen bonds (represented by dots). Peptide group 4 makes hydrogen bonds to two other groups: to group 1, three units behind in the chain, and to group 7, three units ahead in the chain. (For the sake of clarity, hydrogen atoms not involved in hydrogen bonds and some of the peptide groups are not shown.)

◇ What important feature concerning the side chains of the amino acids in a single α-helix is discernible from Figure 3.14?

◆ All the side chains stick out from the helix; this means that they do not participate in its formation or contribute to its stability.

In the absence of interactions other than the hydrogen bonding just described, the α-helix is the preferred (i.e. the most stable) form of a polypeptide chain; in this structure, all the amino acid residues are in identical orientation and each one forms the same hydrogen bonds as every other. Polyglycine, for example, which is a synthetically produced polypeptide composed of only one type of amino acid (glycine) is always an α-helix.

If the amino acid composition of a natural protein is such that the helical structure is long, the protein will be somewhat rigid and fibrous (although not all rigid fibrous proteins are α-helical). This feature is common in many structural proteins, for example, keratin of hair and horn, whose folding pattern is entirely α-helical (discussed later in Section 3.4.4). In contrast, most compact (globular) proteins contain only short stretches of α-helix.

β-structures

A second commonly encountered hydrogen-bonded configuration is the **β-structure**. In this alternative form of secondary structure, the molecule is almost completely extended and hydrogen bonds form between peptide groups of segments of the polypeptide chains lying adjacent to one another (see Figure 3.15a). The side chains lie alternatively above and below the main chain.

Two segments of the same polypeptide chain (or two separate polypeptide chains) can form two types of β-structure, which depend on the relative orientations of the segments: if both segments are aligned in the N-terminal to C-terminal direction, or in the C-terminal to N-terminal direction, the β-structure is said to be parallel; if, however, one segment is N-terminal to C-terminal while the other is C-terminal to N-terminal, the β-structure is antiparallel (Figure 3.15b). When many polypeptides interact in an anti-parallel manner, a pleated structure results called the β-pleated sheet. These sheets can be stacked and held together in large arrays by hydrophobic interactions. Most globular proteins contain only small areas of β-structure.

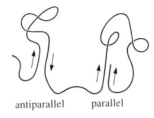

(a)

(b)

Figure 3.15 β-structures. (a) Two regions of extended chains interact via hydrogen bonds (dots) in an antiparallel fashion. The side chains (R groups) are alternately up and down. (b) Antiparallel and parallel β-structures in a single molecule. The arrows in each case point in the N-terminal to C-terminal direction.

In contrast, the fibrous protein fibroin, which is the major component of silk, adopts the β-structure conformation along most of its length. In silk, the structure is stabilized not only by hydrogen bonds but by hydrophobic interactions of the side chains, which project alternately on either side of the β-pleated sheet. The amino acid sequence of silk is predominantly -Gly-Ser-Gly-Ala-Gly-Ala-, repeated over and over again. All the glycine side chains (−H) project on one side of the sheet, and all the serine (−CH$_2$OH) and alanine (−CH$_3$) side chains on the other, allowing the sheets to be packed close together, one on top of the other.

Silk polypeptide chains are almost fully extended and as a result the fibre or sheet has little capacity for stretching, although the sheets themselves—held together by weak bonds—can bend, giving a degree of flexibility. This is a clear example of the primary structure determining the secondary structure and hence the final features of the material.

The one protein building block you would not expect to find in either of the regular secondary structures that we have discussed is proline.

◇ Look back at Table 3.2. What is the unusual feature of proline compared with the other amino acids shown?

◆ Proline has an >NH in place of −NH$_2$ (and is therefore strictly speaking an *imino* rather than an amino acid). Absence of the extra hydrogen atom means that a proline residue cannot participate in hydrogen bond formation with other residues further along the chain.

In addition, the rigid ring structure of proline forces the polypeptide chain to kink; it tends, for example, not to fit into an α-helix and it is therefore described as a *helix breaker residue*. Proline is often found in globular proteins at the end of the regular sections, where the polypeptide chain bends back on itself.

3.4.3 Tertiary structure

Few proteins are pure α-helix or β-structures: usually proteins contain small regions with these regular structures. Since these structures are rigid, a protein in which most of the chain has one of these forms is usually long and thin and is called a **fibrous protein**. In contrast many proteins are much more compact and are called **globular proteins**, in which α-helices and β-structures are short and interspersed with irregularly folded areas.

Fibrous proteins typically have structural roles, for example collagen, the protein of tendon, cartilage and bone. In contrast, the catalytic and regulatory functions of cells are performed by the globular proteins. Of these, the enzymes have been the most extensively studied, as you will learn in the next chapter.

Globular proteins are compact molecules roughly spherical or ellipsoidal in overall shape. However, the molecule is extensively bent and folded. Usually, the stiffer α-helices alternate with very flexible irregularly folded regions which permit bending of the chain without excessive mechanical strain. Numerous segments of the chain form short parallel or antiparallel β-structures (see Figure 3.15b): these are also responsible in part for the folding of the backbone. This extensive folding of the backbone gives rise to the tertiary structure of the protein, which is therefore a mixture of secondary structure and irregularly folded regions.

A distinction can be made between secondary and tertiary structure, namely that the former results from hydrogen bonding between peptide groups, whereas tertiary structure is formed from secondary structures and several different side chain interactions. The most important interactions responsible for tertiary structure are as follows:

☐ Ionic bonds between oppositely charged groups on amino acids, such as positively charged lysine and negatively charged glutamic acid:

$$-NH_3^+ \cdots\cdots {}^-OOC-$$

☐ Hydrogen bonds between an hydroxyl group ($-OH$), e.g. in tyrosine, and an ionized carboxyl group of aspartic or glutamic acid:

$$-OH \cdots\cdots {}^-OOC-$$

☐ Hydrophobic interactions between the hydrocarbon side chains in phenylalanine, leucine, isoleucine and valine.

☐ Disulphide bridges (bonds) between cysteine residues; these are covalent bonds and are therefore more difficult to break than the other interactions listed above (see later).

The sum of all these side chain interactions results in a stable and biologically active tertiary structure for each protein. Figure 3.16 shows the side chain interactions in a hypothetical protein; this should be studied carefully, for it indicates the role of different features of a protein in determining the overall shape of the molecule.

Of the four types of interaction listed above, hydrophobic interactions are especially important. The hydrophobic ('water-hating') groups tend to cluster together in the interior of proteins away from the external aqueous environment. Here they interact through weak bonds to stabilize higher-order structure, as explained in Section 3.3.2. Indeed hydrophobic interactions are the most important stabilizing feature of protein tertiary structure.

◇ What general comment could be made about the role of hydrophobic interactions, ionic bonds, hydrogen-bonding and disulphide bridge formation in determining the tertiary structure of a protein?

◆ As you can see from Figure 3.16, each of these different types of interaction bring together amino acids which are distant from each other in the primary structure of the polypeptide chain, and so aid in making the overall molecule more compact or globular. (An additional point worth making here is that side chains with the *same* charge, i.e. positively charged *or* negatively charged, would have the opposite effect—these would keep the amino acids apart; see Figure 3.11.)

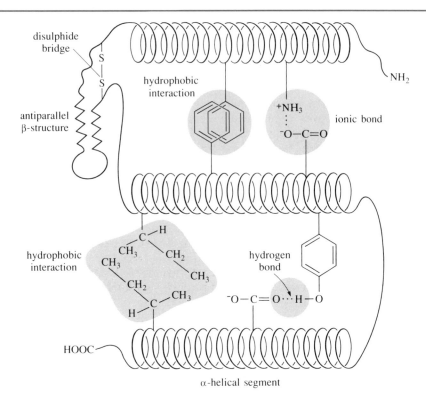

Figure 3.16 Tertiary structure of a hypothetical globular protein which is stabilized by several types of side chain interactions. (Note that in order to emphasize the interactions this figure is not drawn to scale. In reality the helical regions are much larger than the individual interactions shown.)

◇ What effect does the secondary structure (α-helices and β-structures) have on the tertiary structure of the hypothetical protein shown in Figure 3.16?

◆ These give a certain degree of rigidity to regions of the molecule, but note that these are interspersed by irregularly folded regions which provide some flexibility.

To illustrate some of the features of protein molecules described so far, we have chosen the enzyme bovine pancreatic ribonuclease (abbreviated from here on as RNAase), which is shown in Figures 3.17 to 3.19. The primary structure is represented by the sequence of the 124 amino acids shown in Figure 3.17. Notice that there are four disulphide bridges (links between cysteine residues) shown in this figure: we shall return to this feature a little later. The tertiary structure of the molecule is formed by folding of the structure shown in Figure 3.17. It is difficult to represent a three-dimensional object in two dimensions, so to simplify matters, the molecule is shown as a ribbon drawing in Figure 3.18. Here the primary structure of the polypeptide chain is shown as a ribbon and the β-structures are highlighted with blue shading.

◇ What other features of secondary structure in RNAase are visible in Figure 3.18?

◆ There are three α-helices, labelled A, B and C.

◇ The final representation of this molecule is shown in Figure 3.19. What overall impression do you gain from this illustration of the tertiary structure, in comparison to the primary structure?

◆ The tertiary structure is relatively compact, i.e. it is a globular protein.

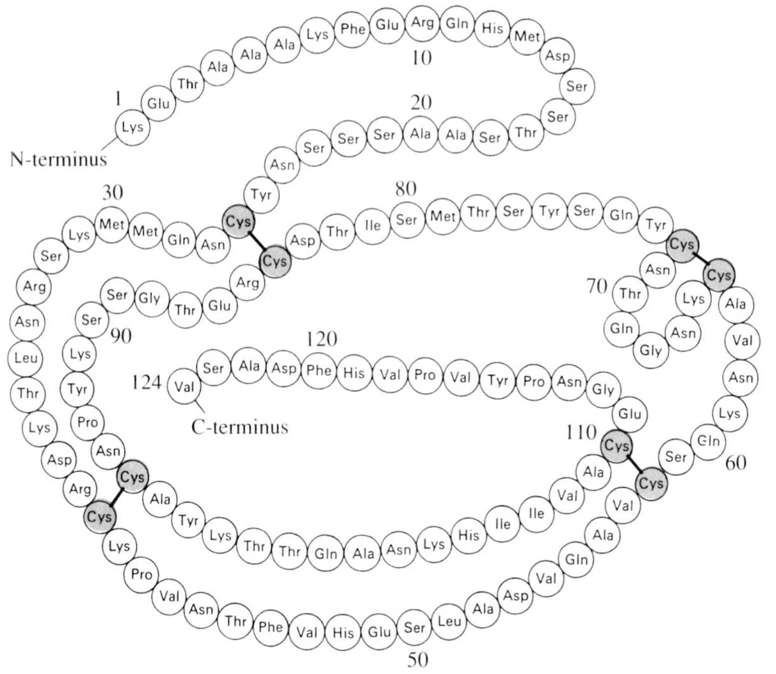

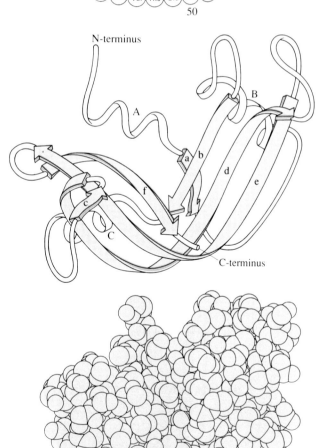

Figure 3.17 The primary structure of RNAase showing the sequence of 124 amino acids. (Eight cysteine residues interact to form four disulphide bridges, shown shaded blue).

Figure 3.18 The tertiary structure of RNAase shown as a 'ribbon' drawing. β-structures are shaded blue and labelled a–f; the arrow-heads indicate the N-terminus to C-terminus orientation of the chain.

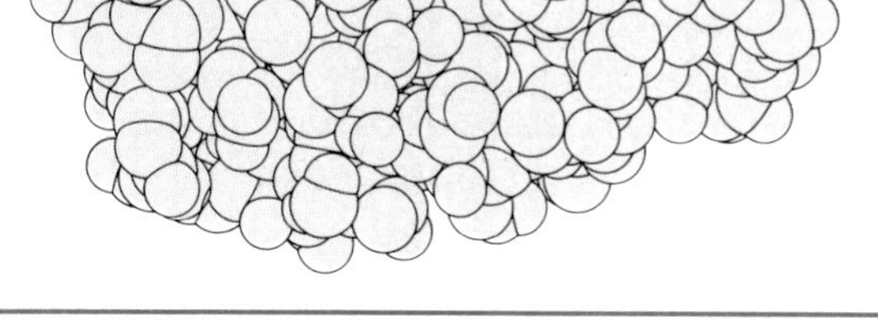

Figure 3.19 The tertiary structure of RNAase shown as a space-filling representation, i.e. the volume of space occupied by each atom is indicated by a sphere.

The three-dimensional shape of a protein is very precise and is intimately related to its function. Much effort therefore has been expended on determining these shapes, by using principally the technique of X-ray diffraction as explained in Box 3.1. As you will see in the next chapter, this compact three-dimensional tertiary structure of proteins is vitally important when considering how enzymes act as catalysts.

Box 3.1 *X-ray diffraction*

X-ray diffraction analysis has been a powerful tool in studying the three-dimensional structures of proteins (and nucleic acids). The basis of the technique is that the precise three-dimensional positions of most of the atoms in a macromolecule can be detected, analysis of which enables the shape of the protein to be determined. A basic requirement of the technique is a preparation in which molecules are regularly arranged, preferably in three dimensions (as in a *crystal*), since this can ultimately give information on the position of individual atoms. (Regularity in two dimensions, however, can still give some structural information, since here the molecules lie in parallel, as in a DNA *fibre*). Crystals of a protein are usually obtained by adding the macromolecule to a concentrated salt solution. Under such conditions solubility of the protein is reduced (as explained in Section 3.2.1) and crystals can form. This stage has often been a stumbling block in X-ray diffraction studies, as a number of macromolecules simply refuse to crystallize!

The principal components used in an X-ray diffraction study are shown in Figure 3.20a, namely a source of X-rays, the protein crystal and a detector. A narrow beam of X-rays is fired at the crystal. Part of the beam goes straight through the crystal; the rest is *scattered* (*diffracted*) in various directions. The diffracted beams are detected by X-ray film, the degree of blackening of the emulsion being proportional to the intensity of the diffracted X-ray beam.

Various physical principles underlie X-ray diffraction which we will not consider here. However, one important concept to note is that the degree of scattering of the X-rays is determined by the number of electrons in an atom. A carbon atom, for example, scatters six times as strongly as a hydrogen atom. This differential scattering is important in detecting the different atoms in the final X-ray diffraction patterns obtained.

The protein crystal is positioned in a precise orientation to the X-ray beam and a whole series of photographs is taken with the crystal at different orientations. Diffraction of X-rays is detected as a large series of spots on the film. The intensity of the spots varies and is proportional to the degree of regularity in the preparation; there is reinforcement by scattering from identically placed atoms leading to greater intensity of the spots. It is these intensities that form the basic data of X-ray diffraction analysis. Electron-density maps are then constructed from the spot intensities, which give the density of electrons at a large number of regularly spaced points in the crystal. This three-dimensional electron-density distribution is represented by a series of parallel sections stacked on top of each other. Each section is a transparent plastic sheet (or a layer in a computer image) on which the electron-density distribution is represented by contour lines (Figure 3.20b), somewhat analogous to geographical maps with height above sea-level represented by contour lines. Interpretation of

(b)

Figure 3.20 (a) The principal components of an X-ray diffraction experiment: a source of X-rays, a crystal and a detector. (b) Part of the electron-density map of the protein myoglobin showing the iron-containing haem group. The 'peak' of electron density in the centre, i.e. closely spaced contours, corresponds to the position of the iron atom.

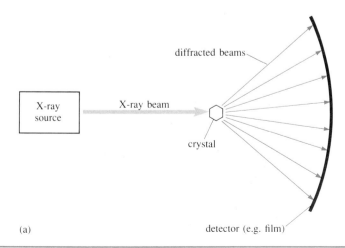

(a)

diffracted beams

X-ray source

X-ray beam

crystal

detector (e.g. film)

the electron-density maps requires complex mathematics and a powerful computer. The position of each atom located represents an average in the various X-ray diffraction pictures initially taken.

The technique of X-ray diffraction analysis is, as you can appreciate, highly involved, and the mathematics are quite staggering in their complexity, but can be handled rapidly by modern computers. All you need to appreciate is that this technique has enabled precise three-dimensional structures to be worked out for many proteins and nucleic acids. The tertiary structure of RNAase (Figures 3.18 and 3.19) and the double helix of DNA (Figure 3.38) are just two examples of the overall pictures of macromolecules produced using X-ray diffraction. A rather static view of macromolecular structure has emerged from such studies, giving the impression that a particular molecule has a *single*, rigid structure. In fact the real situation is far more complex, as you will see from Box 3.3.

RNAase was deliberately chosen as an example to show you tertiary structure because this protein was used in a classic study by Christian Anfinsen in the early 1960s to demonstrate an important concept, namely the relationship between primary structure and three-dimensional shape. RNAase consists of a single polypeptide chain with four disulphide bridges that play a prominent role in maintaining the tertiary structure. A **disulphide bridge** (or disulphide bond) is a covalent bond that cross-links either different parts of the same chain (as in the case of RNAase) or two different chains. It occurs when two cysteine (pronounced *sistayeen*) residues with their reactive sulphydryl groups (−SH) are oxidized (that is lose hydrogen) to form one cystine (*sisteen*) residue. The formation of a disulphide bridge is shown in Figure 3.21.

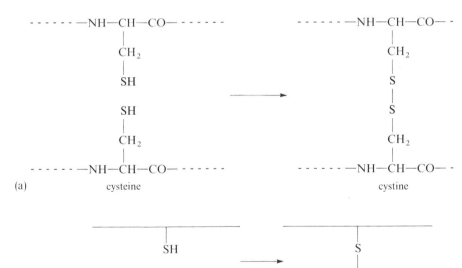

Figure 3.21 (a) Disulphide bridge formation between two cysteine residues forms a cystine, thus linking two adjacent lengths of polypeptide chain. (b) Simplified representation of disulphide bridge formation.

RNAase has eight cysteine residues which are connected into four specific disulphide-bridge pairs. Anfinsen treated this protein with two reagents: mercaptoethanol (a sulphur-containing alcohol, which breaks −S−S− bonds), and urea (which disrupts hydrogen bonds).

◇ Predict the effect of mercaptoethanol and urea on the structure of RNAase.

◈ Disulphide bridges cross-link parts of the polypeptide chain; hydrogen bonds stabilize secondary and tertiary structures. If these bonds are disrupted the protein structure should 'fall apart'.

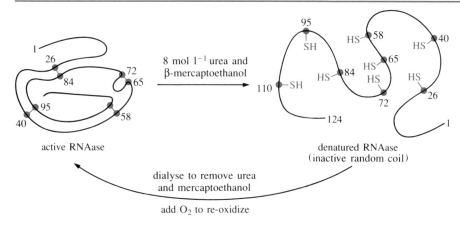

active RNAase

8 mol l^{-1} urea and
β-mercaptoethanol →

denatured RNAase
(inactive random coil)

dialyse to remove urea
and mercaptoethanol

add O$_2$ to re-oxidize

Figure 3.22 Reversible denaturation of RNAase.

The result was indeed loss of three-dimensional tertiary structure and a **random coil** was produced, that is a molecule with no defined structure (Figure 3.22).

The protein is said to have been *denatured* by this treatment. Not only was the tertiary structure lost, but the enzyme activity disappeared too. However, the small molecules of mercaptoethanol and urea can be separated from the relatively larger protein molecules by a process known as dialysis (see Box 3.2). When Anfinsen treated the denatured protein in this way (and oxygen was added to reform the disulphide bridges), 95% of the original biological activity of the enzyme was recovered. This process is termed **reversible denaturation**, for the obvious reason that the original structure and activity is regained after the procedure (Figure 3.22). (This contrasts with protein denaturation that occurs when, for example, an egg is boiled; the structure of the egg white proteins is lost during the cooking process and the denaturation is irreversible in practice, although—as for any chemical reaction—not in principle.)

This was the first direct evidence for one of the basic tenets of molecular biology, namely that *primary structure determines higher-order structure (and hence biological function)*. In other words, a given sequence of amino acids will always fold to give a particular three-dimensional shape, so that folding of the chain depends solely on the primary structure.

Box 3.2 Separation of molecules by dialysis

Proteins and other macromolecules can be separated from small molecules such as salts by the process termed dialysis. A dialysis bag is prepared from a cellulose membrane or similar material which has pores in it, small enough to permit small molecules and ions to pass through, but too small for macromolecules to pass through. The pore diameter is therefore important in determining the degree of separation of the molecules. The mixture of macromolecules and small molecules is added to the dialysis bag, which is sealed so that the solution cannot leak out. The bag is placed in a large volume of water or aqueous solution, as shown in Figure 3.23. In the case of Anfinsen's experiment, for example, the

dialysis bag would contain a solution of RNAase, mercaptoethanol and urea.

The experimental set-up is allowed to stand for several hours, often conveniently overnight, with the surrounding fluid being continuously stirred and replaced at intervals to avoid build-up of small molecules outside the bag. The small molecules, such as mercaptoethanol and urea, cross the membrane through the pores and thus pass out of the bag into the surrounding fluid, leaving the larger macromolecules in the bag, so separation of the molecules is achieved. The bag is later taken out of the bath of liquid, cut open and the macromolecular solution removed for further study.

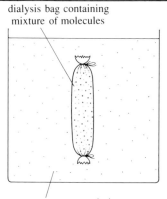

dialysis bag containing
mixture of molecules

water or aqueous solution

Figure 3.23 The experimental set-up for dialysis.

Anfinsen obtained quite a different result when denatured RNAase was re-oxidized while still in the presence of urea, which was then subsequently removed by dialysis (see Box 3.2). RNAase re-oxidized in this way had only 1% of the original enzyme activity. How can this be explained? It seems that under these conditions the wrong disulphide bridges were formed. There are 105 different ways of pairing the eight cysteines to form four disulphide bridges, and only one of these gives a biologically active structure. The 104 wrong pairings have been termed 'scrambled' RNAase (Figure 3.24). Anfinsen found that 'scrambled' RNAase was spontaneously (albeit slowly) converted to fully active RNAase when trace amounts of mercaptoethanol were added to the solution of the re-oxidized protein. This catalysed the rearrangement of the disulphide bridges until the fully active structure was regained, which took about ten hours. When you consider that none of the large number of other RNAase folding patterns that could be stabilized with mismatched pairs of cysteines show any biological activity, it is clear that the most stable folding pattern is indeed the one that is biologically active.

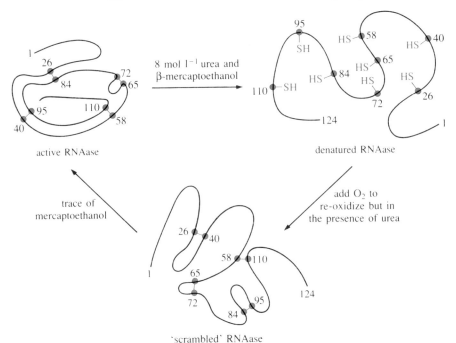

Figure 3.24 'Scrambled' RNAase (one of 104 possible ones is shown) is formed when denatured RNAase is re-oxidized in the presence of urea. (Note that any traces of mercaptoethanol are also oxidized.) Fully active RNAase is reformed from 'scrambled' RNAase over a period of hours in after the addition of a trace of mercaptoethanol.

◇ What feature of Anfinsen's experiment described above was important in detecting the change from tertiary structure to random coil formation?

◈ Loss of biological function. In the case of an enzyme this is relatively easy to measure. Note, however, that this is more difficult for structural proteins, for example, which have a less easily measured biological activity.

Not all disulphide bridge-containing proteins can be reversibly denatured in the same way as RNAase. In these cases, it appears that disulphide bridges hold the polypeptide chain in a comparatively unstable shape, which has been generated by folding of a **precursor protein**, an inactive form which differs from the final biologically active form of the protein; examples include insulin, a hormone, and chymotrypsin, a digestive enzyme. Both of these are produced as inactive precursor proteins—pro-insulin and chymotrypsinogen, respectively—which are converted to the biologically active forms by chopping off (cleaving) particular parts of the polypeptide chain. The pro-insulin–

insulin example is shown in Figure 3.25. Pro-insulin is the inactive precursor, which is activated in the animal cells producing it, by removal of the connecting peptide to give the active hormone insulin. This is released into the bloodstream, and circulates around the body to the target cells.

Pro-insulin, the precursor protein, can be subjected to reversible denaturation in the same way as RNAase. Disruption of the disulphide bridges in the presence of mercaptoethanol and urea results in the conversion of tertiary structure to random coil.

◇ What would you expect to happen if the mercaptoethanol and urea were removed by dialysis?

◆ The disulphide bridges would reform as the random coil is converted back to the tertiary structure.

In contrast, the active form of the protein, insulin, is *irreversibly* denatured by treatment with mercaptoethanol and urea. This is because cleavage of the precursor protein and removal of the connecting peptide results in loss of some amino acids which are necessary for the denatured protein (random coil) to refold into the active tertiary structure.

3.4.4 Quaternary structure

So far we have considered protein structure in terms of a single polypeptide chain and have shown how this can be variously arranged and folded to give higher-order structures. Indeed some proteins comprise only a single such polypeptide chain. However, many proteins comprise more than one polypeptide chain. Such proteins are said to have quaternary structure. There is a huge diversity of such structures, so we will consider just a few examples here; we will start by looking at globular proteins.

A polypeptide chain usually folds so that the hydrophobic (non-polar) side chains are internal, i.e. isolated from water. However, it is rarely possible for a polypeptide chain to fold in such a way that all non-polar groups are internal. So it is often the case that these non-polar amino acids form clusters or hydrophobic patches on the surface, giving minimal contact with water.

◇ Consider two identical polypeptide chains folded in such a way that hydrophobic amino acids on the surface form a hydrophobic patch. How might these two chains interact to bury the patches and separate them from water?

◆ The two patches might 'stick' together, as shown in Figure 3.26.

When two polypeptide chains interact in such a way the protein is said to be a dimer, which in the hypothetical case considered above, consists of two identical **subunits.** So, a protein consisting of more than one subunit, i.e. more than one separate polypeptide chain, has quaternary structure. In the hypothetical example we have just considered, the dimeric protein contains two identical subunits. This is in fact very common, with two, three, four and six subunits occurring most frequently. A multi-subunit (multimeric) protein may also contain non-identical subunits, and this too is quite common. Ionic interactions, hydrogen bonding and disulphide bridges, as well as interactions via hydrophobic patches, may also participate in holding subunits of a multimeric protein together. Generally the weak interactions readily reform when the separated subunits are mixed together, so quaternary structure, like tertiary, can be reversibly broken down and reformed.

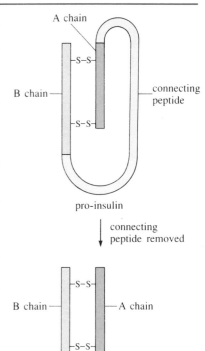

Figure 3.25 The conversion of the inactive precursor protein pro-insulin to the active hormone insulin by removal of the connecting peptide.

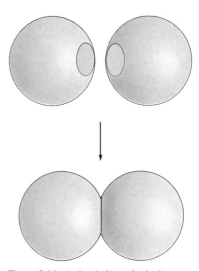

Figure 3.26 A simple hypothetical example of quaternary structure. Hydrophobic patches on two folded polypeptide chains interact to form a dimer.

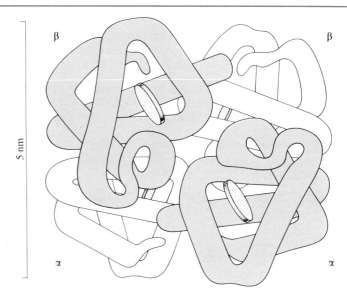

Figure 3.27 The quaternary structure of haemoglobin consists of two α subunits and two β subunits. The resultant protein is therefore tetrameric. (The grey discs are the oxygen-binding haem groups.)

Haemoglobin, the oxygen-carrying protein in blood, is an example of a multimeric protein, which shows how aggregation of subunits alters their properties as individuals. Haemoglobin consists of four polypeptide chains, two called α and two β (not to be confused with the α- and β-structures discussed earlier) (Figure 3.27). The sequences of the two types of chains are only slightly different, and so, as you might expect, their folding patterns are very similar. Each individual subunit of haemoglobin binds a single molecule of oxygen, but in tetrameric haemoglobin the subunits cooperate, and binding of oxygen to one subunit makes subsequent binding of oxygen to the other subunits much easier. This means that oxygen binding to tetrameric haemoglobin can be more precisely tailored to physiological demands than could binding to the single subunit.

Another situation where multimeric complexes sometimes occur is where several enzymes catalyse successive reactions in a chain. The product of one enzyme is the substrate for the next, and if the polypeptides are assembled into a multi-enzyme complex, intermediates can be passed directly between them, thus avoiding wastage of precious materials into the surrounding fluid. An example of this that you will meet later in Chapter 6, is the pyruvate dehydrogenase complex ($M_r = 4\,470\,000$) which consists of 42 subunits.

The immunoglobulins

The immunoglobulins are good examples illustrating the features of protein quaternary structure, because here there is a very clear relationship between structure and biological function. The immunoglobulins (or antibodies) are proteins of the immune system. They react with specific foreign molecules (antigens) and thereby render these inactive. There are various different classes of immunoglobulins, but the best understood one is immunoglobulin G (abbreviated to IgG).

◇ The IgG molecule contains disulphide bridges (Figure 3.28a). If these were disrupted in the same manner as that described in Section 3.4.3 for RNAase, how many subunits would be released or separated?

◆ Four (two H and two L).

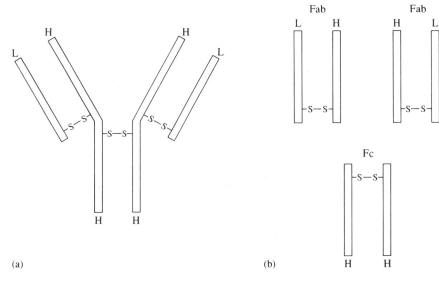

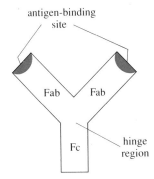

Figure 3.28 (a) The subunit structure of immunoglobulin G. There are two L subunits and two H subunits. Disulphide bridges join each L chain to an H chain and join the two H chains to one another. The polypeptides of the H chains can be cleaved to break specific peptide bonds; the products are shown in (b).

The M_r values of the subunits are about 25 000 and 50 000. The former lighter polypeptide is the L chain; the heavier one is the H chain. IgG is therefore a tetramer, containing two identical L chains and two identical H chains.

IgG can be cleaved experimentally in a different way, by using a technique that actually breaks some specific peptide bonds in the component polypeptide chains. This allows the disulphide bridges to remain intact, but this time causes each of the H chains to break, as shown in Figure 3.28b, thus producing three separate molecules. The two molecules that consist of an L chain and a fragment of the H chain are called Fab fragments. The third molecule, consisting of two equal segments of the H chains, is called the Fc fragment.

An important concept derived from this study of IgG is that different areas of the molecule have different functions. The two Fab fragments can bind to antigen (hence 'ab' from '*a*ntigen-*b*inding'), whereas the Fc portion is not involved in antigen-binding. Proteins which have different functions associated with specific areas of the molecule are said to be composed of **domains**.

IgG has at least two different domains: the antigen-binding domains and the Fc domain. Another important structural feature of IgG is that it contains a hinge region (Figure 3.29). The molecule has an overall 'Y' shape; the hinge region allows a certain amount of movement of the two 'arms'. Here the hinge region therefore contributes to another significant property of proteins, namely **flexibility**; this can accommodate changes in shape of the molecule induced by binding to other molecules, in this case, to the antigen. Box 3.3 gives further details on the current view of flexibility of proteins—a rapidly moving field which is changing the way we consider protein structure.

Figure 3.29 Immunoglobulin G is a 'Y'-shaped molecule. There is a hinge region which imparts flexibility.

Box 3.3 Flexibility of proteins

The classic view of protein structure is of unique, rigid, static form. This view has given way gradually to a more flexible, dynamic one. In fact the various parts of a protein molecule are in constant state of motion. The motion can be of several types, varying in the size of movements involved and the time-scale over which they take place.

At one end of the spectrum are *atomic fluctuations*, such as individual vibrations of atoms. These are random, *extremely* rapid (10^{-15} to 10^{-11} seconds) and over quite small distances (usually under 0.05 nm). Somewhat slower are *collective motions* (10^{-12} to 10^{-3} seconds) where groups of covalently linked atoms (e.g. in an amino acid side chain) move as a unit.

The distances moved can be quite large, in protein terms at least—up to 0.5 nm or more. Finally, there are *triggered conformational changes* where groups of atoms, even whole large sections of proteins (e.g. entire domains), move in response to some specific stimulus—for example, the binding of a substrate or inhibitor (to an enzyme) (Chapters 4 and 5). These changes can be large (from 0.05 to over 1 nm) and relatively slow (10^{-9} to 10^3 seconds).

Given all this movement, all this flexibility, in protein structure, why did the classic rigid, static view arise? The main reason is that most of our information on the precise three-dimensional structure of proteins has been acquired from X-ray diffraction of protein crystals (Box 3.1).

This is not to say that protein molecules in crystals are rigid whereas the same molecules in solution are flexible. Unlike small molecules such as sodium chloride, protein molecules are not tightly packed when in crystals. Typically, around 50% of a protein crystal is composed of solvent and there is ample room for atoms in the protein molecules to move. There is indeed evidence that protein molecules in crystals are flexible and are not generally widely different from the same molecules in solution. The classic static view of proteins emanates more from the nature of the technique of X-ray diffraction itself.

To collect sufficient data from X-ray diffraction of a crystal to be able to build up a three-dimensional model of a protein takes about *one week*. Yet, as can be seen above, even the slowest motions within protein molecules take only about 15 minutes (10^3 seconds) and many are in the nanosecond (10^{-9} second) range or even faster! X-ray diffraction is thus much much too slow to 'capture' individual conformations of the dynamic protein. What one gets amounts to a 'fuzzy' set of data, where those positions frequently occupied by an atom show up as a spread of positions and those most frequently occupied show up most prominently. What is then constructed from such data is essentially a model where the average position of each atom in the molecule is indicated—inevitably a unique static model.

However, some triggered conformational changes, though much too fast to follow in detail by X-ray diffraction, result in new structures which are *stable* as long as the stimulus is maintained (e.g. the substrate or inhibitor is bound to the enzyme). By performing X-ray diffraction on crystals of each of the stable structures the net change can be observed, though the detailed route (i.e. any intermediate stages in going from one stable structure to another) cannot. As you will see in Chapters 4 and 5, such net changes can provide useful insights into how some enzymes function.

Over the last 10–15 years our understanding of protein chemistry has been changing. A variety of techniques has confirmed the more dynamic view of protein structure (and, incidentally, of the structure of nucleic acids too). Ironically, X-ray diffraction, the technique that contributed so much to the static

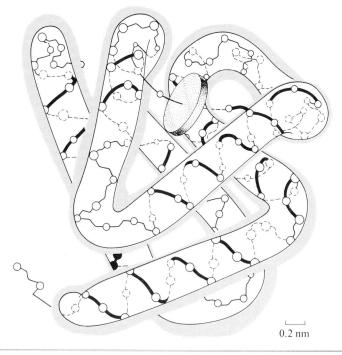

Figure 3.30 Structure of myoglobin, showing the possible positions of the backbone polypeptide chain. The solid outline indicates the path of the backbone chain and the blue shaded area outside indicates the range of positions it can take at any point.

0.2 nm

view, is now contributing dramatically to the dynamic one. For, by subjecting the 'fuzziness' in the data to a series of complex calculations, the range of positions most frequently occupied by each atom can be computed, not just the average position. This then allows a more dynamic, flexible view of the protein to be developed, something illustrated in Figure 3.30.

But even more recently has come what may promise to be a revolution in the way we analyse protein structure. In the late 1980s a different form of X-ray diffraction, called *Laue diffraction*, was applied to a protein such that sufficient data to derive a complete three-dimensional structure could be obtained, not in a week, but in three seconds! As individual diffraction pictures can be obtained in milliseconds, this technique may well open the way to follow step by step changes such as those that occur during triggered conformational changes.

IgG also illustrates another important feature of proteins, that is, **specificity**. A given protein binds to one or at most a few other molecules (ligands). So for example, there are hundreds of thousands of different immunoglobulin molecules, each of which, in order to function, binds to only one antigen. There is therefore a very high degree of specificity in terms of the binding of a protein and the ligand to which it binds. This will become particularly clear when you come to study enzymes in the next chapter (each enzyme is highly specific in terms of the substrate with which it will react). Protein specificity depends on binding sites being of the right size and shape for the ligands to which it binds. The specific binding sites result from precise folding of the polypeptide chain.

Specificity in our IgG example is very well understood. Amino acid sequences of the H and L chains of many IgG molecules are now known. Different regions of these subunits have amino acids which are either variable (V) or constant (C) (Figure 3.31), the variation being amongst the many different IgG molecules. The V and C regions have separate functions.

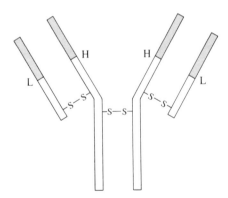

Figure 3.31 The domain structure of immunoglobulin G. Each heavy chain (H) and each light chain (L) has regions in which the amino acid sequence is either variable (V) or constant (C). The V domains are shaded, whereas the C domains are unshaded.

◇ Compare Figures 3.29 and 3.31. What is the significance of the position of the variable (V) regions of the H and L chains?

◆ The variable regions contribute to the antigen-binding sites.

This is how the specificity of each IgG results. Each IgG molecule can only bind one type of antigen molecule because of the nature of the variable regions in its H and L chains. The constant (C) regions are responsible for the overall structure of the molecule and for its recognition by other components of the immune system. Figure 3.32 shows the various features of the quaternary structure of IgG, although as you can appreciate, three-dimensional structures such as this are difficult to represent in two dimensions.

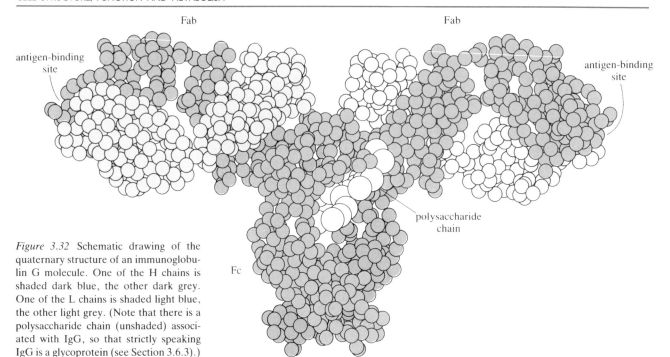

Figure 3.32 Schematic drawing of the quaternary structure of an immunoglobulin G molecule. One of the H chains is shaded dark blue, the other dark grey. One of the L chains is shaded light blue, the other light grey. (Note that there is a polysaccharide chain (unshaded) associated with IgG, so that strictly speaking IgG is a glycoprotein (see Section 3.6.3).)

IgG is now very well characterized and has been useful in illustrating a number of important features of protein quaternary structure. You will now appreciate that proteins are often composed of different domains with different functions. Specificity results from folding of polypeptide chains to give binding sites of very precise size and shape. Biological activity often requires a certain degree of flexibility in the molecule: in the case of IgG this is partly provided for by hinge regions between the domains. Shape changes resulting from biological activity can therefore be accommodated.

Other quaternary structures

Fibrous proteins such as keratin and collagen have a secondary structure consisting only of helices: i.e. no β-structures. Keratin, a major protein component of hair and horn, has a folding pattern which is entirely α-helical. In hair, three of these α-helices (separate polypeptides) are wound around each other to produce a superhelix called a microfibril. Hundreds of these are embedded in a protein matrix, producing a single hair. The ability of hair to stretch depends on the elongation of the α-helices of the individual keratin molecules. These can revert to their original conformations, not through the strength of their hydrogen bonds (which are broken in stretching) but because the helices are covalently cross-linked by disulphide bridges. As with all polymers, the extent of cross-linking determines the rigidity of the product. There are several types of keratin: those with little cystine are soft, i.e. they have little cross-linking, whereas keratins with a lot of cystine are hard. Hair is an example of the former, and horn of the latter. Breaking these disulphide bridges and then reforming them after the fibres (and consequently the keratin molecules) have been distorted is the basis of the permanent waving (or straightening) of hair. Once the molecules have been reset they remain in that conformation, although the 'wave' disappears by growth of the hair.

Collagen is another fibrous structural protein, which is the most abundant protein in mammals, constituting about a quarter of the total. Collagen is the major fibrous element of skin, bone, tendon, cartilage, blood vessels and teeth. It is present to some extent in nearly all organs and serves to hold cells together in discrete units. In addition to its structural role in mature tissue, collagen has a role in developing tissue, helping determine the direction in which cells move. Collagen is distinctive in forming insoluble fibres that have a high tensile strength.

The quaternary structure of collagen depends on intertwined helices, but it is a rather unusual protein in that its regular structure depends on the presence of either proline or its hydroxylated derivative, hydroxyproline. The biological function of collagen depends on its being relatively inert and stiff. These properties are achieved by an elongated helix with entirely different characteristics from the α-helix. This structure is such that at every third position, there is room for only the smallest amino acid, glycine, while the remaining two positions can accommodate proline. It should not surprise you that amino acid analysis of collagen gives about one third glycine, one half proline and one sixth other amino acids, and that its sequence is a repeat of Gly-X-Pro. (Hydroxyproline sometimes replaces Pro, while the X can be any one of a number of amino acids, including Pro.) The primary structure of collagen is therefore remarkably regular. In contrast, globular proteins rarely exhibit regularities in their amino acid sequence.

The quaternary structure consists of three helices (i.e. three polypeptides) which intertwine in the same direction as the α-helix to form a superhelix (Figure 3.33). Hydrogen bonding between the helices stabilizes the superhelix. Yet another level of structure occurs as numbers of superhelices associate to form collagen fibres, which are further stabilized by covalent cross-links. The extent and type of cross-linking varies with the physiological function and age of the tissue. The collagen in the Achilles' tendon of mature rats is highly cross-linked, whereas that of the flexible tail tendon is much less so. Here, then, is another excellent example of the clear link between the detailed molecular structure of a protein and its gross biological function.

We have by no means exhausted the range of protein quaternary structure: you will meet further examples later in this book. In addition, Section 3.8 deals with multimeric complexes containing proteins together with nucleic acids or lipids.

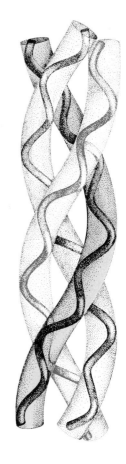

Figure 3.33 Collagen superhelix.

Summary of Sections 3.3 and 3.4

1 Each macromolecule consists of monomers covalently linked together, giving a hierarchy of structure, namely primary, secondary, tertiary and quaternary structures.

2 Various bonds are involved in maintaining the hierarchy of structure: the hydrogen bond, hydrophobic interactions, and the ionic bond.

3 Proteins are polymers based on permutations of different amino acids; each unique primary structure results from the sequence of the amino acids. Peptide bonds join the amino acids together to form the linear polypeptide chain of the protein, which includes at least 20 amino acids. Variations in the number and sequences of the amino acids in these chains give rise to a wide diversity of proteins.

4 The primary structure of a protein is folded to give complex three-dimensional higher-order structures, each of which is unique to a given protein.

5 There are two principal forms of highly regular protein secondary structure: the α-helix and β-structures.

6 The tertiary structure is maintained by non-covalent weak interactions and disulphide bridges between amino acid side chains. Tertiary structure of some

proteins such as RNAase can be reversibly denatured by treatment with mercaptoethanol and urea, which disrupt disulphide bridges and weak interactions, giving rise to a random coil. Removal of the mercaptoethanol and urea enables the precise tertiary structure to reform. Thus, primary structure determines tertiary structure.

7 Some proteins, such as insulin, which are synthesized as precursor proteins, cannot be reversibly denatured owing to loss of polypeptide chains during their activation.

8 Many proteins have quaternary structure, in which several separate polypeptide chains interact to form the active molecule. The separate polypeptide chains are termed subunits, which are held together by non-covalent, weak interactions.

9 Some proteins have different functions associated with specific areas of the molecule known as domains. The classic examples here are the immunoglobulin molecules of the immune system.

10 There are two general shapes of protein molecules: globular proteins, such as enzymes, the biological catalysts, and fibrous proteins which are usually long, thin structural components.

Question 4 (*Objectives 3.1, 3.3 and 3.4*) Below is the amino acid sequence of a polypeptide.

> Met-Ser-Pro-Ala-Ala-Glu-Met-Val-Phe-Leu-Ile-Val-Leu-Pro-Ala-Gly-Ala-Ala-Met-Pro-Arg-Pro-Lys-Ala

Study the sequence and, with reference to Table 3.2, answer the following questions. In each case you should justify your answer.

(a) What is the total number of amino acids in the chain?

(b) Could this polypeptide be described as a protein?

(c) Which is the most common amino acid in this sequence?

(d) Which amino acid is at the carboxyl (C-) terminus?

(e) Which amino acid is at the amino (N-) terminus?

(f) This single polypeptide chain folds to form higher-order structure; could it form both tertiary and quaternary structure?

(g) Could a disulphide bridge form in this polypeptide? Explain your answer.

(h) A small region of the chain forms a hydrophobic core—identify this sequence.

(i) Five amino acids in this polypeptide have the capacity to interact via ionic bonding to stabilize higher-order structure. Identify these amino acids and indicate how ionic bonds might form.

Question 5 (*Objectives 3.1, 3.4, and 3.5*) A strand of hair can be stretched considerably before it breaks. What bond types are involved in resisting stretching?

Question 6 (*Objectives 3.1, 3.3, and 3.4*) Collagen, fibroin (silk protein) and the enzyme RNAase are three different types of protein. Name the types of bonds which stabilize each of the three structures.

Question 7 (*Objectives 3.1, 3.3, 3.4 and 3.5*) Blood clotting occurs at wounds preventing loss of blood and aiding in the healing process. Blood clots consist principally of the protein fibrin, which is a multimeric complex composed of a single type of subunit (monomeric fibrin). Fibrin is synthesized as an inactive precursor protein, fibrinogen, which is activated at the wound by removal of part of the polypeptide chain. Fibrinogen can be reversibly denatured by treatment with urea and mercaptoethanol, whereas monomeric fibrin is irreversibly denatured by this treatment.

(a) Suggest why fibrin is synthesized as an inactive precursor.

(b) Provide an explanation for the observation that fibrinogen can be reversibly denatured, whereas fibrin cannot.

(c) Explain how the conversion of fibrinogen to fibrin might enable the aggregation of monomeric fibrin to multimeric fibrin to occur.

(d) 'Fibrin has quaternary structure, whereas fibrinogen does not.' Explain this statement.

Question 8 (*Objectives 3.1, 3.2, 3.3 and 3.4*) A protein subjected to PAGE (Section 3.2) gives a single band with $M_r = 100\,000$. The same protein, when treated with SDS and then subjected to electrophoresis (i.e. SDS–PAGE) gave two distinct bands on the gel of $M_r = 60\,000$ and $40\,000$. What could you conclude from these results?

Question 9 (*Objectives 3.1 and 3.3*) 'All proteins are polypeptides, and all polypeptides are proteins.' Is this statement true or false? Explain your answer.

3.5 NUCLEIC ACIDS

The nucleic acids deoxyribonucleic acid (DNA) and ribonucleic acid (RNA) are macromolecules involved in the storage and transfer of genetic information. In addition, various forms of RNA have a structural role in ribosomes. The last three decades have seen remarkable advances in our understanding of this class of macromolecules, so that today there is a vast body of information relating to their structure and function. Here we can only skim the surface, and indeed we are really only concerned with one question, namely: Can the concept of hierarchy of structure outlined in Section 3.3 be applied to nucleic acids? As you will see, the answer is that indeed it can, but only to a certain extent, for the various hierarchies are not always applicable to all nucleic acids.

3.5.1 Primary structure

A **nucleic acid** is a **polynucleotide**, that is, a polymer consisting of nucleotides. Each **nucleotide** has three components (Figure 3.34).

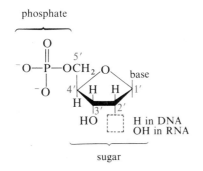

Figure 3.34 Structure of a nucleotide. The carbon atoms in the sugar ring are numbered. DNA contains the sugar deoxyribose (−H at the 2'-carbon atom), while RNA has ribose (−OH at C-2' instead of −H).

1 A five-carbon sugar. This is ribose in the case of RNA, and 2′-deoxyribose in DNA. The difference in the structure of ribose and 2′-deoxyribose is shown in Figure 3.34.

2 A purine or pyrimidine base attached to the 1′-carbon atom of the sugar. The bases, which are shown in Figure 3.35 are the purines, adenine (A) and guanine (G), and the pyrimidines, cytosine (C), thymine (T) and uracil (U). Both DNA and RNA contain A, G and C; however, T is found principally in DNA and U is found only in RNA. (Note that 'base' here does not mean 'base' as in acid/base.)

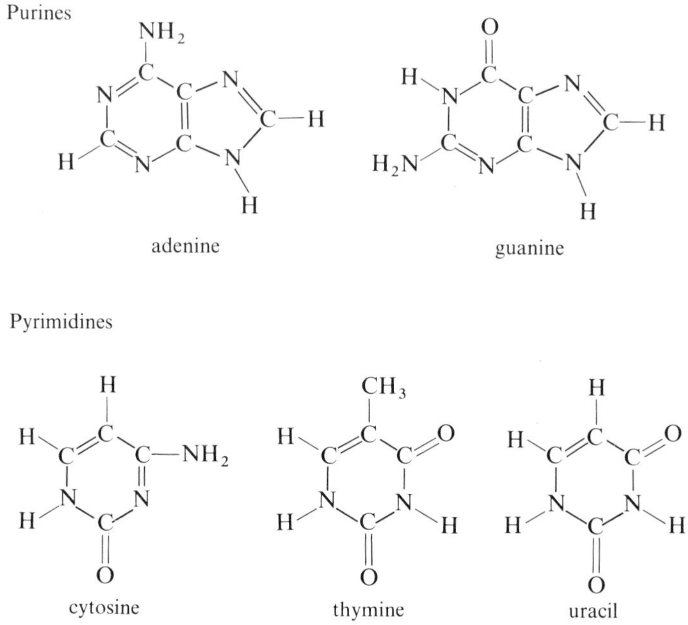

Figure 3.35 The bases found in nucleic acids.

3 A phosphate attached to the 5′-carbon of the sugar by a phosphate–ester linkage. The phosphate is responsible for the strong negative charge on both nucleotides and nucleic acids.

A base linked to a sugar is called a *nucleoside*; a nucleotide is thus a nucleoside phosphate. The nucleotides in nucleic acids are covalently linked by a linkage that joins the 5′-phosphate of one nucleotide and the 3′-OH group of the adjacent nucleotide (Figure 3.36). The phosphate is therefore joined (esterified) to both the 3′- and the 5′-carbon atoms; this unit is often called a **phosphodiester group**. Chains of nucleotides are formed as several of these groups are joined together.

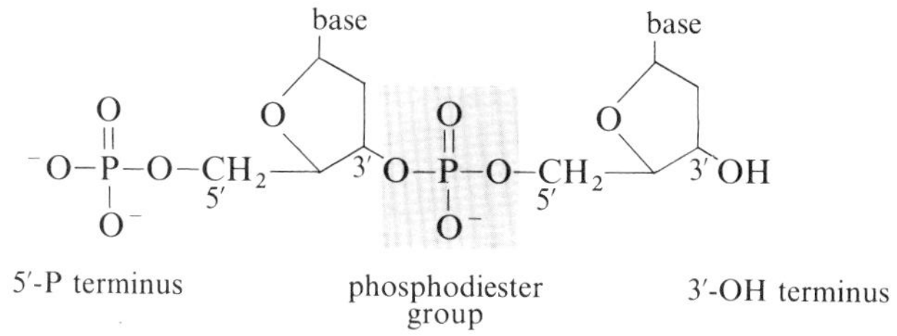

Figure 3.36 The structure of a dinucleotide. A polynucleotide would consist of many nucleotides linked together by phosphodiester groups. Note that all such molecules have a 5′ terminus (on the left) and a 3′ terminus (on the right).

To summarize the key points before we move on. All nucleic acids have a primary structure, i.e. a linear sequence of nucleotides which are covalently linked together, as shown in Figure 3.37. The order of the bases can be used as a shorthand form to denote the sequence of the polynucleotide.

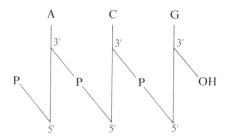

Figure 3.37 The simplified structure of a polynucleotide. A vertical line denotes the sugar to which the appropriate letter is added to indicate the base. The phosphate group that links the sugars is shown by a diagonal with a P.

◇ What is the shorthand sequence of the polynucleotide shown in Figure 3.37?

◆ ACG. Reading from left to right (from the 5′ end to the 3′ end), as is the convention for nucleotide sequences (as it is for amino acid sequences).

The primary structure can often consist of a very long sequence of nucleotides. DNA usually contains many thousands of nucleotides whereas, in contrast, the sequence of transfer RNA (tRNA) is relatively short. The sequences of bases in a nucleic acid can now be determined quite rapidly using modern techniques based on vertical slab gel electrophoresis as described in Section 3.2.3.

3.5.2 Secondary structure

Secondary structures in nucleic acids are based on the formation of helical regions. We shall consider first the double helix of DNA, and then look briefly at secondary structures in RNA.

The DNA double helix

Early studies of DNA indicated that the molecule is an extended chain having a regular structure. The most important technique used in structure determination was X-ray diffraction (see Box 3.1), which provided information about the arrangement and dimensions of various parts of the molecule. The most significant observations were that the molecule is helical and that the bases of the nucleotides are stacked and separated by a spacing of 0.34 nm.

Chemical analysis of the content of each of the four bases (referred to as the base composition) adenine, thymine, guanine and cytosine in DNA isolated from a range of different organisms provided the important relationship that [A] = [T] and [G] = [C]. Here, [ ] denotes the molar concentration of base in a solution of hydrolysed DNA.

Using this information, James Watson and Francis Crick, in their now famous study, suggested a model in which two helical strands are present in DNA, with the two strands coiled around one another to form a double-stranded helix (Figure 3.38). In this model the sugar-phosphate backbones follow a helical path at the edge of the molecule, while the bases form a helical array at the core. The bases of one strand are hydrogen-bonded to those of the other strand to form purine-to-pyrimidine base pairs A—T and G—C (Figure 3.39). This complementary **base-pairing** accounts for the relationships in base composition outlined above.

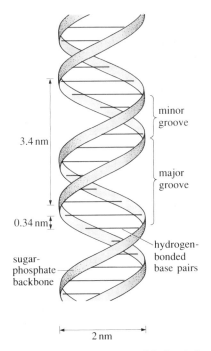

Figure 3.38 The common, right-handed form of the DNA double helix. The ribbon represents the two sugar-phosphate backbones. The parallel blue lines are the stacked, hydrogen-bonded base pairs.

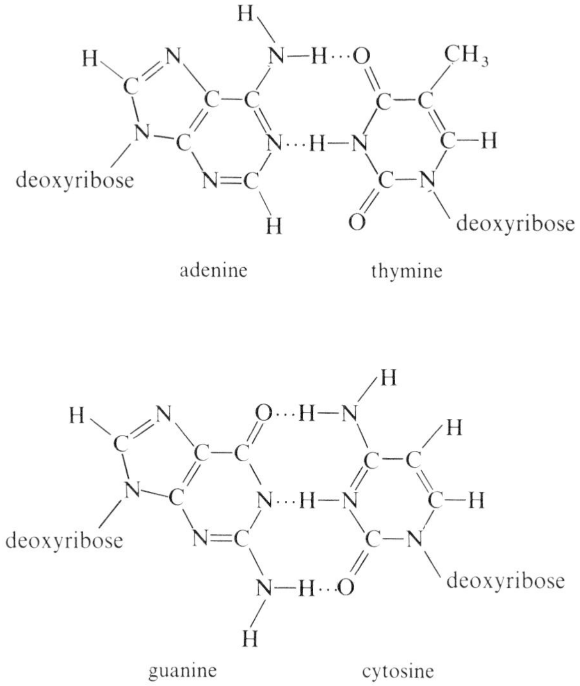

Figure 3.39 The two common base pairs of DNA showing the hydrogen bonds (dotted lines).

◇ What can you deduce from this about the relationship between the content of the purine and pyrimidine bases?

◈ Because [A] = [T] and [G] = [C] it follows that [A + G] = [T + C]. In other words, [purines] = [pyrimidines].

◇ Hydrogen bonds are clearly important in maintaining this secondary structure of DNA. What other type of weak bond forms non-covalent interactions which stabilize the structure?

◈ You can see from Figure 3.38 that the core of the molecule consists of the base pairs which are shown parallel. In fact the bases are stacked, as explained in Section 3.3. Hydrophobic interactions between the non-polar bases are important in maintaining this stacked arrangement.

There are a few additional features of the double helix worthy of note. The form shown in Figure 3.38 is a right-handed helix. This means that when you look down the axis of the helix, each strand follows a clockwise path as it moves away from you. You should also note that DNA can adopt other arrangements, such as left-handed helices, but the one just described appears to be encountered most commonly.

In terms of function it is relevant to point out that the helix of Figure 3.38 has two grooves: a deep one (the major groove) and a shallower one (the minor groove). Both of these are large enough for other molecules, notably proteins, to gain access to the core of the molecule.

◇ Can you suggest why this might be important?

◈ The function of DNA requires interaction with other molecules, notably proteins, hence the need for contact between them.

The base-pairing arrangements shown in Figure 3.39 are an important feature of DNA structure, because it means that the base sequences of the two strands are complementary. The final feature of the DNA helix is that the two strands are antiparallel, they run in opposite directions (i.e. 3′ to 5′ and 5′ to 3′) (Figure 3.40). So that the 3′-OH terminus of one strand is adjacent to the 5′-P (5′-phosphate) terminus of the other. In the linear double helix, therefore, there is one 3′-OH and one 5′-P at each end of the helix.

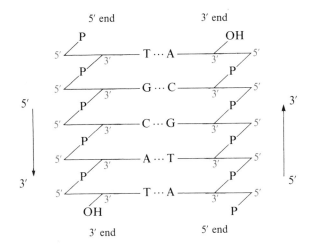

Figure 3.40 A stylized drawing of a segment of a DNA double helix showing the antiparallel orientation of the complementary strands. The arrows indicate the 5′ to 3′ direction of each strand. The phosphates (P) join the 3′-carbon to the 5′-carbon of the adjacent deoxyribose.

To sum up then, the most common secondary structure of DNA is a right-handed double helix consisting of two complementary polynucleotide strands. This is stabilized by base-stacking and by hydrogen bonds between the complementary base pairs.

RNA secondary structures

◇ Secondary structures in RNA are also dependent upon base-pairing. How might base pairs form in RNA molecules?

◆ Here the relationship is that, as in DNA, C pairs with G, but U replaces T, so A pairs with U.

Very few RNA molecules form a very regular helix equivalent to the double helix of DNA. Rather, regular helical regions are interspersed by non-helical sections. The postulated secondary structure of one RNA molecule is shown in Figure 3.41. This is in fact one of the RNA molecules from the ribosome of the bacterium *E. coli* (hence the 'r' of rRNA). Notice how complicated this secondary structure is. You should appreciate that in reality this structure will be highly folded to give a complicated three-dimensional (i.e. tertiary) structure.

◇ What is the primary structure of the molecule shown in Figure 3.41?

◆ The linear sequence of (1 542) nucleotides. Notice that there is just *one* polynucleotide chain, which folds to give the secondary structure. (This contrasts with DNA where there are *two* polynucleotide chains.)

◇ What weak bonds would you expect to be involved in stabilizing the secondary structure of the rRNA molecule?

◆ Hydrogen bonds between the paired bases, hydrophobic interactions between the non-polar bases, in particular base-stacking.

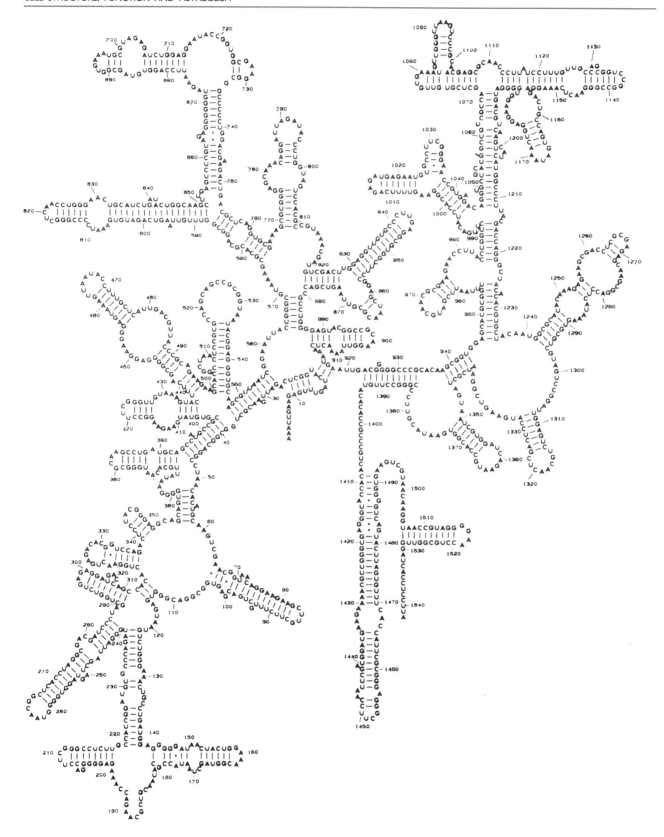

Figure 3.41 The secondary structure of an rRNA molecule from the *E. coli* ribosome.

The secondary structure consists of a mixture of many short regular helical sections which result from base-pairing, together with various circles and loops.

◇ How do the loops and circles come about?

◆ Base-pairing does not occur in these regions.

For the complex rRNA molecule shown in Figure 3.41 it is not yet possible to relate the secondary structure to its functional activity, but this can be done for a smaller molecule, tRNA (Figure 3.42). The tRNA molecules are small, single-stranded nucleic acids ranging in size from 73 to 93 nucleotides. They play a role in protein synthesis, in carrying the amino acids to the sites of synthesis (the ribosomes), each tRNA being associated with a particular amino acid. At the ribosome the tRNA binds to an mRNA ('m' stands for messenger) codon via another three-nucleotide sequence termed the anticodon. This anticodon–codon binding depends on complementary base-pairing. Details here are not important, but you should realize that tRNA binds specifically to both an amino acid and mRNA and so has at least two binding sites. With this in mind, look at the structure of a tRNA molecule (Figure 3.42).

◇ The secondary structure of a tRNA is shown in Figure 3.42. What key features can you see?

◆ There are four short base-paired regions—in three dimensions these are actually helical. These result in three small loops and the acceptor stem without a loop.

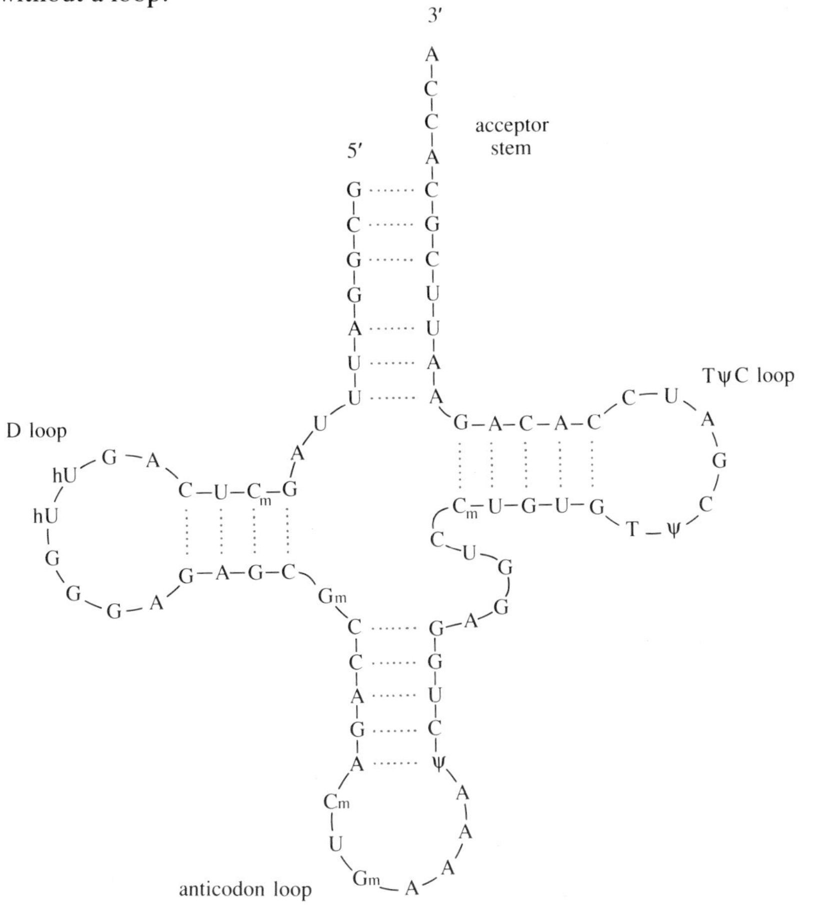

Figure 3.42 Clover leaf secondary structure of tRNA. The dotted lines indicate base-pairing (by hydrogen bonding). Note that Cm, Gm, hU and ψ (pronounced psi) are all modified bases (i.e. variants on C, G and U).

The structure shown in Figure 3.42 is usually referred to as the 'clover leaf' structure of tRNA, for obvious reasons. The loops of this structure are important in terms of its biological function.

◇ Which three nucleotides of the anticodon loop are most likely to interact with mRNA during protein synthesis?

◆ GmAA. These are the most exposed, and furthermore, their bases are not hydrogen-bonded as are those in the stem of the loop. These exposed bases can therefore become hydrogen-bonded to mRNA.

The other important functional feature of the tRNA secondary structure is that the acceptor stem is the region where the specific amino acid is attached.

Nucleic acids, then, have secondary structure maintained by hydrogen bonding and base-stacking. As in proteins, therefore, secondary structure is again stabilized by weak bonds, so forming characteristic regular features.

3.4.3 Tertiary and quaternary structure

Can the regular secondary structures of nucleic acids be folded further to give higher orders of structure as in proteins? Let us consider first tRNA, since the tertiary structure of this group of nucleic acids has been relatively well understood for several years now.

Figure 3.43 shows the tertiary structure of a tRNA, which has a unique structure and, just as in some proteins, the primary structure is sufficient to reassemble the functional molecule after unfolding.

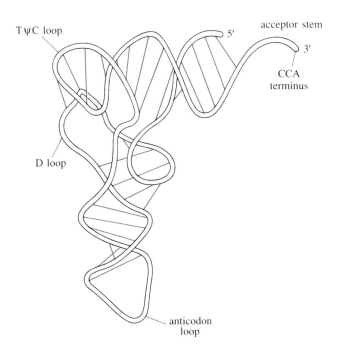

Figure 3.43 Schematic diagram of the three-dimensional (tertiary) structure of a tRNA molecule.

◇ What is the term for this process of unfolding and refolding?

◆ Reversible denaturation.

◇ Predict which types of bonds are broken and reformed during the process.

◆ Hydrogen bonds between the base pairs, and hydrophobic interactions between the stacked bases.

Like globular proteins, tRNA molecules can be crystallized (itself an indication of a specific structure) and studied by X-ray diffraction (see Box 3.1). Most tRNAs have compact 'L'-shaped structures with the anticodon loop and amino acid binding site situated at opposite ends of the molecule (as does the tRNA shown in Figure 3.43). The remaining loops of the 'clover leaf' are twisted so that they lie parallel to the rest of the molecule and are held in place mainly by hydrogen bonds between bases not involved in the helices. The loops appear to be associated with recognition by the enzymes which join the amino acid to the tRNA, so the surface must be such that it is specific for a given enzyme, so enabling correct binding to occur.

The more complicated rRNA shown in Figure 3.41 has, undoubtedly, a specific folding pattern since it combines precisely with other RNAs and proteins to form a ribosome. However, at present we cannot give you any further information on this. It seems likely, though, that the biologically active tertiary structure adopted by this rRNA is determined in some manner by its association with other molecules.

The regular secondary structure of DNA, i.e. the double helix, can be folded further to form higher-order structures. The simplest example is found in most prokaryotes and many viruses, where each DNA molecule forms a circle.

◇ How might a DNA double helix form a circular molecule?

◆ The ends could be joined together.

In fact the two ends are covalently joined together to form a covalently closed circle. This circle in turn can be further coiled (the molecule is said to be 'supercoiled') as shown in Figure 3.44.

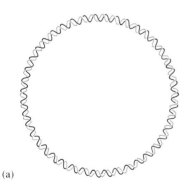
(a)

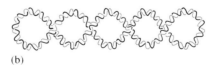
(b)

Figure 3.44 DNA from a prokaryotic organism: (a) covalently closed circle; (b) supercoiled covalently closed circle.

◇ What effect does supercoiling have on the DNA double helix?

◆ It makes the molecule more compact.

Supercoiling is important in relation to the packaging of DNA in a cell. The largest circular DNA molecule in the bacterium *E. coli* has a length of about 1 mm, and yet the cell into which it fits is less than 1 μm across. Clearly the DNA must undergo considerable condensation to fit into the cell, but this process appears to involve other molecules, so the resulting structure is best considered as a macromolecular assembly. The condensing of DNA molecules is also important in eukaryotic cells, but it is clear that this involves interaction between DNA and protein molecules. This results in chromosomes, the macromolecular assemblies we shall look at in Section 3.8.1.

Strictly speaking, relatively few nucleic acids have tertiary structure, in which a single polynucleotide chain folds to form a unique three-dimensional structure—the tRNAs are the only clear examples. Nucleic acids will not generally form quaternary structures because their higher-order structure is dependent on the interaction with *other* macromolecules. This contrasts with proteins where separate polypeptide chains can interact to form a protein quaternary structure.

Summary of Section 3.5

1 Nucleic acids (DNA and RNA) are involved in the storage and transfer of genetic information.

2 Each nucleic acid is a polymer of nucleotides, i.e. a polynucleotide.

3 Secondary structures in nucleic acids are based on the formation of helical regions, notably the classic double helix of DNA, which depends on complementary base-pairing and base-stacking. RNAs seldom have regular secondary structure.

4 The tRNAs have well-defined tertiary structures, somewhat analogous to those of globular proteins. Other nucleic acids (on their own) do not have precisely defined tertiary structure.

Question 10 (*Objectives 3.1, 3.3, 3.4 and 3.5*) DNA can be reversibly denatured by gently heating a solution of this macromolecule and then cooling it. The molecules are double helices at the start of the experiment and return to this conformation after cooling. In the denatured state random coils are formed. Explain these observations. What can you deduce from this regarding the relationship between the primary and secondary structure of DNA?

Question 11 (*Objectives 3.1 and 3.4*) Figure 3.35 shows the structures of the bases in nucleic acids and Figure 3.39 illustrates how base-pairing can occur. Use this information to calculate the number of hydrogen bonds in the clover leaf structure of tRNA shown in Figure 3.42. For the purposes of this question assume that the hydrogen-bonding potential of the unusual bases is the same as that of the parent base.

Question 12 (*Objectives 3.1 and 3.4*) An RNA molecule contains the following base sequence:

A-G-U-G-C-G-A-U-A-C-G-C

Draw out and arrange this to form a stem with a loop, so that the maximum number of hydrogen bonds can form.

Question 13 (*Objectives 3.1 and 3.3*) An unusual deep-ocean prokaryote has recently been isolated and its properties have been described in a scientific journal. The authors of the report proposed a model which suggested that some of the RNA from this organism consisted of four supercoiled molecules which associated to give a tetrameric complex, which formed in the absence of protein. From your knowledge of nucleic acid structure, criticize this model and suggest why it is unlikely.

3.6 POLYSACCHARIDES

Wood, paper, cotton, rayon, bread and potatoes are all everyday materials whose major components (at least as far as macromolecules are concerned) are **polysaccharides**, polymers of sugar molecules. Polysaccharides have been rather neglected because they do not play particularly glamorous roles. They do not have catalytic functions, nor do polysaccharides carry genetic information (although, as sugars are involved—ribose and deoxyribose—you could argue that the backbone of DNA and RNA is a *sort* of polysaccharide). It is not surprising therefore that less attention has been paid to these 'mundane' macromolecules than to the more 'exciting' nucleic acids and proteins. One particular polysaccharide, cellulose, is the most abundant organic molecule on earth, as a consequence of its major structural role in plants. Chitin is another abundant polysaccharide. It is found in the cell walls of most fungi and in the exoskeletons of insects and crustaceans, where it is produced in extremely large quantities (but is degraded more rapidly than cellulose). Other polysaccharides form the food stores of plants and animals. In addition, some polysaccharides are responsible for lubricating skeletal joints and for initiating the reactions that lead to rejection of blood transfusions or transplants, and these roles have encouraged more extensive study of polysaccharides in general.

3.6.1 Primary structure

◇ Based on your knowledge of the polymeric macromolecules we have looked at so far, suggest two features of the primary structure of a polysaccharide which would be important in determining its structure.

◆ The type of monomer (building block) and the nature of the bonds which link these together to form the polymer.

Hydrolysis of polysaccharides releases the component sugars, a few of which are shown in Figure 3.45. Unlike proteins and nucleic acids, many polysaccharides contain only a single type of monomer. Glycogen, starch and cellulose fall into this category—these are all polymers of glucose and yet they represent the two extremes of polysaccharide function, food storage and rigid support. Here is a clear case where the type of bond between component building blocks (i.e. the primary structure) has a marked effect on the biological activity of the macromolecule.

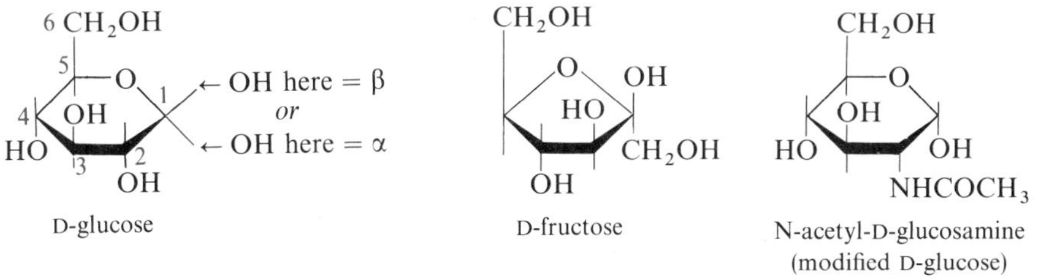

Figure 3.45 Some examples of simple sugars. (Note the differences in the orientation of the —OH group at C-1 on glucose, which gives rise to α- and β-glycosidic bonds).

Sugars are joined together into a chain through links known as **glycosidic bonds** (as shown in Figure 3.46). If you look at the ring structure of glucose (Figure 3.45), you will be able to work out why the nature of this link between sugars is as important in primary structure as the nature of the sugars themselves. Of the six hydroxyl groups, only that attached to C-1 (which is also bonded to the ring oxygen) forms a glycosidic bond by linking to one of

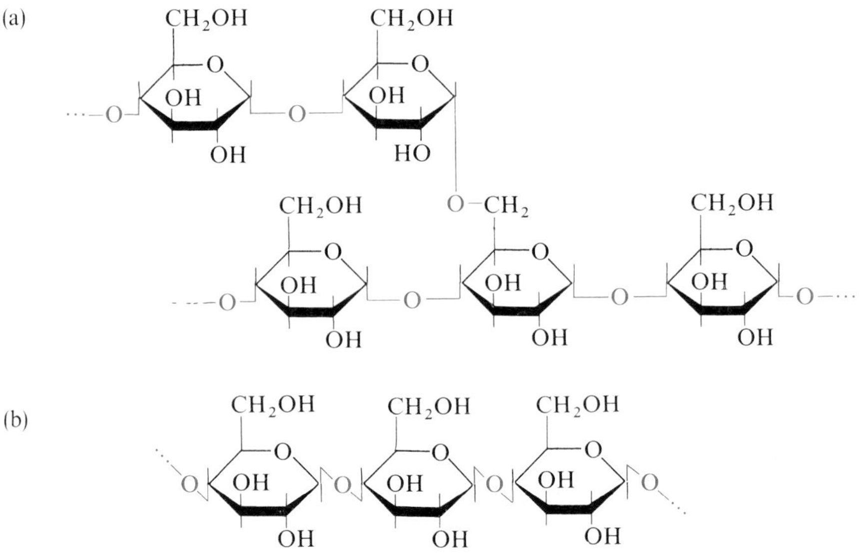

(a)

(b)

Figure 3.46 Primary structures of (a) glycogen and (b) cellulose. The glycosidic bonds are shown in blue.

the five hydroxyls on the next glucose monomer in the chain. The chain can also become branched if more than two hydroxyl groups on the same sugar are involved in bonding.

◇ What sort of molecule would a branched polysaccharide be, and how would this compare with a globular protein?

◈ It would be a globular polysaccharide; these are generally bush-like molecules formed by branching, in contrast to globular proteins, which are formed by folding up a single polypeptide chain, or in multimeric proteins by several such folded chains associating together.

◇ How might you estimate the amount of branching in a globular polysaccharide?

◈ There are two possibilities: you might either 'count' the number of branching points, or measure the number of ends of chains.

Many sugars other than glucose are found in polysaccharides, some with exotic names like abequose, rhamnose or hamamelose, whose structures need not concern us here. They are all variations on the theme of glucose or fructose: a 5- or 6-membered ring containing an oxygen atom and one particularly reactive hydroxyl group. The variations are in the orientation of the hydroxyl groups and the nature of the functional (reactive) groups; for example, $-OH$ may be replaced by $-NH_2$ or $-NHCOCH_3$ and $-CH_2OH$ by $-H$, $-CH_3$ or $-SO_3^-$. Just one of these variants is shown in Figure 3.45.

The primary structure of polysaccharides varies, therefore, in terms of the sugar composition and the way in which these monomers are joined via glycosidic bonds. There is a wide diversity of polysaccharides, but here we shall consider mainly glycogen and cellulose; they illustrate different functions of polysaccharides and their primary structures are shown in Figure 3.46.

◇ What is the principal difference between these two primary structures?

◈ Glycogen is a branched structure, whereas cellulose is an unbranched, linear molecule.

3.6.2 Higher-order structure

Glycogen is an example of a polysaccharide with a food storage role in animals. It is a compact globular molecule and can therefore be stored economically within liver and muscle cells where it functions as an energy reserve. Glycogen has a variable arrangement of bonds, so it does not have one fixed shape, i.e. there is no regular higher-order structure, in contrast to other macromolecules which we have examined so far. Instead, glycogen has an overall bushy shape, well fitted to its function. Large amounts of glucose can be quickly mobilized from the liver when energy is required (and transported into the blood) by the removal (by enzymes) of a few residues off the end of each of the many branches.

A second key role played by polysaccharides is that of support. These structural molecules are found principally in bacteria, plants and invertebrates.

◇ Recall the structural proteins such as collagen and keratin—what features have these in common with structural polysaccharides?

◆ They are fibrous, with a linear chain of repeat units, folded into a regular higher-order structure, and stabilized by weak bonds along and between the chains.

Cellulose is one example of a fibrous, structural polysaccharide. The primary structure (Figure 3.46b) is formed from long chains of β-1,4-linked glucose residues, each of which is related to the next by a rotation of 180 degrees. This forms a zigzag ribbon with the ring oxygen atom of one glucose residue forming a hydrogen bond to the 3-OH group of the next, i.e. *intra*chain (within the chain) hydrogen bonds form (Figure 3.47). *Inter*chain (between different chains) hydrogen bonds can also form between adjacent chains via the $-CH_2OH$ groups. Fibres are formed by parallel chains. The cell wall of plants consists of many such fibres stacked together in such a way that the chain directions are different for successive layers. Hydrogen bonding between the layers results in a strong, protective multilayered coat for the plant cell. Further stability comes from covalent bonding to non-cellulose cell wall polymers.

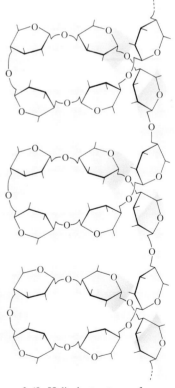

Figure 3.48 Helical structure of amylose. The blue tone is used to show the shape of the helix more clearly.

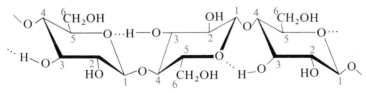

Figure 3.47 The structure of cellulose showing a single chain with (intrachain) hydrogen bonds.

You can now see why the physical properties of silk (a fibrous protein) and cotton (made from a fibrous polysaccharide) are so similar. In fact the cotton fibre used for sewing is a rather unusual adaptation of the cotton plant, in which cellulose, whose normal role is as a support molecule, acts as an aid for seed dispersal, each seed having a tuft of fibres which acts like a parachute.

Our final polysaccharide is amylose, a component of the plant food reserve starch. It has a helical structure, unlike the other two polysaccharides we have considered. Amylose is a polymer of glucose joined by α-1,4-glycosidic bonds. The α-linkage allows for greater freedom of rotation and the chain forms a stable helix with six glucose monomers per turn (Figure 3.48).

Incidentally, the iodine molecule fits inside the central core of this helix almost exactly, and the complex thus formed is responsible for the colour change in the iodine test used to detect starch.

3.6.3 Glycoproteins

Many polysaccharides, especially those from animals, are not 'pure' but are **glycoproteins**—hybrid macromolecules, in which polysaccharide chains are covalently linked to a protein or polypeptide. (Note that hybrid molecules differ from the macromolecular assembles you will meet in Section 3.8, where *weak* bonds, i.e. non-covalent interactions, hold *different* macromolecules together.) Although the sugar chains may be several hundred residues long, glycoproteins come in many different forms and even short chains of sugar may have important roles. For example, proteins such as digestive enzymes and antibodies (immunoglobulins), which operate outside the cell, are often found to carry short polysaccharide chains (see Figure 3.32).

On their outside surfaces, all cells carry polysaccharide chains of precisely defined sugar sequence. These are anchored in position by a protein or a lipid tail, which is embedded in the surface of the cell membrane. (If the polysaccharide is attached to lipid rather than to protein, then the whole is called a glycolipid, not a glycoprotein.) The sugar chains appear to act as cellular identification labels and, incidentally, form the basis of tissue typing for suitable transplant patients. An individual's immune system will destroy transplanted cells if these carry 'foreign' sugar identification labels. Some cell surface markers whose structures are known in detail are those that determine the A and B blood groups. Other sugar identification labels are involved in a range of other processes, such as the ordered interaction of cells during development, recognition of the roots of legume plants (peas, beans, etc.) by symbiotic nitrogen-fixing bacteria, and the interaction of disease-causing organisms with their hosts.

3.7 LIPIDS

Lipids are the final class of biological molecules to be considered in this chapter. **Lipids** are water-insoluble organic molecules found in all living cells. Strictly speaking lipids are not macromolecules, because their average relative molecular mass (M_r) is less than a thousand, which contrasts with the true macromolecules whose M_r may be up to several millions. Lipids tend to be classed with the others because lipid molecules show a strong tendency to form high M_r aggregates, in which individual weak bonds reinforce each other to give a large, fairly stable structure. Note, however, that lipids do not conform to the general principles of structural hierarchy (Section 3.3) which we have outlined for proteins, nucleic acids and polysaccharides.

There are many families of lipids but they all derive their distinctive behaviour from the hydrocarbon structures which form a major part of each molecule. The biological functions of lipids fall into three groups: 1, food storage; 2, structural and recognition components of membranes; and 3, carriers of biological signals (as hormones). The food storage lipids will be the most familiar to you, so we shall start with these.

3.7.1 Food storage lipids

If asked to give an example of a lipid, commonly referred to as fats and oils, you might well think of butter, lard, or the vegetable oil used for cooking. These are all neutral fats (i.e. the molecules are uncharged), and the naturally occurring ones function as food reserves. Most of the neutral fats are triglycerides, esters composed of glycerol (an alcohol) and three fatty acid molecules. The **fatty acids** involved are long hydrocarbon chains each with a terminal carboxyl ($-COOH$) group. The triglyceride is formed between glycerol $-OH$ and fatty acid $-COOH$ groups (Figure 3.49).

$$
\begin{array}{c}
CH_2OH \\
| \\
CHOH \\
| \\
CH_2OH
\end{array}
\;+\;
\begin{array}{c}
\overset{\displaystyle O}{\overset{\|}{HO-C-R^1}} \\[6pt]
\overset{\displaystyle O}{\overset{\|}{HO-C-R^2}} \\[6pt]
\overset{\displaystyle O}{\overset{\|}{HO-C-R^3}}
\end{array}
\;\longrightarrow\;
\begin{array}{c}
\overset{\displaystyle O}{\overset{\|}{CH_2OC-R^1}} \\[6pt]
\overset{\displaystyle O}{\overset{\|}{CHOC-R^2}} \\[6pt]
\overset{\displaystyle O}{\overset{\|}{CH_2OC-R^3}}
\end{array}
\;+\; 3H_2O
$$

glycerol   fatty acid   triglyceride   water

Figure 3.49 Formation of a triglyceride from glycerol and three molecules of fatty acid. R^1, R^2 and R^3 represent the fatty acid hydrocarbon chains, which can be either the same or different.

The bulk of the neutral fat molecule therefore consists of long fatty acid chains of non-polar $-CH_2-$ groups.

◇ How could such long non-polar chains interact?

◆ By hydrophobic interactions.

Indeed aggregates of neutral fat molecules tend to become spherical, the shape that gives the smallest surface area in contact with the surrounding watery cell contents. (This is what happens to oil droplets in water.) In this way, lipids can be stored separately from the rest of the cell contents. Although it takes longer to mobilize stored fat than glycogen (and fat is more difficult to transport in the bloodstream than sugar because of its insolubility), adipose (fat) tissue is the major form of food storage in many animals because it is more economical: the breakdown of a gram of fat yields more than twice as much energy as the breakdown of a gram of glycogen. This high energy yield is one explanation for the abundance of fats as a food source in milk.

Animal fats tend to be more solid than the vegetable oils. This is a direct consequence of the different structures of their fatty acid components, which may be saturated (without double bonds: $-CH_2-CH_2-$) or unsaturated (with a double bond or bonds: $-CH=CH-$), as shown in Table 3.3 and Figure 3.50. As you can see from Figure 3.50, a double bond can kink the chain, producing a more open fluid structure when the molecules are packed together. This fluidity is important in terms of membrane structure and functions as you will see later.

Animal fats such as those found in butter, lard, etc. are mostly saturated (40–60%) mixed with mono-unsaturated fats (30–50%), i.e. those with a single double bond such as oleic acid, and traces of polyunsaturated fats (those with two or more double bonds, for example, linoleic acid). Plant fats, such as those found in seeds, are more strictly oils and generally contain lower amounts (10–20%) of saturated fatty acids and high (80–90%) levels of

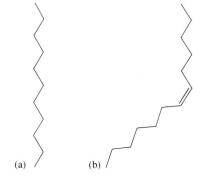

Figure 3.50 Shapes of fatty acid molecules: (a) saturated; (b) (mono-) unsaturated.

Table 3.3 Fatty acid components of neutral fats and phospholipids

Common name	Number of carbon atoms	Number of double bonds	Formula
saturated fatty acids			
palmitic	16	0	$CH_3(CH_2)_{14}COOH$
stearic	18	0	$CH_3(CH_2)_{16}COOH$
arachidic	20	0	$CH_3(CH_2)_{18}COOH$
unsaturated fatty acids			
oleic	18	1	$CH_3(CH_2)_7CH{=}CH(CH_2)_7COOH$
linoleic	18	2	$CH_3(CH_2)_4(CH{=}CHCH_2)_2(CH_2)_6COOH$

unsaturated fats. The exact composition of the unsaturated fraction varies from plant to plant, for example, olive oil is 79% oleic acid, whereas sunflower seed oil is 75% linoleic acid. Margarine is in fact made by chemically reducing the double bonds in the natural oil, either completely to give a hard, butter-like fat, or partially to give a soft variety 'rich in polyunsaturates'.

Linoleic acid, the major constituent of sunflower oil, is actually one of two essential fatty acids in our diet, since our bodies are unable to synthesize it. Essential fatty acids can be obtained only from dietary sources, and when they are absent from the diet a severe pathological condition, characterized by scaly skin, loss of hair and low growth rate, will develop.

The neutral fats are non-polar lipids. The other groups of lipids found in cells are relatively polar molecules (i.e. contain charged groups) and have interesting properties that will become apparent later when considering how these polar lipids are assembled, along with other macromolecules, into cell structures. The polar lipids are principally involved in membrane structure and function.

3.7.2 Membrane lipids

The majority of lipids found in cell membranes are **amphipathic lipids**, which have both a long hydrocarbon tail region with little affinity for water (i.e. hydrophobic) and a hydrophilic part, commonly referred to as the *polar head group*. The peculiar ability of these compounds to form both hydrophilic and hydrophobic interactions makes them ideally suited to the formation of membranes. This property can be illustrated by an experiment in which the lipid is suspended in aqueous solution. In the presence of a large excess of water over lipid, the lipids form *micelles* (Figure 3.51) in which the hydrophobic tails associate in a spherical droplet with the polar head groups exposed at the surface to the polar solvent, water.

This amphipathic nature of the membrane lipids is due to their 'tuning-fork'-like shape (Figure 3.52a). Unlike the triglycerides we met earlier, where all three hydroxyl groups of glycerol are replaced by long fatty acid chains, molecules of membrane lipids have only two fatty acid chains, together with a polar group (Figure 3.52). The 'tuning-fork' arrangement results from the polar group forming the 'handle' and the two fatty acid chains the 'prongs'.

There is a large range of membrane lipids, which differ in the identities of their fatty acid chains, or of the polar groups, or both. The major lipids of cell membranes are phospholipids, which are similar to triglycerides in that they are based on glycerol bound to fatty acid chains, but they have a phosphate-

Figure 3.51 Amphipathic lipid molecules in water form a micelle, with the hydrophobic tails internal, and the polar head groups exposed to the water.

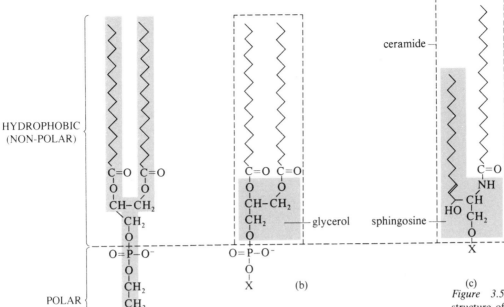

HYDROPHOBIC
(NON-POLAR)

POLAR

ceramide —

glycerol

sphingosine —

(a)

(b)

(c)

Figure 3.52 The basic 'tuning-fork' structure of membrane lipids. (a) Phosphatidyl choline, a common phospholipid. (b) The general structure of phospholipids. (c) The general structure of sphingolipids. X represents alternative end groups.

containing group and one of a number of alternative end groups (X in Figure 3.52). Phosphatidyl choline (Figure 3.52a) is the predominant phospholipid of most mammalian cell membranes. Frequently a nitrogen-containing base is attached to the phosphate group to form a doubly charged (+ and −) head group.

Another major class of membrane lipids is the sphingolipids. These have the basic 'tuning-fork' design, but differ from the phospholipids in several ways (Figure 3.52). The hydrophobic portion of the molecule is ceramide (a derivative of sphingosine, a long chain amino alcohol) and the polar group may be either a phosphate-containing group or some sugar residues.

The final group is known as the *simple lipids*, which do not contain fatty acids, represented in membranes by **cholesterol** which is amphipathic, like the phospholipids and sphingolipids, but not based on glycerol and fatty acids. The polar hydroxyl group is attached to a system of hydrophobic carbon rings and a hydrocarbon tail (Figure 3.53). Cholesterol plays a major role in the structure and function of many cell membranes; for example, it represents about 26% of all lipids in the mammalian red blood cell membrane. In contrast, it rarely occurs in plant and prokaryotic cell membranes.

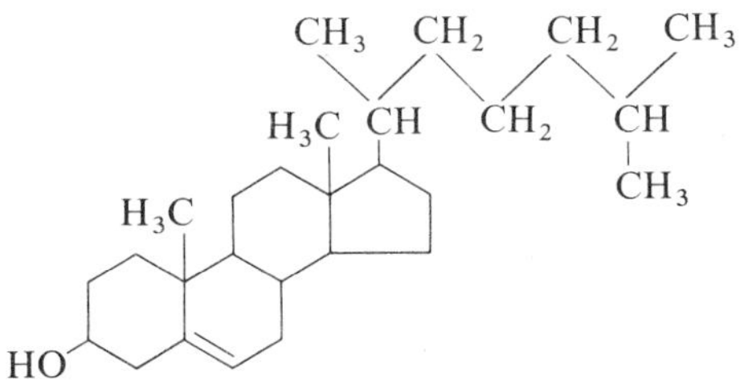

Figure 3.53 The structure of cholesterol: a polar (−OH) group with hydrophobic rings and side chain.

Cholesterol is equally important in other roles that it performs in living organisms. The simple lipids are a heterogenous collection of molecules, including fat-soluble vitamins and steroid hormones. Cholesterol is the molecule from which all steroid hormones in animals are synthesized (e.g. the male sex hormone testosterone and the female sex hormone oestrogen). It is also the major component of the bile salts which, after secretion by the liver, move to the small intestine where they are mixed with ingested fats and so aid their absorption into the bloodstream.

3.8 MACROMOLECULAR ASSEMBLIES

So far we have considered the structure of individual macromolecules. We have seen that proteins often contain a number of separate subunits, which may or may not be identical, which interact to form the biologically active molecule. Individual lipid molecules also interact to form large aggregates or micelles. Some macromolecules, such as the glycoproteins, for example, are hybrids between different types of molecules which are covalently linked together. Indeed, interactions between macromolecules are the rule rather than the exception.

The structural components of cells that you examined in Chapter 2 are invariably assemblies of macromolecules, which are held together by weak (*non-covalent*) interactions. Nucleic acids are often associated with proteins, for example, in eukaryotic chromosomes which are DNA–protein complexes, in ribosomes which are RNA–protein complexes, and in viruses which consist of nucleic acid cores encased in protein coats. Proteins can also interact with lipids to produce membranes, such as those that separate the contents of a cell from the environment and those that separate different intracellular organelles from one another. Membrane proteins are also often *covalently* linked to polysaccharides (i.e. glycoproteins, see Section 3.6.3). Finally, polysaccharides form extraordinarily complex structures such as the cell walls of bacteria and of plants.

Some of these macromolecular assemblies are almost completely understood, while the structures of others await further elucidation. Here we will consider two examples which illustrate the general principles of macromolecular assembly: eukaryotic chromosomes and membranes.

3.8.1 Eukaryotic chromosomes

The DNA of all eukaryotes is organized into morphologically distinct units called chromosomes, each of which contains only a single enormous DNA molecule. For example, the *single* DNA molecule in a chromosome of the fruit-fly *Drosophila* has an M_r value greater than 10^{10} and is a double helix with a length of 1.2 cm.

This is an extraordinary value and indeed these DNA molecules are much too long to be seen in their entirety by electron microscopy, because the field is only about 0.01 cm in diameter.

◇ From the information given in Figure 3.38, what is the ratio of length to width of *Drosophila* DNA?

◈ The width of all DNA molecules is 2 nm or 2×10^{-7} cm. The ratio of length to width is therefore $1.2 \text{ cm}/2 \times 10^{-7} \text{ cm} = 6\,000\,000 : 1$.

A chromosome is therefore very much more compact than the DNA molecule that it contains, but DNA cannot fold spontaneously to form such a compact structure because the molecule would be strained enormously, as a consequence of the large number of negative charges it carries. Instead, DNA is made compact by a hierarchy of different types of folding, each of which is mediated by one or more protein molecules.

The DNA molecule in a eukaryotic chromosome is bound to basic proteins called **histones**, and the complex of DNA and histones is called **chromatin**. Histones have an unusual amino acid composition in that they are extremely rich in the amino acids lysine and arginine.

◇ Look back at Table 3.2. What is the similarity between these two amino acids?

◈ They are both positively charged at physiological pH.

The positive charge on histones is why they are described as 'basic proteins', and it is one of the major features of the molecules.

◇ Why is this feature significant in terms of DNA–histone interactions?

◈ Recall from Section 3.5.1 that nucleic acids have a strong negative charge because of the phosphate groups.

The ionic interactions between positively charged histones and the negatively charged DNA are apparently the most important stabilizing force in chromatin. This can be demonstrated by placing chromatin in solutions of high salt (NaCl) concentration which disrupts the ionic attractions. In these conditions chromatin breaks down to yield free histones and DNA; such chromatin is described as being completely dissociated (see left-hand side of Figure 3.55).

◇ From your knowledge of the reversible denaturation experiment on RNAase, how might chromatin be reconstituted in the above experiment?

◈ The NaCl could be removed by dialysis (as explained in Box 3.2).

Chromatin reforms under these conditions, showing that no other components are required for its formation. Thus weak interactions between two different types of macromolecule (DNA and protein) stabilize a cellular component.

Chromosomes have been isolated from cells, *gradually* dissociated, and examined by electron microscopy. The basic unit of structure is chromatin, a fibre 11 nm wide, which appears like a string of beads. The structural hierarchy of a chromosome, which has been deduced from a variety of studies, is shown in Figure 3.54. The higher orders of structure are still somewhat speculative but a great deal of information is known about the 'beads' themselves, each of which is an ordered aggregate of DNA and histones.

Treatment of chromatin with enzymes which digest it has revealed information about the composition of the 'beads'. The enzyme DNAase I (which digests 'free' DNA but which cannot attack DNA that is in contact with protein) chops chromatin into a collection of small particles each containing DNA and histones (Figure 3.55). Removal of the histones reveals that each fragment of DNA is 200 base pairs long. The 'bead' is called a nucleosome, each of which contains the 200 base pairs of DNA, one molecule of histone H1 and two molecules each of four other histones (i.e. an octamer). The

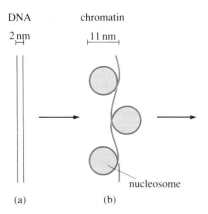

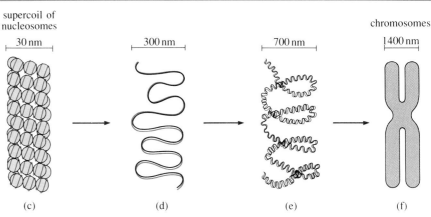

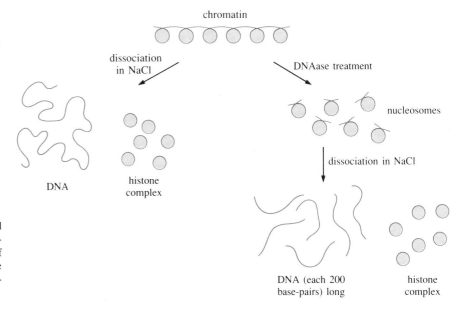

Figure 3.54 Various stages in the condensation of (a) DNA and (b–e) chromatin to form (f) a chromosome. The dimensions indicate known widths of intermediates, but the detailed structures are somewhat speculative.

Figure 3.55 Chromatin can be disrupted to give DNA and histones either by dissociation in NaCl or by a combination of DNAase treatment and dissociation. The DNAase-digestion method alone produces separate nucleosomes.

nucleosome can be digested further with a second enzyme to remove a short section of DNA and histone H1, to give a core particle of a DNA molecule with 140 base pairs and the histone octamer (Figure 3.56).

A nucleosome therefore consists of a core particle (the histone octamer around which the 140-base-pair segment of DNA is wrapped like a ribbon, Figure 3.56), additional linker DNA and one molecule of histone H1. The H1 binds to the histone octamer and to the linker DNA, causing the linkers to draw closer to the histone core.

The overall structure of the chromatin fibre is probably a zigzag as shown in Figure 3.54b. Assembly of DNA and histones into chromatin is the first stage of shortening of the DNA molecule in a chromosome—this is a sevenfold reduction in length of the DNA and the formation of a beaded flexible fibre 11 nm wide, roughly five times the width of a DNA molecule.

The binding of DNA and histones in chromatin has been studied. About 80% of the amino acids in the histones are in α-helices with many of these structures sitting in the major groove of the DNA helix. The complex is stabilized by ionic attraction between the positively charged lysines and arginines of the histones and the negatively charged phosphates of the DNA.

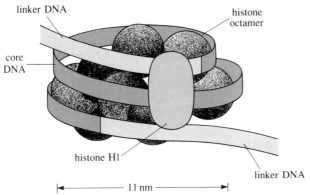

core
DNA

linker DNA

histone
octamer

histone H1

linker DNA

|← 11 nm →|

Figure 3.56 Schematic diagram of the nucleosome core particle, plus linker DNA and histone H1. The DNA molecule makes almost two full turns around a histone octamer. Histone H1 is bound to the linker DNA.

Compact chromosomes are formed by folding and refolding of the 11 nm nucleosome fibre as shown in Figure 3.54. The second level of folding shortens the 11 nm fibre to form a supercoil with six nucleosomes per turn—called the 30 nm fibre, which has been isolated and is well characterized. The remaining levels of organization—folding of the 30 nm fibre—shown in Figure 3.54d–f are less well understood. Electron micrographs of isolated chromosomes from which histones have been removed indicate that the partially unfolded DNA has the form of an enormous number of loops that seem to extend from a central core composed of non-histone proteins.

The compaction of DNA and protein into chromatin and ultimately into a chromosome greatly facilitates the distribution of the genetic material during cell division: DNA strands in the chromosomes divide at an early stage when the structures are less condensed, but final separation is of the compact structures. Some unfolding and refolding of chromatin is probably involved in gene activity (expression).

3.8.2 Membranes

Our second example of a macromolecular assembly is the cell membrane. The structures of all membranes have many common features, but differences exist which accommodate the varied functions of different membranes.

◇ From Chapter 2, what is the most notable feature of membrane structure viewed with the electron microscope?

◈ It consists of two layers called a bilayer.

This structure is a consequence of the chemical nature of the lipids that form the membrane.

◇ From Section 3.7.2 summarize the main feature of a membrane lipid.

◈ Membrane lipids are amphipathic, with long hydrophobic fatty acid tails and a polar head group.

There are many membrane lipids, of which the most common are the phospholipids, consisting of a polar phosphate-containing head group and two long non-polar hydrocarbon chains. Such lipids can therefore participate in both hydrophobic and hydrophilic interactions. As we saw in Section 3.7.2, when isolated and purified from cells these membrane lipids tend to take up positions in aqueous solution so that the hydrocarbon tails aggregate via hydrophobic interactions in such a way that the polar head groups are in contact with the watery medium.

The second major group of membrane components is the proteins, which are contained wholly or in part within the membrane. Numerous physical measurements have shown that the membrane proteins can diffuse freely laterally throughout the lipid bilayer. The degree of motion is determined by the fluidity of the lipid layer, which in turn is a function of the hydrophobic interactions between the hydrocarbon tails of the membrane lipids. This phenomenon was first noted by S. Jonathan Singer and Garth Nicholson and led to the fluid-mosaic model of membrane structure shown in Figure 3.57.

◇ What interactions are involved in stabilizing this macromolecular assembly?

◈ Hydrophobic interactions between the hydrocarbon tails of the membrane lipids and also between these and the hydrophobic regions of the membrane proteins. Hydrophilic interactions between the polar head groups of the lipids and the polar regions of the proteins and with the watery medium.

A feature of the arrangement of the proteins within the lipid bilayer is that these molecules do not spontaneously rotate (flip over) from one side of the membrane to the other. This is because the membrane proteins have polar and non-polar regions as shown in Figure 3.57. The external region is polar and the internal region is non-polar, so rotation would require the polar region to pass through the non-polar centre of the bilayer.

Figure 3.57 The structure of the cell membrane according to the fluid-mosaic model. Four proteins (a–d) are embedded to various degrees in a lipid bilayer such that the hydrophobic surface of each protein (heavy lines) is buried in the membrane and the polar regions are external.

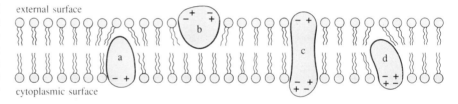

◇ What is the consequence of this in terms of the two surfaces of the membrane?

◈ If certain proteins protrude from one surface but not the other, then the two surfaces are different.

Different proteins protrude from the two surfaces of the membrane, so the membrane is an asymmetric structure with an inside and an outside. The inner surface is referred to as the cytoplasmic surface, while the side exposed to the watery medium outside the cell is called the external surface. Notice also from Figure 3.57 that some proteins go all the way through the lipid bilayer (c, for example, in the figure). As you will see in Chapter 7, these proteins are important in transport across membranes.

The final important macromolecular components of membranes are the polysaccharides, which are always covalently linked to proteins (as glycoproteins), to lipids (as glycolipids) or to both lipids and proteins (as glycolipoproteins).

◇ What roles do glycoproteins play?

◈ They are informational molecules, important in cell recognition in a variety of processes.

Structure and function of membranes will be considered in greater detail in Chapter 7. For the present it is sufficient to realize that membranes are complex assemblies of a wide variety of macromolecules, whose structures are stabilized by different sorts of interactions involving weak bonds.

Summary of Sections 3.6 to 3.8

1 Polysaccharides have structural or food-storage roles. They are polymers of sugars linked by glycosidic bonds. The primary structure of polysaccharides varies in terms of sugar composition and in the way in which these monomers are joined via the glycosidic bonds.

2 Higher-order structure of polysaccharides varies according to function. Glycogen, a food store, has a globular tertiary structure, whereas cellulose, a structural component of plants, forms long rigid fibres.

3 Lipids are water-insoluble organic molecules with a variety of functions: food storage; structural and recognition components of membranes; and carriers of signals (as hormones).

4 Triglycerides are the main food-storage lipids, and consist of glycerol and long chain fatty acids.

5 Membrane lipids are amphipathic molecules, with hydrophobic hydrocarbon tails and polar head groups. There are two main types: phospholipids and cholesterol.

6 Different classes of macromolecules can interact via non-covalent weak interactions to form macromolecular assemblies, of which eukaryotic chromosomes and membranes are two examples.

7 DNA is condensed in a eukaryotic cell by interaction with histone proteins to form chromatin, which is further compacted by folding and refolding to form the chromosome.

8 Membranes are macromolecular assemblies of phospholipids and proteins. Hydrophobic and hydrophilic regions are arranged to give the characteristic bilayer, with hydrophobic fatty acid hydrocarbon chains 'inside' the membrane and the polar head groups in contact with the external watery medium. Proteins protrude out of the bilayer and interact with various components on both sides of the membrane.

Question 14 (*Objectives 3.1, 3.4 and 3.5*) If you pull out a length of sewing cotton, which consists mainly of cellulose, it will resist stretching and then suddenly break. What bond types are involved in resisting stretching?

Question 15 (*Objectives 3.1 and 3.6*) Explain what is meant when a diet is described as being 'high in polyunsaturates'.

Question 16 (*Objectives 3.1 and 3.6*) The following chemical structure is that of a particular lipid molecule. Carefully study this structure and then choose two accurate descriptions of it from the list overleaf.

(a) It is a fully saturated lipid.

(b) The presence of sphingosine indicates that this is a sphingolipid.

(c) The presence of a phosphate group indicates that it is a phospholipid.

(d) It is a triglyceride.

(e) The lipid is amphipathic with fatty acid chains and a polar head group.

(f) The structure is that of cholesterol.

Question 17 (*Objectives 3.1, 3.4 and 3.7*) A DNA–protein complex has recently been isolated from an unusual species of marine alga. It has been proposed that the DNA and protein in this complex are covalently linked. Which of the following experimental results provide information about the proposed covalent linkage?

(a) Treatment of the complex with concentrated NaCl does not dissociate the complex.

(b) Treatment of the complex with a reagent that disrupts hydrophobic interactions does not dissociate the complex.

(c) Treatment of the complex with a protease (an enzyme which breaks down protein) removes all protein from the DNA.

SUMMARY OF CHAPTER 3

1 The contents of a cell are subdivided by membranes into various compartments, each with its characteristic macromolecular components. The main classes of macromolecules are the proteins, nucleic acids and polysaccharides; and lipids are usually grouped with them, although they are not strictly macromolecules.

2 Macromolecules can be isolated and purified on the basis of differences in solubility, charge, size and/or specificity. Electrophoresis is a widely used technique for separation on the basis of charge or size. Gel filtration separates on size differences. Separation using affinity chromatography is based on differences in specificity.

3 Each macromolecule consists of monomers covalently linked together, giving a hierarchy of structure, namely primary, secondary, tertiary and quaternary structures.

4 Various bonds are involved in maintaining the hierarchy of structure: the hydrogen bond, hydrophobic interactions, and the ionic bond.

5 Proteins are polymers based on permutations of 20 different amino acids; each unique primary structure results from the sequence of the amino acids. Peptide bonds join the amino acids together to form the linear polypeptide chain of the protein, which includes at least 20 amino acids. Variations in the number and sequences of the amino acids in these chains give rise to a wide diversity of proteins.

6 The primary structure of a protein is folded to give complex three-dimensional higher-order structures, each of which is unique to a given protein.

7 There are two principal forms of highly regular protein secondary structure: the α-helix and β-structures.

8 The tertiary structure is maintained by non-covalent weak interactions and disulphide bridges between amino acid side chains. Tertiary structure of some proteins such as RNAase can be reversibly denatured by treatment with mercaptoethanol and urea, which disrupt disulphide bridges and weak

interactions, giving rise to a random coil. Removal of the mercaptoethanol and urea enables the precise tertiary structure to reform. Thus, primary structure determines tertiary structure.

9 Some proteins, such as insulin, which are synthesized as precursor proteins, cannot be reversibly denatured owing to loss of polypeptide chains during their activation.

10 Many proteins have quaternary structure, in which several separate polypeptide chains interact to form the active molecule. The separate polypeptide chains are termed subunits, which are held together by non-covalent, weak interactions.

11 Some proteins have different functions associated with specific areas of the molecule known as domains. The classic examples here are the immuno-globulin molecules of the immune system.

12 There are two general shapes of protein molecules: globular proteins, such as enzymes, the biological catalysts; and fibrous proteins which are usually long, thin structural components.

13 Nucleic acids (DNA and RNA) are involved in the storage and transfer of genetic information.

14 Each nucleic acid is a polymer of nucleotides, i.e. a polynucleotide.

15 Secondary structures in nucleic acids are based on the formation of helical regions, notably the classic double helix of DNA, which depends on complementary base-pairing and base-stacking. RNAs seldom have regular secondary structure.

16 The tRNAs have well defined tertiary structures, somewhat analogous to those of globular proteins. Other nucleic acids (on their own) do not have precisely defined tertiary structure.

17 Polysaccharides have structural or food-storage roles. They are polymers of sugars linked by glycosidic bonds. The primary structure of polysaccharides varies in terms of sugar composition and in the way in which these monomers are joined via the glycosidic bonds.

18 Higher-order structure of polysaccharides varies according to function. Glycogen, a food store, has a globular tertiary structure, whereas cellulose, a structural component of plants, forms long rigid fibres.

19 Lipids are water-insoluble organic molecules with a variety of functions: food storage; structural and recognition components of membranes; and carriers of signals (as hormones).

20 Triglycerides are the main food-storage lipids, and consist of glycerol and long chain fatty acids.

21 Membrane lipids are amphipathic molecules, with hydrophobic hydrocar-bon tails and polar head groups. There are two main types: phospholipids and cholesterol.

22 Different classes of macromolecules can interact via non-covalent weak interactions to form macromolecular assemblies, of which eukaryotic chromosomes and membranes are two examples.

23 DNA is condensed in a eukaryotic cell by interaction with histone proteins to form chromatin, which is further compacted by folding and refolding to form the chromosome.

24 Membranes are macromolecular assemblies of phospholipids and proteins. Hydrophobic and hydrophilic regions are arranged to give the characteristic bilayer, with hydrophobic fatty acid hydrocarbon chains 'inside' the membrane, and the polar head groups in contact with the external watery medium. Proteins protrude out of the bilayer and interact with various components on both sides of the membrane.

OBJECTIVES FOR CHAPTER 3

Now that you have completed this chapter, you should be able to:

3.1 Define and use, or recognize definitions and applications of, each of the terms printed in **bold** in the text.

3.2 Describe methods for separating macromolecules on the basis of solubility, charge, size and/or specificity. (*Questions 1, 2, 3 and 8*)

3.3 Explain using simple diagrams and named examples, how the folding patterns of proteins, nucleic acids and polysaccharides may be described in terms of a hierarchy of structure, namely primary, secondary, tertiary and quaternary structures. (*Questions 4, 6, 7, 8, 9, 10 and 13*)

3.4 Give examples to illustrate the role of weak bonding in higher-order macromolecular structure. (*Questions 4, 5, 6, 7, 8, 10, 11, 12, 14 and 17*)

3.5 Illustrate, with named examples from a variety of macromolecules, the relationship between structure and function. (*Questions 5, 7, 10 and 14*)

3.6 Describe the relationship between structure and function in neutral and membrane lipids. (*Questions 15 and 16*)

3.7 Describe the macromolecular assemblies that are chromosomes and membranes. (*Question 17*)

ENZYMES: STRUCTURE, SPECIFICITY AND CATALYTIC POWER   ◆ CHAPTER 4 ◆

4.1 ENZYMES: THE KEY TO UNDERSTANDING METABOLISM

So far we have largely built up a picture of the cell as a somewhat static entity, though an undoubtedly intricate one (Chapters 1–3). Given that we have concentrated on techniques designed to obtain a *physical* description of the cell, we should not be too surprised that that is just what we have obtained! Yet, as stated in the Introduction, cells are far from static. Probably the most dramatic proof of the cell as a dynamic entity is its capacity for growth and division, to engage in the **cell cycle** outlined in Figure 4.1. It is worth considering briefly what this cell cycle must entail.

In the growth phase, prior to cell division, the cell must double its mass. This is no mere 'swelling', no simple taking in of water. In fact it involves the cell in a massive feat of chemical synthesis. It must double its quantities of all the many thousands of different molecules it contains. This it does largely by chemical synthesis from the relatively simple mix of substances it takes in from its surrounding environment. But synthesis at the molecular level is only part of the story. Various molecules, notably macromolecules, must be assembled together to fashion the organelles characteristic of the type of cell in question (Chapter 2), for these too must double in number. Finally, when the doubling in mass has been achieved, the cell divides. Like growth, division is a complex process. Most obviously, it must be accurate to ensure that each of the two resulting **daughter cells** receives its appropriate inheritance of organelles, macromolecules (including, of course, DNA) and so on, so that each daughter cell is identical to the other and to the parent cell from which they both derive.

As if this were not enough evidence of activity, the cell often carries out the mammoth demands of the cell cycle in a matter of hours or less. In some prokaryotes the complete cell cycle takes just tens of minutes, with only a very frugal input from the cell's surrounding environment. For example, given just an aqueous solution of ions of magnesium, sodium, potassium, chloride, phosphate, sulphate, ammonium, a few traces of some other ions, a source of carbon such as glucose, a supply of oxygen and a temperature of

Figure 4.1 The cell cycle of growth and division. Each cell grows till double its initial mass, then divides to give two daughter cells. These grow and divide, and so on. Each doubling of mass plus a division constitutes one cell cycle.

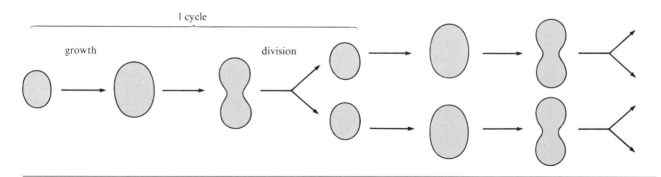

about 35 °C, the common intestinal bacterium *E. coli* can go through the complete cell cycle of growth and division, producing two cells from one, in 30–40 minutes. A highly automated car plant turning out say 20 000 Metros, Bluebirds, Sierras or what you will, *plus an additional car plant*, every hour on the hour, starting with only the basics such as rubber latex, iron ore, crude oil and sand would seem child's play by comparison.

It is this capacity for producing rapidly a whole range of molecules from simple inputs that is the epitome of the dynamic cell. And though this capacity is most readily demonstrated in growing cells, it is one that is shared to some degree or other by all cells, growing or not. For even in non-growing cells, various components, from molecules to organelles, turn over—they 'wear out' and are continuously maintained or replaced. This requires chemical synthesis, and chemical synthesis requires energy. Energy itself comes from yet more chemical activity. Thus a non-growing cell, like its growing counterparts, is a hive of chemical activity. Even 'to stand still' it must be in a kind of dynamic equilibrium, both internally and with its surrounding environment—hardly a static structure!

The sum total of the cell's chemical activity is called **metabolism**. One aspect of metabolism is how the cell handles small molecules—the sugars, fatty acids, nucleotides, amino acids and so on. Most of these the cell synthesizes from the relatively narrow range of substances available to it from its environment. This synthesis involves a number of series of stepwise chemical reactions. Each series is called a **metabolic pathway**. Each pathway leads to one or more specific small molecule products. Synthesis requires energy, and, except for cells that can trap the energy of sunlight, this too is generated chemically, also via metabolic pathways. In this case the pathways are generally specific degradative (catabolic) ones also involving small molecules. Taken all together, the synthesis and breakdown of small molecules in the cell involves many hundreds or thousands of different individual chemical reactions and intermediate chemical substances (i.e. intermediate on the pathways to particular products or energy production). This small molecule industry necessitates a large number of distinct metabolic pathways, yet many with interconnections. This whole network of pathways is lumped together under the name of **intermediary metabolism**. Intermediary metabolism is the key phase in metabolism as a whole.

◇ Intermediary metabolism is how cells produce energy and small molecules from what they take in from the environment. How can it also help to explain the production of macromolecules and organelles?

◈ Macromolecules such as polysaccharides, proteins and nucleic acids are polymers composed of small molecule monomers covalently linked together (Chapter 3). Forming the links takes energy. Synthesis of macromolecules thus utilizes some of the *products* of intermediary metabolism—the small molecule monomers *and* energy. Organelles in turn are largely composed of macromolecules.

A recent chart showing just the main metabolic pathways of intermediary metabolism has many hundreds of chemical compounds and many hundreds of reaction arrows. It is far too complicated to reproduce here. A chemist looking at such a chart for the first time might be very impressed by the cell's considerable expertise as an organic chemist and equally *de*pressed about our ever understanding how the cell achieves what it does. The chemist might, however, also be suitably impressed by how much we apparently *do* already know about intermediary metabolism. Such charts are the briefest of summaries of one of the greatest scientific achievements of the 20th century; they represent the results of the industrious hands and able minds of many thousands of biochemists over the 90 or so years since **biochemistry** (the study of the chemistry of living organisms) proper was born.

In Chapter 6 we will examine just a small portion of the reactions detailed in such metabolic charts—those of some pathways central to intermediary metabolism, to energy production in particular, and common to very many different types of living cell. For now, we are not concerned with detail or indeed any particular pathway. We just wish to try to 'make some sense' of intermediary metabolism, to see if any general principles emerge from among the great mass and variety of reactions, to attempt to understand *how* it operates. With this achieved, the more specific examination in Chapter 6 should fall into place. In trying to answer how intermediary metabolism operates, three more specific 'how' questions come to mind:

1 To try to carry out the large range of reactions that are typical of intermediary metabolism, an organic chemist in a laboratory would probably use a large range of temperatures, pressures and pH, and undoubtedly a lot of time. The cell achieves all its chemical reactions very rapidly and under very mild and relatively very constant conditions of temperature, pressure and pH—*how?*

2 The cell is by definition a single entity, a single 'pot', yet it can carry out a large number of chemical reactions and apparently not 'get confused'—*how?* For example, as Figure 4.2 shows, pyruvate can have any one of four fates inside a cell. How does the cell perform each reaction to the 'required' degree? Put another way, can the cell 'direct' how much pyruvate is converted into each product and, if so, how?

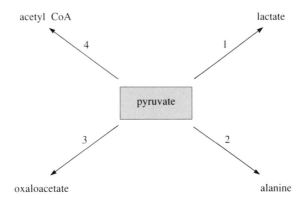

Figure 4.2 Four alternative fates of pyruvate.

◇ Can you suggest one means by which the cell can carry out a number of quite different reactions in the 'same pot'?

◆ As Chapter 2 made clear, in many senses a cell is not a 'single pot', it is compartmentalized into several 'pots', the different organelles. This 'division of labour' within a cell undoubtedly goes some way towards avoiding conflicting reactions occurring in the same space. The cell is perhaps more properly visualized as a 'laboratory' or 'kitchen' with many 'pots'.

Though compartmentation may help explain how the cell copes with a multitude of different reactions at the same time, it cannot be the whole explanation.

◇ Why not? (Think of how many types of organelle there are.)

◆ Put simply, there are just a few types of compartments but hundreds or thousands of different reactions.

In fact compartmentation does have a role to play in the fate of pyruvate—two of the reactions (1 and 2) occur within the cytosol and two (3 and 4) within the mitochondria (Figure 4.2). But, as we implied, this is only a part solution to our question, for can the cell discriminate between the two reactions in each compartment (between 1 and 2; and 3 and 4)?

3 Asking whether the cell can direct the fate of a particular substance (such as our example, pyruvate) is in a sense an instance of a wider issue—can the cell *control* certain reactions and regulate their outputs relative to others? On the larger scale, the whole of intermediary metabolism, the instinctive answer is *yes*. For without some degree of control over the various interconnected pathways we would expect the result to be chaos. That this is not the case is obvious when we recall that the cell can reproduce itself accurately and rapidly, cycle after cycle. Thus it seems that the cell *must* be capable of controlling intermediary metabolism.

As the cell is to some degree also in dynamic equilibrium with its surroundings it must also be capable of responding to external changes, because external changes could affect inputs to the cell and these in turn could affect intermediary metabolism. To maintain its internal dynamic equilibrium, and not be merely at the mercy of external changes, the cell must be able to regulate the extent of its various reactions to compensate for both internal and external changes. To regulate reactions appropriately it once again seems likely that the cell must be able to control them. *How* is this regulation and control of intermediary metabolism achieved?

Consider the above three questions carefully. In answering them, we shall develop a powerful unifying concept: namely, the key to the cell's ability to carry out intermediary metabolism effectively, and the key to *our* understanding *how*, is that the cell has a whole array of sophisticated catalysts, the **enzymes**.

Though many different enzymes exist, they all have two general features in common. One is by definition: they are all biological catalysts. The other is chemical: they are all proteins.

Like other, non-enzyme, proteins, enzymes can vary quite widely in size and structure (Chapter 3). Some enzymes are composed of a single subunit, others have several. And though most enzymes have relative molecular masses (M_r) of a few tens of thousands, the overall range is from as low as about 10 000 up to several hundred thousand. It is of course the fact that proteins are polymers that permits such a wide range of sizes and structures. Ultimately, each different protein, and hence enzyme, reflects a particular number and sequence of amino acids in the polypeptide chain(s) that constitute(s) its primary structure (Chapter 3). And, as we shall see, the particular amino acid sequence of each enzyme is critical to its function in metabolism. The key to understanding how any particular enzyme works is to unravel the link between its *structure* and its metabolic *function*. Each enzyme is unique and there are many enzymes. Nevertheless there are some general aspects of how *all* enzymes work.

We can begin to examine these general aspects by considering three properties of enzymes, ones that are central to the 'how' questions about intermediary metabolism that we posed above:

☐ Enzymes have considerable catalytic power.

☐ Enzymes are highly specific.

☐ The rates at which some enzymes operate are often subject to regulation and control.

What catalytic power and specificity entail will become evident as this chapter proceeds, as will, to some extent, what it is about enzyme structure that endows enzymes with these properties. The regulation and control of metabolic pathways, via regulation and control of enzymes, is more the subject of Chapters 5 and 6. It is however useful first to define briefly what these three aspects of enzyme function mean.

Put simply, the **catalytic power** of an enzyme is the degree to which it speeds up a reaction. Frequently enzymes can effect increases in reaction rate of the order of 10^9 to 10^{12} times that of the uncatalysed reaction. This alone goes a long way towards explaining how a cell can carry out reactions rapidly under mild conditions. For example, the efficient transportation of carbon dioxide from the tissues to the lungs depends on converting the gas into bicarbonate ions which are carried in the blood plasma: But the actual formation of bicarbonate ions occurs in the red blood cells from where they diffuse into the plasma:

$$CO_2 + H_2O \rightleftharpoons H^+ + HCO_3^-$$

A mammal's ability to do this rapidly is largely due to the presence in red blood cells of the enzyme carbonic anhydrase. This catalyses the formation of bicarbonate ions, increasing the rate of reaction by about 10^7 times over the uncatalysed reaction.

As an aside, you might like to note how this knowledge of the catalytic power of an enzyme, carbonic anhydrase, helps provide both an explanation at the *molecular* level of the reaction *and* at the *physiological* level. Without the rapid enzyme-catalysed reaction of carbon dioxide, the tissues would not be able to get rid of this potentially toxic waste product quickly enough.

Catalysts, even ones with considerable power, are not unique to living cells—non-biological catalysts are part of the day-to-day armoury of chemists. Though enzymes do often possess greater catalytic power than their synthetic counterparts, it is their **specificity** that really marks enzymes out. That is, each type of enzyme will only catalyse one particular chemical reaction or a set of closely related reactions. The reactions must be closely related because enzymes are specific as to the **substrate** (the reactant in an enzyme-catalysed reaction) that they will act on. In contrast to most non-biological catalysts, which are relatively promiscuous in their use of substrates, enzymes are very specific. Indeed some enzymes are totally specific: they will only accept one particular substance, and one alone, as a substrate for the reaction. In so doing, some enzymes can even discriminate between substances that are chemically almost identical. For example, the enzyme lactate dehydrogenase (LDH) obtained from certain bacteria is able to distinguish between optical isomers. It acts on (+) lactate to catalyse its oxidation to pyruvate, but has no effect on (−) lactate:

$(+)$ lactate   pyruvate $(-)$ lactate

◇ What in structural terms is the difference between (+) and (−) lactate?

◈ The relative orientation of an −OH and an −H about the central (chiral) carbon atom.

LDH from some other organisms has the opposite specificity; that is, it will only catalyse the reaction involving $(-)$ lactate, not the $(+)$ isomer.

(For convenience we speak of an enzyme's specificity for its substrate, but enzyme-catalysed reactions often involve more than one reactant/substrate; say, reactions of the type $A + B \rightleftharpoons X + Y$. The enzyme will be specific for each of its substrates; only accepting as substrates A or substances closely related to A, and B or substances closely related to B. A and B need not be at all related chemically; the enzyme is independently specific for each substrate, here A and B.)

In addition to being very selective about a substrate, an enzyme is very precise about what reaction it catalyses, what chemical conversion that substrate undergoes. Thus as well as having considerable *substrate* specificity, enzymes also exhibit **product specificity**. Whereas one type of enzyme may help convert a substance to a particular product, another type of enzyme may catalyse the conversion of the same substance to some other product. Each enzyme, by lowering the activation energy for *its* particular reaction rather than for the other, exerts its own specific *directive effect*. Once each product is made in the cell it is further converted (via another different enzyme-catalysed reaction) into the next product in the stepwise sequence of the particular metabolic pathway of which it is an intermediate. Thus different enzymes can effectively direct or funnel the same substance into different metabolic pathways. You can now begin to see how a cell *might* be able to discriminate between fates for the same substance (Figure 4.3) even when present in the same compartment in the cell.

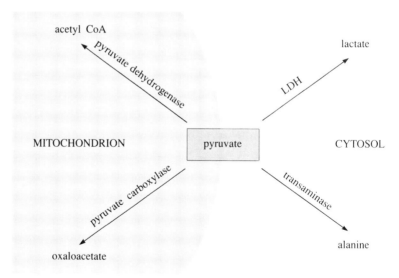

Figure 4.3 Alternative fates of pyruvate, the enzymes involved, and where the reactions take place.

Different enzymes could 'compete' for their common substrate (pyruvate, for example). Their relative competitiveness would then depend on their relative catalytic power and the relative amount of each enzyme in the cell.

◇ Are the enzymes actually needed for the reactions (of, say, pyruvate) to occur?

◈ Strictly speaking, *no*. As a catalyst, an enzyme speeds up a reaction by lowering the activation energy. It does not alter the equilibrium position of the reaction and it cannot 'cause' an otherwise impossible reaction to occur. In practice, however, only those reactions which are enzyme-catalysed will occur at any appreciable rate in the living cell. Hence those pathways, of which the enzymes present are members, will effectively 'mop up' any common substrate long before other, non-catalysed, reactions get a real look in.

In this way enzymes might exert a degree of control, influencing the rates of flow of substances into, and conversion via, particular metabolic pathways. However, in living cells, both the catalytic power and the actual physical amount of certain enzymes can change in response to local conditions, such as the presence or absence of some specific substances. Altering these aspects of a particular enzyme inevitably in turn affects *its* influence on a metabolic pathway. As you will see in Chapters 5 and 6, systems capable of altering the amount or catalytic power of enzymes form the fundamental basis of **regulation and control** of metabolic pathways.

We have now introduced the aspects of enzymes that we examine later in more detail—*catalytic power* and *specificity* in this chapter, and *regulation and control* in the next two. We have also shown briefly why greater understanding of these aspects of enzymes is the key towards an understanding of intermediary metabolism. However, a logical starting point for a more detailed discussion of enzymes is to consider how we actually study them in the laboratory.

As you are no doubt all too aware by now, a common feature of scientific study is naming things—'if you don't know what it is, name it'! Less cynically, when studying anything, giving it a distinct, and hopefully informative, name is perfectly sensible. Enzymes are no exception. So first we shall briefly consider the logical basis behind the naming of enzymes (Section 4.2). Then we shall look at what is the first fundamental of any enzyme study: how to detect and measure the amount of an enzyme, that is how to *assay* it (Section 4.3).

4.2   THE NAMING OF ENZYMES

By definition an enzyme is a biological catalyst that catalyses a specific chemical reaction. Thus, logically, enzymes can be named after the specific reactions that they catalyse; indeed this is the convention adopted, with the additional feature that all enzymes thus named have the suffix *-ase*. Nowadays, just as there is a systematic method for naming and classifying living organisms (taxonomy), so too is there one for enzymes.

Thus mucopeptide *N*-acetylmuramoylhydrolase together with its internationally agreed Enzyme Commission (EC) number 3.2.1.17 identifies an enzyme unambiguously as one that catalyses 'hydrolysis of β-1,4 linkages between *N*-acetylmuramic acid and 2-acetamido-2-deoxy-D-glucose residues in a mucopolysaccharide or mucopeptide'. The hierarchy of numbers in 3.2.1.17 identifies its membership of various classes and sub-classes of enzymes. The first number (in our example, 3) places the enzyme in one of six broad classes of enzymes according to the general type of reaction that it catalyses, as shown in Table 4.1; the second and third numbers (here, 2 and 1) further define the type of reaction, placing it in a sub-class and sub-sub-class; the final number (here, 17) is the serial number for the enzyme in its sub-sub-class. Over 2 000 enzymes so far have been sufficiently well studied to have acquired systematic names and numbers.

A particular enzyme, i.e. one catalysing a particular reaction, generally bears the same systematic name and number irrespective of where the enzyme comes from (which species of organism or which tissue). As you will see in

Table 4.1 Enzyme nomenclature: the six broad classes of enzymes

Class	Type of reaction catalysed	Example (EC number and *common* name)
1 Oxidoreductases	oxidation and reduction (removal or addition of hydrogen atoms)	1.1.1.27 lactate dehydrogenase (LDH)
2 Transferases	transfer of chemical group (e.g. phosphate, methyl) from one substance to another	2.7.12 glucokinase
3 Hydrolases	hydrolysis (splitting of bonds by water)	3.2.1.22 β-D-galactosidase
4 Lyases	removal of chemical groups from substances	4.2.1.1 carbonic anhydrase
5 Isomerases	structural rearrangements of atoms within one molecule	5.3.1.9 glucose 6-phosphate isomerase
6 Ligases	joining together of two molecules utilizing energy (such as from ATP)	6.2.1.1 acetyl CoA synthetase

Section 4.8, this does not mean that the same enzyme isolated from different species or tissues is necessarily identical chemically. It just means that it catalyses essentially the same reaction.

Many enzymes also have other, more common, names. These are less scientifically descriptive, but infinitely more manageable and were often acquired before the systematic method was adopted. Thus mucopeptide *N*-acetylmuramoylhydrolase (EC 3.2.1.17) is better known as lysozyme (not even an *-ase* suffix!), L-lactate:NAD^+ oxidoreductase (EC 1.1.1.27) as lactate dehydrogenase or LDH, and carbonate hydro-lyase (EC 4.2.1.1) as carbonic anhydrase (Section 4.1). We will use the less systematic, more common, names for the enzymes that we have occasion to mention; often these are in fact ones recommended by the Enzyme Commission, in preference to the full cumbersome systematic ones.

4.3 THE ASSAYING OF ENZYMES

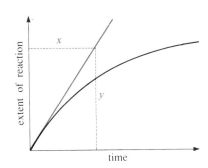

Figure 4.4 Reaction–time plot for an enzyme-catalysed reaction and how to calculate initial rate from the tangent at time zero (blue line).

To measure the amount of enzyme present in a particular preparation we perform an **enzyme assay**; the term *assay* means simply measurement. We do not measure enzyme *mass*: what we actually measure is the rate of the reaction catalysed by a preparation of the enzyme in question. We take a measured sample of the preparation, add the components needed for the chemical reaction (substrates) under appropriate conditions (pH, temperature, etc.) and follow the reaction with time. The extent of reaction (appearance of products *or* disappearance of substrates) per unit time is, by definition, the rate of reaction. As the rate of reaction is directly proportional to the amount of enzyme present (e.g. double the enzyme, double the rate), the rate of reaction gives a measure of the amount of enzyme present. To calculate the rate of reaction we can plot a graph of the measured extent of reaction against time, a **reaction–time plot**. This tends to give a curve like the black line in Figure 4.4.

◇ How can we be sure that the reaction observed is actually being catalysed by the enzyme preparation and not just occurring at this rate anyhow?

◆ This is a serious point. Enzymes as catalysts only speed up reactions, they do not 'cause' them to occur. Therefore there must be an uncatalysed rate of reaction. This is taken into account by including *controls* in the experiment. One control comprises a so-called *enzyme blank*, an assay performed under identical conditions but leaving out the enzyme preparation. The measured values of any reaction that occurs in the blank can be subtracted from those occurring in the sample containing the enzyme preparation. The net result must be due to catalysis by the enzyme and it is these net values that are plotted in the reaction–time plot, as in Figure 4.4. In practice most blanks react so slowly as compared with the enzyme-containing sample that there is little or no measurable reaction in the blank within the time-span of the experiment (i.e. the time needed to get sufficient reaction in the presence of enzyme to enable a reaction–time plot to be constructed).

◇ Why does the enzyme-catalysed reaction shown in Figure 4.4 tail off with time?

◆ There are several possible contributing reasons, including: the enzyme becoming inactivated during the reaction as a result of its thermal instability; possible inhibition of the enzyme by the products of the reaction; depletion of substrate, causing the reaction rate to drop (Section 4.4.1); the build-up of products causing the reverse reaction to become more significant (all chemical reactions are, in principle, reversible).

Because of this tailing off, the assessment of the rate of reaction depends on what part of the curve you look at. As the reaction is less likely to be affected in its early stages by any of the factors mentioned above, the rate of reaction is defined as being equal to the **initial rate** of the reaction. So the terms *rate* and *initial rate* are generally used synonymously. (The initial rate is also frequently called the *initial velocity*.) The initial rate is obtained by constructing the tangent to the curve at time zero (blue line in Figure 4.4) and calculating the rate based on this tangent (i.e. y/x). By convention, initial rate is expressed as μmoles of substrate converted (to product(s)) per minute.

Knowing initial rate we can then express the amount of enzyme present as enzyme activity (i.e. the catalytic activity) in a particular preparation in terms of the initial rate per amount of preparation. By convention the amount of preparation is expressed in terms of the amount of protein (in milligrams, mg) contained in the preparation (protein can be measured chemically). The initial rate per mg of protein is known as the **specific activity** (of enzyme) of a preparation. So:

specific activity = μmoles substrate converted per minute
per mg protein (i.e. $\mu mol\, min^{-1}\, mg^{-1}$)

The specific activity of a particular preparation of an enzyme is a good indicator of the purity of the preparation relative to other preparations of the same enzyme.

◇ Can you see why this should be the case? Consider, for example, two preparations of the same enzyme; one, *a*, in which the enzyme is 25% pure and another, *b*, in which it is 50% pure. What will be the relative specific activities of the two preparations? (Assume the impurities are other proteins.)

◆ Specific activity is expressed per mg protein. So to compare the specific activities of the two preparations we would have to compare their initial rates for the same mass of protein (i.e. per mg). For each mg of protein in *a* only 25% is actually the enzyme in question, and therefore only 25% of the protein contributes to catalysis and hence to the measured initial rate. In *b* 50% of the protein is the enzyme in question and therefore each mg of protein used in the assay would contribute twice as much to the initial rate as in *a*. Therefore *b* will have twice the specific activity of *a*.

Thus even if we never knew the actual purities of the two preparations, on measuring their specific activities (i.e. initial rates per mg protein) we would find that of *b* was twice that of *a* and hence preparation *b* must be twice as pure as preparation *a*, irrespective of the actual purity values.

When actually assaying an enzyme preparation in order to calculate its specific activity, it is very important to choose optimal conditions (i.e. those yielding maximum activity) and to note these carefully. This is because factors such as temperature, pH and the composition of the solution can all profoundly affect the activity of an enzyme. To make the assay reproducible it is vital to know what these factors are and to make others aware of them if the results are published. Using specific activities of two preparations as a measure of relative enzyme purity is of little meaning if they were assayed under widely different conditions, as factors other than the relative purities of the two enzymes may well play a significant part.

We have now seen how the raw data gathered from following an enzyme-catalysed reaction can, via a reaction–time plot, be used to calculate the initial rate of a reaction and hence the specific activity of an enzyme preparation. This does however beg the not insignificant question of how, in the first place, we can follow a reaction! The method used depends on the reaction in question but the principle is universal. For in a reaction where some reactants, say A and B, are converted to products, say C and D:

$$A + B \rightleftharpoons C + D$$

the reaction can be followed by observing either the appearance of one or both of the products (C and D) or the disappearance of one or both of the reactants (A and B). To do this, any feature that distinguishes C or D from A or B can be used. So, for example, in the reaction catalysed by the enzyme urease:

$$\begin{array}{c} H_2N \\ \diagdown \\ \diagup \\ H_2N \end{array} C{=}O \ + \ H_2O \ \underset{urease}{\rightleftharpoons} \ CO_2 \ + \ 2\,NH_3$$

urea   water   carbon   ammonia
                              dioxide

the enzyme 'could be' detected by the production of the pungent smell of ammonia, something distinguishing it from the other components. But how much smell? How can you measure it? Obviously some quantifiable entity must be used if it is to form the basis for measuring a rate of reaction. Smell is not readily quantifiable. We will now consider some standard assay techniques based on characteristics that are quantifiable and accurately measurable.

4.3.1 Absorption of light

Most organic compounds absorb light. The wavelengths of the light they absorb depend on their chemical structures, and the amounts they absorb on their concentrations. If the substrates and products of a reaction absorb light of different wavelengths, then by observing the change in the intensity of absorption at one such wavelength, the reaction can be followed (see Figure 4.5, later). Where the wavelength in question is in the visible region of the spectrum, the reaction will be accompanied by a colour change and this can be measured by a **colorimeter**, an instrument capable of measuring the intensity of colour. Frequently, enzyme-catalysed reactions do not involve coloured substrates or products, but it is sometimes possible to replace the natural substrate of the enzyme by a related substance that is also a substrate and which on reaction yields a coloured product. For example, phosphatases are enzymes that occur very widely in living cells. They are generally not highly specific (unlike LDH, for example) and will catalyse the removal of phosphate groups from a variety of molecules. Most such reactions involve no colour change. However, instead of using a natural substrate, phosphatases can be assayed with *para*-nitrophenyl phosphate as substrate, because on reaction, this colourless compound yields *para*-nitrophenol, which is intensely yellow at neutral to alkaline pH:

para-nitrophenyl phosphate (colourless)   *para*-nitrophenol (yellow)   phosphate (ion) (colourless)

Another trick is to measure the substrates remaining or products produced at various times by some *secondary* reaction that gives a colour change. Thus using natural substrates that give *colourless* products, phosphatase can be assayed by taking samples from the reaction mixture after various times and measuring the phosphate ions produced by adding a substance that reacts with phosphate ions to give a blue compound. The intensity of this blue colour is measured with a colorimeter.

Instruments are also available for measuring absorption in non-visible parts of the spectrum, such as the ultraviolet or infrared regions. So reactions involving changes in absorption in these regions can be readily assayed. One example involves the **coenzyme** (of which more in Section 4.5.1) **nicotinamide adenine dinucleotide (NAD$^+$)**, which acts as a **hydrogen acceptor** in certain enzyme-catalysed reactions. That is, it participates in reactions of the type:

$$AH_2 + NAD^+ \xrightleftharpoons[\text{}]{\text{enzyme}} A + NADH + H^+$$

One such reaction is in fact that catalysed by lactate dehydrogenase (LDH) mentioned in Section 4.1. The reaction is more properly represented by:

lactate   pyruvate

As NADH absorbs ultraviolet light at 340 nm and NAD^+ does not, reactions of the type shown above can be followed at this wavelength (see Figure 4.5) with an ultraviolet **spectrophotometer**, an instrument that can measure the intensity of ultraviolet light.

4.3.2 Measurement of pH

Some enzyme-catalysed reactions result in a production of hydrogen ions (H^+), or as biochemists more often call them **protons**.

◇ What will this do to the *acidity* and the *pH* of the solution in which the reaction is occurring?

◈ The acidity of a solution depends on the concentration of protons, $[H^+]$, where [] denotes concentration; the greater $[H^+]$, the greater the acidity. pH is a measure of $[H^+]$. Thus if the reaction results in an increase in $[H^+]$, the acidity will *rise* and, as $pH = -\log_{10}[H^+]$, the pH will *fall*.

In such reactions the decrease in pH, related to the increase in proton concentration, can be measured and from it the extent of reaction calculated. (Of course, the technique is applicable equally to reactions where $[H^+]$ decreases, and hence pH rises.) The measurement can be done with a **pH meter**. This essentially consists of two electrodes attached to a meter calibrated in pH units. One electrode is a standard reference electrode, the other an electrode sensitive to protons (H^+). When these electrodes are put into the solution under study, the meter measures the potential difference between the sensitive electrode and the reference electrode, and the magnitude of this difference depends on the concentration of H^+, that is $[H^+]$. The difference is registered as $-\log_{10}[H^+]$, that is, pH.

However, in using pH changes to follow an enzyme-catalysed reaction, complications can arise.

All enzymes are sensitive to pH, each enzyme having a particular **optimum pH** at which its catalytic activity is maximum (Section 4.5.3). Thus when assaying an enzyme the composition of the solution is chosen so as to be at the optimum pH of that enzyme. And to ensure that this pH is kept steady during the reaction the solution is *buffered* (as, for example, the reaction itself may be one that involves a change in $[H^+]$). A buffered solution (i.e. one that contains a **buffer**) is one that resists changes in pH; a relatively large change in $[H^+]$ produces little change in pH.

◇ But what if we are assaying a reaction of a type that itself involves an increase in $[H^+]$ and, as here, we are actually trying to use this change, detected by change in pH, as a measure of the extent of reaction? What will a buffer do to our assay?

◈ A buffer will 'swamp' the increase in $[H^+]$, show relatively little change in pH and render the assay insensitive.

We could leave out the buffer, but then the change in pH due to the reaction might affect drastically the very enzyme activity that we want to measure! A way round this dilemma is to omit buffer but keep the pH of the reaction mixture steady by adding alkali continuously to neutralize the acid as it is produced. The rate of addition of alkali gives a measure of the rate of reaction. An instrument called a **pH-stat** can be used to keep the pH constant automatically and measure the alkali added to do so.

An analogous approach could be used for following reactions where protons are used up in the reaction (i.e. $[H^+]$ decreases and pH rises).

Electrode techniques are not limited to the measurement of hydrogen ions, because, in principle, it is possible to construct a wide variety of electrodes, each specifically sensitive to a particular ion. Electrodes made of special glasses can measure ammonium ions, for example. So reactions involving changes in the concentration of this ion could be followed too. Other electrodes, depending on somewhat different principles, measure the concentration of oxygen. Still others measure carbon dioxide or hydrogen.

4.3.3 Coupled enzyme assays

Where no convenient techniques exist for following directly an enzyme-catalysed reaction, it is sometimes possible to use an additional enzyme-catalysed reaction to measure the one that we wish to study. Say, for example, we wish to follow the rate of the enzyme-catalysed reaction $A + B \rightleftharpoons C + D$, which we can call reaction (i), and no direct technique is available. Let us suppose that one of the products of reaction (i), C, is capable of participating in another enzyme-catalysed reaction $C + X \rightleftharpoons Y + Z$, which we can call reaction (ii). Let us further suppose that the amount of Z can be readily measured. Then by *coupling* reaction (i) to reaction (ii), the rate of reaction (i) can be determined.

For example, suppose the enzyme catalysing reaction (ii) is supplied with an excess of X plus a sample taken after a measured time from reaction (i). Provided that X is in great excess, the amount of Z produced will depend on the amount of C in the sample. Hence measure the amounts of Z produced in a series of reaction (ii)s supplied with different samples (i.e. taken at different times) from reaction (i) and this in essence will allow calculation of the rate of production of C in reaction (i). In practice, coupled assays are often done continuously, that is, by mixing the components of (i) plus X and *both* enzymes, all together. As C is produced in reaction (i), it is consumed by reaction (ii), yielding Z (which is measured) continuously.

The following actual example may help clarify the principles underlying **coupled enzyme assays**. It shows how the enzyme hexokinase, an important enzyme in the metabolism of glucose (Chapter 6), can be assayed by such an assay. Hexokinase catalyses the production of glucose 6-phosphate from glucose and ATP, and so its activity can be estimated by measuring the rate of production of glucose 6-phosphate. This can be done using glucose 6-phosphate as a substrate in a *second* enzyme-catalysed reaction, that involving the enzyme glucose 6-phosphate:$NADP^+$ oxidoreductase. As this long but explicit name suggests (one of the advantages of systematic names in action!), this enzyme catalyses the oxidation of glucose 6-phosphate using $NADP^+$ (a substance closely related to NAD^+) as coenzyme. For each molecule of glucose 6-phosphate produced in the *first* (the hexokinase-catalysed) reaction, one molecule of $NADP^+$ is converted to NADPH in the *second* (the glucose 6-phosphate:$NADP^+$ oxidoreductase-catalysed) reaction. Therefore, the amount of NADPH produced, which like NADH can be estimated by its absorption at 340 nm (Figure 4.5), gives a measure of the extent of the hexokinase reaction:

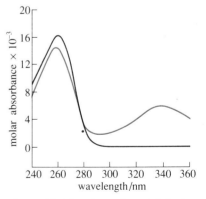

Figure 4.5 The absorption of light by NAD^+ (black line) and NADH (blue line). As NADH has a peak at 340 nm and NAD^+ does not, measuring absorption at this wavelength will reveal the amount of NADH present in a sample.

$$\text{glucose} + \text{ATP} \underset{\text{hexokinase}}{\rightleftharpoons} \text{glucose 6-phosphate} + \text{ADP}$$

$$\text{glucose 6-phosphate} + NADP^+ \underset{NADP^+ \text{ oxidoreductase}}{\overset{\text{glucose 6-phosphate:}}{\rightleftharpoons}}$$

$$\text{6-phosphogluconate} + \text{NADPH} + H^+$$
$$\text{(measure at 340 nm)}$$

◇ Can you suggest an alternative way of following the second (glucose 6-phosphate : NADP$^+$ oxidoreductase) reaction.

◈ One alternative is to measure the rate of production of H$^+$ (Section 4.3.2).

4.3.4 Choosing an assay technique

You have now read about the principles of some of the techniques used for assaying enzymes. The survey has been by no means exhaustive nor have we dealt with all the qualifying complications underlying each technique. Nevertheless we have outlined at least some of the techniques with wide application. But how in practice do we decide which technique to use in any particular case? First, from the physico-chemical properties of the substrates and products of the reaction, we can decide, on paper at least, what technique might be suitable. There is, however, a limit to 'paper predictions'. The enzyme may prove unstable at the temperature chosen, the buffer may interfere with the reaction, and so on. There must also be appropriate controls to ensure that the changes observed are actually dependent on enzyme catalysis (see Section 4.3). These and many other possible factors can only be observed by trial and error; so the 'paper decisions' are only the first step in developing a suitable enzyme assay. Success often entails much work and is dependent also on influences such as the cost of chemicals, the availability of the necessary equipment, and the time involved. All are important; the ideal enzyme assay is, once designed, quick, accurate and cheap.

Summary of Sections 4.1 to 4.3

1 The efficient operation of the hundreds of chemical reactions that constitute intermediary metabolism is due largely to enzymes.

2 Enzymes owe their key role in intermediary metabolism to their great catalytic power, their specificity and the sensitivity of certain enzymes to specific regulation and control.

3 Enzymes are named after the reactions that they catalyse. By convention enzymes are classified into six broad classes.

4 The amount of an enzyme present in a preparation can be assayed by the initial rate of reaction catalysed by a sample of the preparation.

5 An enzyme assay depends on finding some feature that distinguishes the reactants from the products of the reaction.

6 Absorption of light and pH are among features used to assay enzymes. Some enzymes are assayed by coupling the reactions that they catalyse to other enzyme-catalysed reactions.

Question 1 (*Objectives 4.1, 4.2, 4.3 and 4.5*) In which of the aspects listed below do enzymes generally differ from non-biological catalysts?

(a) Enzymes, coming from living organisms, are cheap to obtain.

(b) Enzymes have a smaller M_r than non-biological catalysts.

(c) Given a range of potential substrates, an enzyme affects fewer than a non-biological catalyst might.

(d) Enzymes can increase the rate of reaction by up to about 10^{12} times.

(e) Enzymes are not affected by temperature.

(f) The systematic names of enzymes end in *-ase*.

Question 2 (*Objectives 4.1 and 4.4*) Here is an opportunity for you to do some 'paper chemistry'. In Table 4.2, some details of the properties of a number of substances, A–Z, are given, and in Table 4.3, a number of enzyme-catalysed reactions involving these compounds are listed. For each reaction in Table 4.3, select the principles on which to base an assay for that reaction from List I, and suitable apparatus from List II. Only if no direct assay is available, should you choose a coupled assay; in such cases indicate which reactions are to be coupled and choose the appropriate items for those reactions from Lists I and II. Where more than one answer can be found for a reaction, you should give alternatives.

List I

Ia   Coupled assay

Ib   Reaction of substrate to give red colour

Ic   Reaction of product to give red colour

Id   Colour of substrate

Ie   Colour of product

If   Ultraviolet (u.v.) absorption of substrate

Ig   u.v. absorption of product

Ih   Measurement of $[H^+]$

List II

IIa   pH meter

IIb   Colorimeter

IIc   u.v. spectrophotometer

IId   Centrifuge

IIe   Filter paper

IIf   pH stat

Table 4.2 The properties of substances A–Z

Substance	Solution in water coloured yellow	Solution absorbs u.v. at 340 nm	Reacts with dye in solution to give red colour
A			
B		+	
D			
F			
G		+	
L			+
N	+		
Q			
T			
R			
S			
Z			+

Table 4.3 Some reactions involving substances A–Z

	Reaction
(i)	$B \rightleftharpoons T$
(ii)	$B + F \rightleftharpoons G + L$
(iii)	$N \rightleftharpoons L$
(iv)	$B + A \rightleftharpoons G + R + H^+$
(v)	$B + S \rightleftharpoons L + R + H^+$ (where enzyme is inactive at acid pH)
(vi)	$L + G \rightleftharpoons Z + F$
(vii)	$Q \rightleftharpoons T$

All the reactions shown in Table 4.3 occur in aqueous solution where all the substances shown above are readily soluble. Only where indicated by a '+' sign does a substance either give a yellow solution, absorb at 340 nm or react with dye to give a red colour. L and Z react with the dye to give an identical red colour, i.e. absorbing at the same wavelength.

All reactions are enzyme-catalysed and reversible, but substrates are regarded as those substances on the left side, products on the right, of the equations. H^+ is a proton.

4.4 THE MECHANISM OF ACTION OF ENZYMES IN GENERAL

Having considered how we can study the rate of an enzyme-catalysed reaction we can now return to the most intriguing of all questions about enzymes, and one central to this chapter: *how* do they work?

4.4.1 The enzyme–substrate complex

By definition enzymes are catalysts. So when we ask how they work, it is as well to recall that we are in fact asking 'how do they operate as catalysts?' To what do they owe their great catalytic power and specificity (Section 4.1)? A logical starting point in answering these questions is to compare an enzyme-catalysed reaction with its non-catalysed counterpart. In particular we shall examine how the concentration of reactants (substrates) affects the rate of the reaction.

Consider the simplest possible hypothetical reaction, where a single substance, S, is converted to a single product, P, i.e. S→P. We could carry out the reaction several times, each time varying the initial concentration of S and for each measuring the rate of reaction.

◇ How could we measure the rate of reaction?

◆ By techniques such as those described in Section 4.3. The basic *ways* of following a reaction so as to plot a reaction–time plot are the same, irrespective of whether the reaction is enzyme-catalysed or not.

If we did this for the uncatalysed reaction, we would find that the greater the concentration of S, the greater the rate. In fact the rate of such a reaction, v ('v' from velocity, the alternative name for rate), is dependent on the initial concentration of S, such that $v = k[S]$, where [S] is the shorthand for concentration of S and k is what is known as the **rate constant** for the forward reaction, S→P. This equation has the form of a straight line; increase [S] and the rate of reaction increases linearly. This linear relationship would be revealed by a graph where each measured reaction rate, v, is plotted against its corresponding concentration of reactant, [S], as in Figure 4.6.

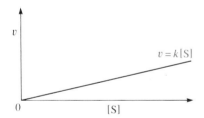

Figure 4.6 The relation between rate of reaction, v, and concentration of reactant, [S], for an uncatalysed reaction.

Now suppose that the reaction S→P can be catalysed by an enzyme and that we repeated the procedure for the enzyme-catalysed reaction; that is, we again carried out the reaction several times, in the presence of a similar amount of enzyme each time, but each time at a different concentration of S. For each we measured the initial rate of reaction (by the procedure shown in Figure 4.4). Once again we would find that as [S] increases so does the rate of reaction, v. But on plotting a graph of v against [S] we would obtain a different result from that for the uncatalysed reaction (that in Figure 4.6). The sort of graph we would generally get is shown in Figure 4.7.

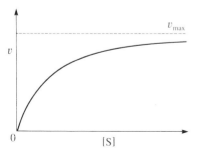

Figure 4.7 The relation between rate of reaction, v, and concentration of substrate, [S], for an enzyme-catalysed reaction.

Compare Figures 4.6 and 4.7. Without any actual numbers one dramatic difference *cannot* be seen by comparing the graphs—the enzyme-catalysed reaction is much much faster than its uncatalysed counterpart. But the difference in the shape of the curves *is* readily seen. Whereas v for the uncatalysed reaction increases linearly with increasing [S], for the enzyme-catalysed reaction it traces a curve in which the increasing rate gradually tails off towards a maximum value (Figure 4.7), a value termed the **maximum rate**, v_{max}. What accounts for this difference between the uncatalysed and enzyme-catalysed reactions?

The answer comes from appreciating that though enzymes do have remarkable powers, they do not actually operate by 'magic'! In common with all catalysts, enzymes do not act at-a-distance; they themselves participate in the reaction. However, unlike the other reactants, at the end of the reaction the catalyst/enzyme is regenerated, unchanged.

We can thus *assume* that the first step in an enzyme-catalysed reaction is the interaction of enzyme, E, and substrate, S, to form a complex, termed an **enzyme–substrate complex** or **ES complex**. The ES complex then breaks down to yield the product and free enzyme, E. Thus the enzyme-catalysed version of the reaction can be represented as:

$$E + S \rightleftharpoons ES \longrightarrow P + E$$

Free enzyme, E, is then available to combine with further S, forming ES, and so on.

Assuming this is indeed what happens, consider what varying the concentration of S, that is [S], would do to the rate, v, of this enzyme-catalysed reaction. At low [S] not all the molecules of enzyme present will be occupied at any one time in the form of ES complex; so at any one time some enzyme molecules will effectively be idle and so the rate of measured reaction in the solution, v, will be relatively low. However as [S] is increased to high values, more and more of the molecules of enzyme at any given time will be in the form of ES complex, and more molecules of enzyme being involved simultaneously in catalysis means a *greater rate* of measured reaction, v. Eventually as [S] is increased still further, all the molecules of enzyme at any given time would be occupied as ES complex (the enzyme would be saturated), and hence no more S could be accommodated until some ES complex was broken down. In such a situation the rate of reaction would be maximum, v_{max}. As some ES complex would be breaking down at any given time, no such point could ever be reached in practice. It could however be closely approached and at such [S] the maximum rate of reaction, v_{max}, would be approached.

Look at Figure 4.7. The actual experimental findings of a gradual approach to v_{max} fit this postulated behaviour of an enzyme-catalysed reaction where we *assumed* an ES complex to be a vital early stage in catalysis.

Henri, in 1902, was the first person to appreciate that curves from experimental data like that seen in Figure 4.7 were consistent with the assumed formation of an ES complex. This idea was firmly established by Michaelis and Menten in 1913. By assuming, as had Henri, that the first step in an enzyme-catalysed reaction was the formation of an ES complex they derived *theoretically* an equation describing the dependence of (initial) reaction rate, v, on substrate concentration, [S]:

$$v = \frac{v_{max}[S]}{K_M + [S]} \tag{4.1}$$

where K_M, known as the *Michaelis constant*, and [S] are both measured in moles per litre of solution ($mol\,l^{-1}$).

Equation (4.1) is known as the **Michaelis–Menten equation**. We need not concern ourselves with its mathematical derivation but note what happens when it is plotted as a graph (i.e. v against [S]). The graph obtained is that shown in Figure 4.8.

Compare Figure 4.8 with the plot of v against [S] shown in Figure 4.7. The curves have exactly the same shape (they 'flatten off'), that known mathematically as a rectangular hyperbola. So a graph of v against [S] obtained from a *theoretically derived equation* (Figure 4.8), the Michaelis–Menten equation (Equation 4.1), coincides with that of the curve obtained from *experimental data* (Figure 4.7). This is persuasive evidence that the crucial *assumption* used in the theory, the existence of an ES complex, is essentially correct.

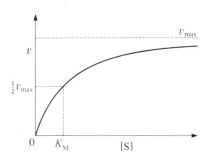

Figure 4.8 Graph of v against [S], as derived from the Michaelis–Menten equation. (The significance of K_M is outlined in Section 4.4.2)

It would be useful to have some direct evidence for the existence of ES complexes; to be able to detect them directly, perhaps even to isolate and purify them and examine their chemical structure. As you might well anticipate, this is very difficult (but see Section 4.5.4, later), because enzyme-catalysed reactions are very rapid and the ES complex is a transient and unstable stage on the path from substrate (S) to product (P). However, in certain cases, there is somewhat more *indirect* evidence for the formation of an ES complex, this evidence depending on some peculiar features of the enzymes concerned.

One instance where a marriage of circumstance and technology allows us a 'glimpse' of an ES complex is in the working of the bacterial enzyme tryptophan synthetase. This enzyme catalyses the synthesis of the amino acid tryptophan from serine and indole:

$$HOCH_2CHCOO^- \atop \qquad |\atop \quad {}^+NH_3} \; + \; \text{indole} \; \underset{\text{synthetase}}{\overset{\text{tryptophan}}{\rightleftharpoons}} \; CH_2-CH-COO^- \atop \qquad\qquad |\atop \qquad\quad {}^+NH_3}$$

serine   indole   tryptophan

In solution, the enzyme exhibits some fluorescence at about 500 nm, which can be measured by a special instrument called a fluorimeter. On addition of one of the two substrates, serine, there is a marked change in fluorescence, as can be seen in Figure 4.9. On subsequent addition of the other substrate, indole, the fluorescence changes again, being reduced to below that of the free enzyme. The data in Figure 4.9 are most easily understood if we assume that serine initially forms an ES complex with tryptophan synthetase, a complex with very high fluorescence compared with the enzyme alone. Then indole reduces this fluorescence, presumably by converting the enzyme–serine complex to an enzyme–serine–indole complex. Thus, this experiment reveals not one, but two, ES complexes, albeit somewhat indirectly.

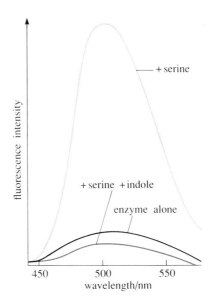

Figure 4.9 The change in fluorescence during interaction between tryptophan synthetase and its substrates, serine and indole.

More direct evidence for the occurrence of ES complexes comes from X-ray diffraction studies of enzymes (Chapter 3, Box 3.1), something we discuss in Section 4.5.4.

4.4.2 The use of the Michaelis–Menten equation

You have seen how the derivation of the Michaelis–Menten equation helped establish the concept of the existence of the ES complex; an event of undoubted importance in the history of enzyme studies. But the equation still has its uses today and it is to these that we now turn briefly.

Look again at the equation:

$$v = \frac{v_{max}[S]}{K_M + [S]} \tag{4.1}$$

We have already described what v_{max} is (Section 4.4.1), but have yet to describe K_M, other than to identify it as the **Michaelis constant**.

Operationally K_M is in fact that concentration of substrate at which the rate of reaction is half the maximum value (see Box 4.1). As a concentration, it is expressed in moles per litre of solution (mol l^{-1}), as is [S].

Box 4.1 More data on the Michaelis–Menten hypothesis

The definition of K_M

The Michaelis–Menten equation is derived from considering the enzyme-catalysed reaction

$$E + S \underset{k_2}{\overset{k_1}{\rightleftharpoons}} ES \overset{k_3}{\longrightarrow} P + E$$

where E is enzyme, S is substrate, ES is enzyme–substrate (ES) complex, P is product and k_1, k_2 and k_3 are the rate constants for the reactions, as shown. It is assumed that almost none of the product, P, reverts to substrate, S; a condition that is true in the initial stage of the reaction when [P] is low. Hence the one-way arrow, ES→P+E.

From these considerations it is possible to show that

$$v = \frac{v_{max}[S]}{(k_2 + k_3)/k_1 + [S]}$$

However it is much more convenient to define a new constant equal to $(k_2 + k_3)/k_1$. This constant is the Michaelis constant, K_M. Substituting K_M for $(k_2 + k_3)/k_1$ we obtain

$$v = \frac{v_{max}[S]}{K_M + [S]} \tag{4.1}$$

the Michaelis–Menten equation.

Though K_M is defined by the relationship $K_M = (k_2 + k_3)/k_1$, it is also equal to that concentration of substrate (mol l^{-1}) at which the rate of the reaction is half the maximum, that is, when $v = \frac{1}{2}v_{max}$.

We can see this to be the case by actually examining the Michaelis–Menten equation when $v = \frac{1}{2}v_{max}$.

That is, putting $\frac{1}{2}v_{max}$ for v in Equation 4.1, we see that

$$\tfrac{1}{2}v_{max} = \frac{v_{max}[S]}{K_M + [S]}$$

which on dividing by v_{max} and rearranging, gives

$$K_M + [S] = 2[S]$$

Thus $K_M = [S]$ (when $v = \frac{1}{2}v_{max}$).

The Lineweaver–Burk plot

One well-known transformation of the Michaelis–Menten equation is the Lineweaver–Burk equation. This is obtained by inverting the Michaelis–Menten equation:

$$v = \frac{v_{max}[S]}{K_M + [S]} \tag{4.1}$$

giving

$$\frac{1}{v} = \frac{K_M + [S]}{v_{max}[S]}$$

that is

$$\frac{1}{v} = \frac{K_M}{v_{max}[S]} + \frac{[S]}{v_{max}[S]}$$

which on cancelling out [S] gives

$$\frac{1}{v} = \frac{K_M}{v_{max}} \frac{1}{[S]} + \frac{1}{v_{max}} \tag{4.2}$$

the Lineweaver–Burk equation.

Equation 4.2 has the form of the general equation for a straight line ($y = mx + c$) which means that when plotting $1/v$ against $1/[S]$ we get a straight line (Figure 4.10). Values of v_{max}

---

and K_M are easily obtained by reading off $1/v_{max}$ and $-1/K_M$ from the intercepts of the line with the vertical (y) and horizontal (x) axes, respectively; that is where $1/[S]$ and $1/v$ are zero, respectively. (If you wish, you can prove this for yourself by alternately putting $1/[S]$ and $1/v$ as zero in Equation 4.2 and rearranging the equation as necessary.)

The Hofstee–Eadie plot

If we take Equation 4.1, the Michaelis–Menten equation

$$v = \frac{v_{max}[S]}{K_M + [S]} \tag{4.1}$$

and multiply both sides of the equation by $K_M + [S]$, we get

$$v(K_M + [S]) = v_{max}[S]$$

that is, $K_M v + [S]v = v_{max}[S]$
Rearranging, $v[S] = v_{max}[S] - K_M v$
whereupon, by dividing by $[S]$,

$$v = v_{max} - K_M \frac{v}{[S]} \tag{4.3}$$

the Hofstee–Eadie equation.

From Equation 4.3, plotting v against $v/[S]$ gives a straight line (as in Figure 4.12), where the intercept on the vertical (y) axis corresponds to v_{max}, and that on the horizontal (x) axis to v_{max}/K_M. (If you wish, you can check this for yourself by putting alternately $v/[S]$ and v as zero in Equation 4.3 and rearranging the equation as necessary.) The slope of the line is equal to $-K_M$.

From experience, it appears that K_M and v_{max} estimated from Hofstee–Eadie plots are less affected by experimental error in measure-ment of v than they are when estimated from Lineweaver–Burk plots. Both plots (and some others) are in current use.

Use of v_{max} and K_M

As well as their use in 'description' of a particular enzyme, v_{max} and K_M have some other uses:

☐ They are useful in distinguishing between different types of enzyme inhibition, a topic dealt with in Section 4.4.3.

☐ Under certain circumstances K_M is a measure of the enzyme's *affinity* for the substrate, how tight the binding is of E to S in the ES complex. This is *only* true for reactions where the dissociation of ES complex to E and substrate S is much more rapid than the formation of E and product P. That is where $k_2 \gg k_3$. For then $K_M = (k_2 + k_3)/k_1$ effectively becomes k_2/k_1; that is $K_M = k_2/k_1$. But k_2/k_1 is the dissociation constant of the ES complex, as this is given by $[E][S]/[ES] = k_2/k_1$.

Thus under these particular circumstances K_M is equal to the dissociation constant of the ES complex, another way of saying that it is a measure of the strength of this complex, a measure of the enzyme's *affinity* for its substrate. The greater the dissociation constant of the ES complex, the weaker the binding of E to S. So in situations where K_M is equal to the dissociation constant of the ES complex, a high K_M indicates weak binding (low affinity) and a low K_M indicates strong binding (high affinity).

The concept of an ES complex has its equivalent in certain non-enzymic systems such as ion transport. Here the parameters K_t and J_{max} have analogous use to K_M and v_{max} (see Chapter 7).

The v_{max} and K_M of a particular enzyme acting on a particular substrate are useful values to measure. (They vary with conditions such as temperature, pH and concentration of ions, and so measurement of v_{max} and K_M for an enzyme should be done under well-defined conditions.) An organic chemist may measure the melting temperature and examine the absorption spectrum of a newly prepared compound as ways of checking whether it fits known descriptions for that compound. So might a biochemist measure the v_{max} and K_M of an enzyme that he or she has just purified. v_{max} and K_M can be 'diagnostics', part of the 'description' of an enzyme.

Of course, where an enzyme has more than one substrate there will be a different K_M for each substrate.

◇ Under what circumstances can an enzyme have more than one substrate?

● There are two distinct circumstances:
1 Where the enzyme catalyses a reaction which involves more than one reactant; e.g. $A + B \rightleftharpoons C + D$. For each reactant (i.e. A and B), the enzyme will have a specific K_M.

2 Where, as is usually the case, the enzyme is not totally specific, it can have alternative substrates which are chemically closely related. For each alternative substrate, the enzyme will have a specific v_{max} and K_M.

(Note that the two circumstances can occur together. That is, an enzyme can catalyse a reaction that has more than one reactant *and* there may also be alternative reactants. For example, the same enzyme might catalyse the reaction $A + B \rightleftharpoons C + D$ and, a chemically related but different one, $L + M \rightleftharpoons N + O$. Such an enzyme would have two v_{max} values, one for each of the two reactions; and four K_M values, one for each of A, B, L and M.)

From the small selection of v_{max} and K_M values given in Table 4.4, you can see that these can vary over a wide range for different enzymes and for the same enzyme with different substrates.

Table 4.4 Some K_M values

Enzyme	Substrate	$K_M/\mu mol\,l^{-1}$
β-galactosidase	lactose	4 000
pyruvate carboxylase	pyruvate	400
penicillinase	benzylpenicillin	50
red blood cell LDH	pyruvate	59
glucose phosphate isomerase	fructose 6-phosphate	120
carbonic anhydrase	carbon dioxide	8 000
yeast hexokinase	{ glucose*	100
	fructose*	1 000

* Alternative substrates of hexokinase.

For *some* enzyme-catalysed reactions the K_M is equivalent to a measure of the enzyme's **affinity** for the substrate: the higher the K_M, the lower the affinity. In such cases, if the enzyme has alternative substrates the K_M values can indicate the relative affinity of the enzyme for those substrates. (Box 4.1 indicates the only circumstances under which the relationship between affinity and K_M holds true.)

But how do we actually calculate v_{max} and K_M? Look again at Figure 4.8, the graphical representation of the Michaelis–Menten equation. The upper dashed line indicates v_{max}, to which the curve approaches. A similar line can be drawn for data actually derived experimentally, as in Figure 4.7. But where exactly do we draw the line? As v_{max} is never actually reached in practice (theoretically it occurs at *infinite* [S]) it is hard to say precisely where it lies. If it is hard to place v_{max} precisely on plots of v against [S], so too is it to calculate K_M from them (see Figures 4.7 and 4.8). Because if we do not know exactly where v_{max} lies, it stands to reason that we cannot know exactly where $\frac{1}{2}v_{max}$ lies, and this we need to know in order to calculate K_M precisely (see Box 4.1).

As explained in Box 4.1, it is possible to perform various algebraic transformations on the Michaelis–Menten equation and some of these help overcome the problem of calculating v_{max} and K_M precisely. One such rearrangement, the **Lineweaver–Burk equation** (Equation 4.2) relates $1/v$ to $1/[S]$:

$$\frac{1}{v} = \frac{K_M}{v_{max}} \frac{1}{[S]} + \frac{1}{v_{max}} \qquad (4.2)$$

When $1/v$ is plotted against $1/[S]$ the result is a straight line, as can be seen in Figure 4.10.

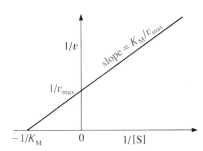

Figure 4.10 Graph of $1/v$ against $1/[S]$; as derived from the Lineweaver–Burk equation. The intercept on the y-axis is $1/v_{max}$. The intercept on the x-axis is $-1/K_M$; the slope of the line is K_M/v_{max}.

123

Where the line intersects the vertical (y) axis $1/v = 1/v_{max}$; where it intersects the horizontal (x) axis $1/[S] = -1/K_M$. Thus by reading off these intercepts we can obtain, easily and *precisely*, values for $1/v_{max}$ and $-1/K_M$, and hence precise values for v_{max} and K_M; something very difficult to do from a direct plot of v against [S] (Figures 4.7 and 4.8).

So in practice, having obtained experimental data by measuring v at different [S], we can plot the reciprocals, $1/v$ and $1/[S]$, that is as a Lineweaver–Burk plot, from which we can then obtain v_{max} and K_M (Figure 4.10). An example may make this clearer.

◇ Given below are data for X-ase, an enzyme that hydrolyses substance X.

Concentration of X/mol l^{-1}	Rate of hydrolysis of X/μmol min^{-1}
0.20×10^{-5}	16.7
0.22×10^{-5}	17.5
0.25×10^{-5}	19.6
0.27×10^{-5}	20.0
0.50×10^{-5}	28.6
0.77×10^{-5}	37.0
1.00×10^{-5}	38.5

Use these data to calculate K_M and v_{max} for X-ase with X as substrate. (We give the answer in stages, so if you wish to save time you can attempt parts of the question only.)

The first step is to identify what the data correspond to and to make sure that they are in the correct units.

The concentration of X is [S], and the rate of hydrolysis of X is v, as needed for calculating v_{max} and K_M. They are already in the right units: mol l^{-1} and μmol min^{-1}, respectively.

To plot the data in the Lineweaver–Burk form we need the reciprocals, that is $1/v$ and $1/[S]$:

[S]	1/[S]	v	1/v
0.20×10^{-5}	5.0×10^{5}	16.7	6.0×10^{-2}
0.22×10^{-5}	4.5×10^{5}	17.5	5.7×10^{-2}
0.25×10^{-5}	4.0×10^{5}	19.6	5.1×10^{-2}
0.27×10^{-5}	3.7×10^{5}	20.0	5.0×10^{-2}
0.50×10^{-5}	2.0×10^{5}	28.6	3.5×10^{-2}
0.77×10^{-5}	1.3×10^{5}	37.0	2.7×10^{-2}
1.00×10^{-5}	1.0×10^{5}	38.5	2.6×10^{-2}

◇ Use the reciprocals to construct a Lineweaver–Burk plot.

When $1/v$ is plotted against $1/[S]$ we obtain the Lineweaver–Burk plot seen in Figure 4.11.

◇ From Figure 4.11 calculate v_{max} and K_M for X-ase with X as substrate.

Remember that the intercept on the y-axis is $1/v_{max}$ (see Figure 4.10). In Figure 4.11 this intercept is at a value for $1/v$ of about 1.75×10^{-2}. Hence v_{max}, the reciprocal of this, is 57.1 μmol min^{-1}.

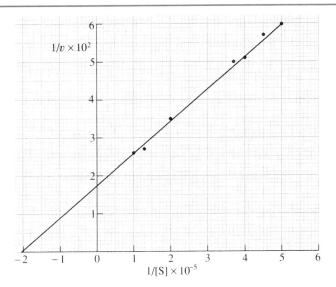

Figure 4.11 Lineweaver–Burk plot of $1/v$ against $1/[S]$ from data for X-ase.

For calculating K_M we can use the intercept on the x-axis which is $-1/K_M$ (Figure 4.10). In Figure 4.11 this intercept is at a value for $1/[S]$ of about -2.05×10^5. Hence K_M, which is minus the reciprocal of this, is 4.9×10^{-6} mol l^{-1}.

The Lineweaver–Burk equation is only one of the possible transformations of the Michaelis–Menten equation. Another useful transformation is that leading to the **Hofstee–Eadie equation**:

$$v = v_{max} - K_M \frac{v}{[S]} \qquad (4.3)$$

Plotting v against $v/[S]$ in Equation 4.3 also takes the form of a straight line, as seen in Figure 4.12.

Thus here too v_{max} and K_M can be accurately obtained. v_{max} is obtained directly as the intercept on the vertical (y) axis, and this can also be used in calculating K_M as the intercept on the horizontal (x) axis corresponds to v_{max}/K_M; K_M can also be calculated from the slope of the line which is equal to $-K_M$ (Box 4.1).

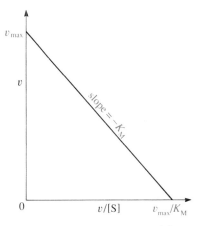

Figure 4.12 Graph of v against $v/[S]$; as derived from the Hofstee–Eadie equation. The intercept on the y-axis is v_{max}. The intercept on the x-axis is v_{max}/K_M; the slope of the line is $-K_M$.

4.4.3 The active site; enzyme inhibitors

As proteins, enzymes are very large molecules (Section 4.1) but only a relatively small portion of an enzyme is in contact with its substrate in the ES complex. This relatively small portion of the enzyme is termed the **active site**. As we discuss later in this chapter, small chemical alterations occurring at the active site can lead to a complete loss of catalytic activity. This supports the view that the active site is an accurately shaped region in which *only* the substrates can fit *and* be acted on.

The accuracy of shaping of the active site is supported by the high specificity of enzymes for their substrates. The active site is somewhat like a lock, the substrate a key. This idea of a rigid 'lock and key' relationship of enzyme to substrate was developed by the great German chemist Emil Fischer in 1894, and has lasted relatively unchallenged until recent times (Section 4.6). Further knowledge about active sites can often be gained from studying substances that inhibit the catalytic activity of enzymes, **enzyme inhibitors**.

When working with enzymes, a fairly frequent finding is that some preparations of an enzyme fail to work, or at least the enzyme's catalytic power is less than expected. Often this is because the enzyme has 'gone off'; being proteins, enzymes have many weak bonds responsible for their higher-order structures (Chapter 3) and so are sensitive to temperature, pH, and high concentrations of ions. These factors are often the cause of loss of enzyme activity.

Sometimes, however, there is another more subtle explanation; some particular substance or substances in the enzyme preparation or the assay mixture are *inhibiting* the enzyme. Though often a nuisance, if such inhibitors can be identified or others produced, they can sometimes provide powerful clues as to how the particular enzyme works. You will come across some examples of this in Section 4.5, but now we merely introduce two broad classes of enzyme inhibitors; *irreversible* and *reversible*.

As the name implies, an **irreversible inhibitor** is a substance that inhibits an enzyme such that the inhibition cannot be reversed merely by removing the inhibitor from the solution; the enzyme is 'poisoned'. Generally this is because the inhibitor has altered some covalent bond(s) in the enzyme, often by part of the inhibitor itself becoming attached to part of the enzyme. Often where the chemical nature and site of these covalent attachments can be determined, we can gain considerable information on how the enzyme works *normally* (Section 4.5.2).

In the case of **reversible inhibitors** the inhibition can be relieved by removal of the inhibitor from the solution. Reversible inhibitors themselves fall into various categories: competitive inhibitors form one category; other categories include so-called non-competitive inhibitors, uncompetitive inhibitors, etc. We shall not deal here with any but *competitive inhibitors* and *non-competitive inhibitors*.

Competitive inhibitors seem to compete with substrate for the *same* site on the enzyme, presumably the active site, and thus tend to impede the formation of ES complex. However, given a fixed concentration of inhibitor, at high enough concentrations of substrate, the effect of the inhibitor can be swamped and v_{max} approached, at least in theory. (In practice, it may not always be possible to approach v_{max} closely because a high enough concentration of substrate may be impossible to provide for other reasons, such as the limited solubility of the substrate.) The K_M is altered, *increased*, in the presence of a competitive inhibitor. As competitive inhibitors seem to compete directly with the substrate for the active site, thus reducing the amount of ES complex formed, you might well expect that a competitive inhibitor for a particular enzyme will be chemically similar to the substrate. This is indeed often the case and can have interesting consequences. For example, the well-known antibacterial agent sulphanilamide is effective because of its ability to act as a competitive inhibitor for an enzyme involved with the specific bacterial growth substance *para*-aminobenzoic acid. As Figure 4.13 shows, the structures (shape and size) of *para*-aminobenzoic acid and sulphanilamide are indeed similar. Thus the exclusively bacterial enzyme is 'fooled' into binding sulphanilamide instead of *para*-aminobenzoic acid. The mode of action of sulphanilamide was not known at the time of its discovery or when first used. More exact knowledge about enzymes and their inhibitors might allow us to design specific therapeutic agents, such as antibiotics, in the future. This 'rational design' is an approach currently much in favour with some drug companies.

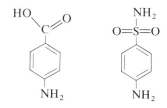

para-aminobenzoic acid   sulphanilamide

Figure 4.13 The structures of *para*-aminobenzoic acid and sulphanilamide.

Non-competitive inhibitors are not necessarily related structurally to substrates and do not appear to prevent the formation of the ES complex. They bind at a separate site on the enzyme from the active site. They do not reduce the binding of substrate but do reduce catalysis. Unlike with competitive inhibitors, there is no concentration of substrate that can totally swamp their effect and v_{max} can never be approached closely, even in theory. K_M, however, is unaffected by non-competitive inhibitors.

As you will see in Section 4.5, it is sometimes necessary to be able to decide whether an enzyme inhibitor is competitive or not.

◇ From your knowledge of the Michaelis–Menten relationship (Section 4.4.2) and our discussion of the nature of competitive and non-competitive inhibitors, can you suggest a possible way of distinguishing between them?

◆ They have different effects on the Michaelis–Menten relationship and in particular as to whether v_{max} can be approached closely. It should be possible to identify an inhibitor as being either competitive or non-competitive by examining the effect of the inhibitor on the relationship between substrate concentration ([S]) and reaction rate (v) (e.g. how it affects graphs like those in Figures 4.7, 4.8, 4.10, 4.11 and 4.12).

◇ Figure 4.14 is a plot for an enzyme in the presence of (blue line) and absence of (black line) an inhibitor. Is the plot of the Lineweaver–Burk or Hofstee–Eadie type? Is the inhibitor competitive or non-competitive? (*Hint*: the type of plot should be evident from Section 4.4.2. Having identified it, recall how to derive K_M and v_{max} from it. You should be able to see if these are affected by the inhibitor without actually having to calculate what the values are. The manner in which they are affected should identify the type of inhibition.)

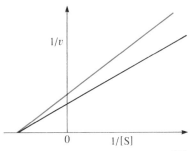

Figure 4.14 Graph of $1/v$ against $1/[S]$ for an enzyme in the presence of (blue line) and absence of (black line) an inhibitor.

◆ Figure 4.14 is a plot of $1/v$ against $1/[S]$ and is therefore of the Lineweaver–Burk type. (Hofstee–Eadie plots involve v against $v/[S]$.) Competitive inhibitors bind at the active site, thus increasing the K_M of the enzyme for the substrate; as these inhibitors can ultimately (in principle) be swamped by substrate, v_{max} is unaltered. In Figure 4.14, K_M is given by $-1/K_M$, the intercept on the x-axis, and v_{max} by $1/v_{max}$, the intercept on the y-axis. As the presence of inhibitor (blue line) alters the intercept on the y-axis (and hence reduces v_{max}) but leaves the intercept on the x-axis (and hence K_M) unaltered, it must be of the non-competitive type. A competitive inhibitor would give a result like that shown in Figure 4.15.

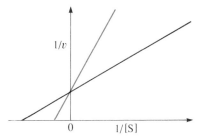

Figure 4.15 Lineweaver–Burk plot for an enzyme in the presence of (blue line) and absence of (black line) a competitive inhibitor.

There is yet another kind of enzyme inhibition of a reversible type, called *allosteric inhibition*; it is of great importance to our third crucial point about the role of enzymes in metabolism—their role in regulation and control—and it will be dealt with at some length in Chapter 5.

Summary of Section 4.4

1 Where enzyme, E, catalyses the conversion of substrate, S, to product, P, the first step in the reaction is the binding of substrate to the active site of the enzyme, forming an enzyme–substrate complex (ES complex).

2 In an enzyme-catalysed reaction, if the initial concentration of substrate, [S], is increased, so does the rate of reaction, v, approaching a maximum rate, v_{max}. This relationship is expressed in the Michaelis–Menten equation,

$$v = \frac{v_{max}[S]}{K_M + [S]}$$

where K_M, the Michaelis constant, is equal to [S] when $v = \frac{1}{2}v_{max}$.

3 For some enzymes, K_M gives a measure of the affinity of the enzyme for its substrate.

4 The Michaelis–Menten equation can be transformed into a Lineweaver–Burk or Hofstee–Eadie plot from which K_M and v_{max} can be estimated readily.

5 Enzymes can be inhibited by reversible or irreversible inhibitors. Reversible inhibitors can be subdivided into various types including competitive and non-competitive ones.

6 Competitive inhibitors compete with the substrate for binding at the active site; they often closely resemble the substrate. They leave v_{max} unaffected but increase K_M.

7 Non-competitive inhibitors do not bind at the active site. They reduce v_{max} but leave K_M unaffected.

Question 3 (*Objectives 4.1 and 4.7*) On addition of increasing amounts of substrate, the rates of all enzyme-catalysed reactions increase until they closely approach a maximum value. Which of (a)–(d) are correct explanations of this fact? (More than one answer may be possible.)

(a) It is because of the Michaelis–Menten equation.

(b) It is because more and more enzyme molecules are operating until all are fully occupied with substrate.

(c) It is because at high concentrations, the substrate is an inhibitor of the enzyme.

(d) It is because all the substrate is used up.

Question 4 (*Objectives 4.1, 4.6 and 4.8*) Figure 4.16a–d gives four graphs that, in some way or other, relate enzyme reaction rates (v) to substrate concentration ([S]) for reactions in the presence of (blue lines) and in the absence of (black lines) inhibitors. Explain which graph(s) correspond(s) to competitive inhibition occurring?

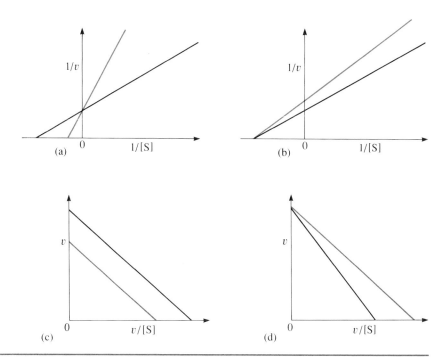

Figure 4.16

Question 5 (*Objectives 4.1 and 4.8*) Plot the data given on page 124 for the enzyme X-ase as a Hofstee–Eadie plot and from this calculate the v_{max} and K_M for the enzyme.

4.5 THE STUDY OF ENZYME MECHANISMS

In attempting to understand how any enzyme works, there are essentially three questions to be answered:

- □ What is the nature of the chemical changes to the substrate(s)?
- □ What is the chemical nature of the active site?
- □ How do the chemical groups present in the active site effect catalysis?

To answer the first, the techniques for investigating the mechanisms of organic reactions are used, and we shall not deal with them here. In this section, we deal with the second question, while the third we consider in Section 4.7.

4.5.1 Coenzymes and vitamins

All enzymes are proteins. Therefore the active site is largely shaped by a three-dimensional array of the amino acid side chains in the site. However, some enzymes also contain one or more non-protein parts vital to enzyme activity. Such substances, which are in effect an integral part of the enzyme as far as its catalytic activity is concerned, are often called **coenzymes**. (Some are more often called *cofactors* or *prosthetic groups*; as the distinctions are not very important, we shall simply refer to all of them as *coenzymes*.)

The role of a coenzyme is to help in catalysis. Sometimes coenzymes are regenerated at the end of the reaction, sometimes they are altered (e.g. NAD^+ to NADH). They often react with the substrates, and are located at the active sites of enzymes. The relationship between active site, coenzyme and substrate is shown schematically in Figure 4.17.

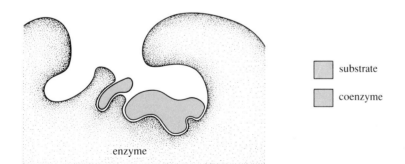

substrate

coenzyme

enzyme

Figure 4.17 Coenzyme and substrate in an enzyme active site.

A coenzyme can often be separated from the protein portion of an enzyme. Sometimes, in the test-tube, such isolated coenzymes can be shown to have *some* catalytic activity of their own (i.e. without the protein component of the enzyme), though in the actual cell this is insignificant compared to the much greater catalytic power of the intact enzyme (i.e. protein component plus coenzyme). Coenzymes are much simpler molecules than intact enzymes. By studying the chemistry of the simpler coenzyme-catalysed reaction we might learn something about the chemistry of the more complex (intact) enzyme-catalysed one. Such systems are **model systems**, as it is supposed that they are models for explaining the more complicated (intact) enzyme systems. For example, enzymes that catalyse the interconversion of the $(+)$ and $(-)$ isomers of amino acids (racemases):

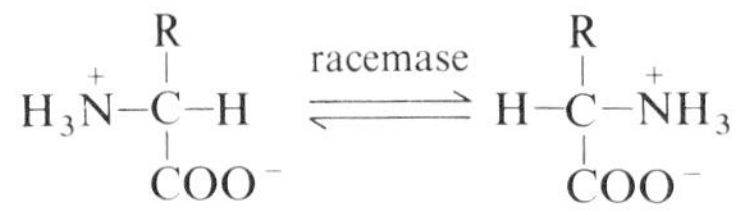

$$\underset{\overset{|}{COO^-}}{\overset{\overset{R}{|}}{H_3\overset{+}{N}-C-H}} \quad \overset{racemase}{\rightleftharpoons} \quad \underset{\overset{|}{COO^-}}{\overset{\overset{R}{|}}{H-C-\overset{+}{N}H_3}}$$

or the transfer of amino groups between amino acids and oxoacids (transaminases), such as glutamate and oxaloacetate:

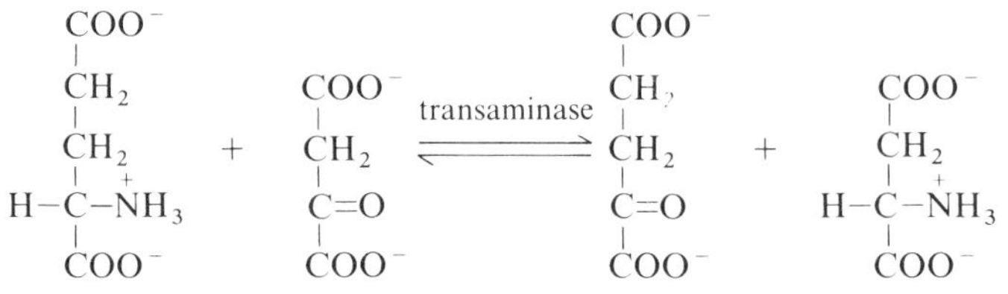

$$\underset{\text{glutamate}}{\underset{\overset{|}{COO^-}}{\overset{\overset{COO^-}{|}}{\underset{|}{\overset{|}{CH_2}}}\overset{|}{\underset{|}{CH_2}}\overset{|}{\underset{|}{H-C-\overset{+}{N}H_3}}}} \quad + \quad \underset{\text{oxaloacetate}}{\underset{\overset{|}{COO^-}}{\overset{\overset{COO^-}{|}}{\underset{|}{\overset{|}{CH_2}}}\overset{|}{C=O}}} \quad \overset{transaminase}{\rightleftharpoons} \quad \underset{\text{2-oxoglutarate}}{\underset{\overset{|}{COO^-}}{\overset{\overset{COO^-}{|}}{\underset{|}{\overset{|}{CH_2}}}\overset{|}{\underset{|}{CH_2}}\overset{|}{C=O}}} \quad + \quad \underset{\text{aspartate}}{\underset{\overset{|}{COO^-}}{\overset{\overset{COO^-}{|}}{\underset{|}{\overset{|}{CH_2}}}\overset{|}{H-C-\overset{+}{N}H_3}}}$$

or the decarboxylation of amino acids (decarboxylases):

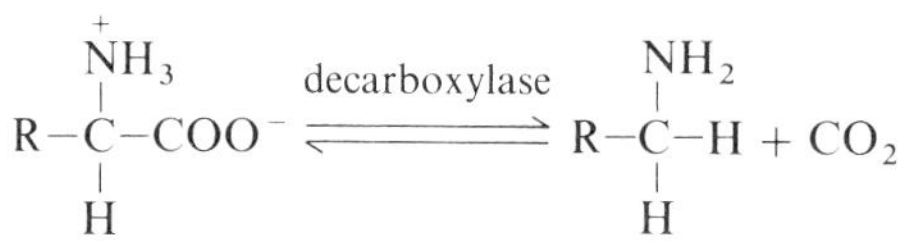

$$\underset{\overset{|}{H}}{\overset{\overset{+}{N}H_3}{R-\overset{|}{C}-COO^-}} \quad \overset{decarboxylase}{\rightleftharpoons} \quad \underset{\overset{|}{H}}{\overset{NH_2}{R-\overset{|}{C}-H}} + CO_2$$

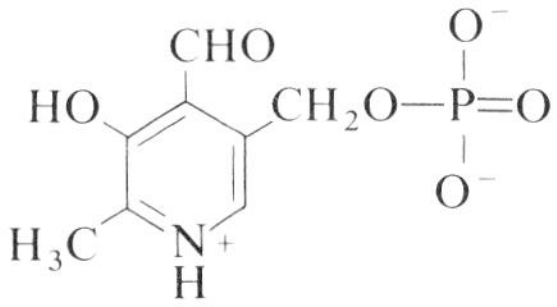

Figure 4.18 The structure of pyridoxal phosphate.

all need pyridoxal phosphate as a coenzyme (Figure 4.18).

It is possible to effect catalysis of some of these three types of reaction by adding pyridoxal phosphate, sometimes plus a metal ion (but without enzyme), to the various substrates shown. This argues strongly for pyridoxal phosphate being actively involved at the active site of the enzyme-catalysed reactions. In these model systems, it is easier to study the exact nature of the complexes formed between the pyridoxal phosphate and the substrates, and thus, by inference, learn more about the enzymic (i.e. intact enzyme-catalysed) reactions. With model systems of pyridoxal phosphate, the rate of catalysis achieved is much less than that achieved by the enzymic systems, which, of course, involve pyridoxal phosphate. (Thus in whole cells too, any catalysis by coenzyme alone would be very insignificant compared to catalysis by the entire enzymic system.) The much greater rate of reaction with coenzyme *plus* enzyme means that influences other than the pyridoxal phosphate are important in the enzymic reactions.

◇ Can you suggest evidence, other than reaction rate, that coenzymes alone (without enzyme) do not act entirely as do the enzymic systems?

The most dramatic difference relates to product specificity (Section 4.1). In our example, the coenzyme 'alone' (sometimes plus a metal ion in the model system) might catalyse more than one of the above reactions (racemization, transamination, or decarboxylation). In contrast when the coenzyme is part of an intact enzyme, *only one or other* of the reactions is catalysed, depending on the enzyme in question. For example, the enzyme catalysing the interconversion of $(+)$ and $(-)$ amino acids will catalyse this rather than decarboxylation. Hence each enzyme imposes its own rapid rate and highly specific directive effect (Section 4.1), over and above the types of catalysis that the coenzyme alone could effect.

So although it may be relatively easy to learn about the chemical interconversions catalysed by some enzymes by studying their coenzymes alone, these data must be treated with caution when applied back to the actual enzymic systems.

Much of what is known about the role of coenzymes ties up with studies from a quite different field—nutrition. Anyone who has read the labels on fruit juice containers or muesli packets or watched TV commercials is familiar with the word **vitamin**. Vitamins are organic substances present in foodstuffs and needed by the body in small amounts in addition to an adequate supply of carbohydrate, protein, fats and mineral salts. Deficiency of any particular vitamin can lead to illness. Deficiency diseases are as old, indeed older, than humanity itself—there is no reason to suppose that *Tyrannosaurus rex* too did not experience vitamin deficiency on occasion! And as for our closer ancestors, skeletons of prehistoric humans reveal signs of vitamin deficiency. The recognition of such diseases is also ancient, as is some notion of cure—blind Tobias cured with fish bile, as mentioned in the Bible, suggests an awareness of night-blindness (caused by a deficiency of vitamin A), and its treatment. By the mid-16th century, oranges and lemons were known as cures for scurvy, and in 1720, the Austrian physician Kramer wrote

'... if you have oranges, lemons, citrons, or their pulp and juice preserved with whey in cask, so that you can make a lemonade, or rather give to the quantity of 3 to 4 ounces of their juice in whey you will, without other assistance, cure this dreadful evil'.

By 1804, issue of lemon juice was compulsory in the British Navy, thus making scurvy a comparatively rare disease among the sailors. (Issue of lemons or limes probably gave rise to the US nickname of 'limeys' for British sailors, later extended to soldiers and then to the British generally.)

In the Japanese Navy, another disease, beriberi (a form of paralysis, due to degeneration of nerves), was very common until in 1885 Takaki, convinced that it arose from the eating of polished rice (rice without husks) as the chief staple food, managed to have some of the rice replaced by a richer diet. The disease virtually vanished in the Navy. It seems that something in the grain husks was needed in the diet. A great step forward was taken in 1887 when Eijkman reported that by feeding polished rice to hens, he had produced experimentally a condition in these birds resembling beriberi. By 1912, Funk believed he had isolated the anti-beriberi factor, which he called the 'beriberi vitamine' (he believed it to be *vital* and an *amine*), thus giving us the word *vitamin*. It is interesting to note that most vitamins are not in fact amines, despite the use of Funk's terminology.

Similar studies on the feeding of controlled diets to animals, notably by Hopkins in England and by Mendel and Osborne in the USA, made the existence of other vitamins clear. Many such vitamins, often known by letters and numbers, have now been isolated, purified and their chemical nature determined, What then do vitamins do?

At first sight this is a puzzling question. Vitamins include a large number of substances which are chemically totally unrelated. Though they are *vital* for good health, they are only needed in small amounts as compared with major dietary components such as proteins, carbohydrates and fats. A substance that is a vitamin for one species of animal is not necessarily so for another. However, on closer examination of the chemical nature of individual vitamins, it is evident that often a particular vitamin is closely related chemically to a particular coenzyme. Thus niacin (a vitamin that prevents pellagra, a disease leading to dry skin and nervous disturbances) is found to be nicotinic acid, which is closely related to NAD^+ and $NADP^+$; the anti-beriberi substance, vitamin B_1, is thiamine, a substance related to a coenzyme known as TPP (thiamine pyrophosphate). We are therefore led to the conclusion that some animals, ourselves included, cannot synthesize all the coenzymes they need from the simple sugars, amino acids and fats that are provided in the diet. In addition, they need from their food some more complex building blocks—vitamins—which form components of the final coenzyme molecules.

◇ Why are vitamins only needed in small amounts compared to the major dietary materials (carbohydrates, etc.)?

◆ Coenzymes are either regenerated unchanged at the end of the reactions in which they participate or else are regenerated subsequently in other reactions. For example, NAD^+ is converted to NADH in a variety of enzyme-catalysed reactions for which it is a coenzyme; see Section 4.3.1, for example. However other reactions require NAD*H* as a coenzyme, converting *it* in the process to NAD^+ (Chapter 6). Coenzymes are thereby regenerated and only a relatively small reservoir of each type of coenzyme is needed by the cell; this reservoir is kept topped up by a small regular dietary intake of vitamins, from which the cell synthesizes the coenzymes. (Note, however, that not all vitamins are needed for cellular production of coenzymes.)

There is as yet insufficient evidence to support the idea that all the symptoms characteristic of a vitamin deficiency disease necessarily arise from a lack of coenzyme(s) resulting from a deficiency of the vitamin in question. Nevertheless, one thing that is clear is that vitamin deficiency diseases can be cured readily if treated early enough, and can be prevented easily, that is, easily from a medical standpoint.

4.5.2 Chemical modification of active sites

You have already seen how the active site of an enzyme is a precisely shaped entity that accepts just its specific substrate(s) or close chemical relative(s). This is largely because the active site consists of a cluster of amino acid side chains orientated to form a specific three-dimensional shape. The amino acid side chains in the active site can be subdivided operationally into three classes, as shown in Figure 4.19.

Figure 4.19 The active site of a hypothetical enzyme. The three classes of amino acid side chain are:

A Those directly involved in the bond breaking and bond making steps in the catalysed reaction. (The substrate is shown in blue as a circle attached to an oblong. The bond to be broken is shown as a blue hatched bar.)

B Those not involved in the catalysed reaction as such, but in contact with the substrates or coenzymes, thus helping provide specificity.

C Those present in the active site region, but not involved in the reaction or substrate or coenzyme binding at all.

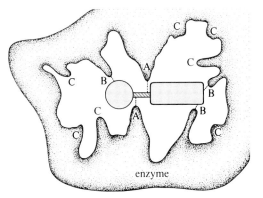

enzyme

One of the first steps towards understanding how any enzyme works is to identify the amino acid side chains in the active site and their orientations to each other and to the substrate(s) and any coenzyme(s). One approach involves **chemical modification of the active site**. A chemical 'label' is attached to the active site, the protein is hydrolysed, and the amino acid side chain(s) carrying the label are identified. But how can one be sure that the label was attached to amino acid side chains actually in the active site and not in some other part of the protein?

Ideally a *substrate* of the enzyme would be used as the label, but in practice, this has rarely been possible because the association between substrate and active site, as an ES complex, is too brief to survive the lengthy and destructive procedure required to isolate a labelled amino acid. On occasions, however, by luck or by design, this or analogous procedures have proved possible.

One historically early and classic case is the interaction of a substance, with the rather grand name of di-isopropylfluorophosphate (or DIFP for short), with a number of different enzymes that share certain common features. All these enzymes are *esterases* (they catalyse the hydrolysis of esters) and *some* of them can also hydrolyse proteins (thus these can also be classified as *proteases*, and indeed usually are). DIFP is an irreversible inhibitor (Section 4.4.3) of these enzymes. After treatment with DIFP, the enzymes can be subjected to partial hydrolysis (using acid), breaking the polypeptide chains into short peptide fragments (Chapter 3) which can be separated from each other. In every case, it was found that the peptide containing the DIFP (in fact, part of it; see Figure 4.20) had it bound to a serine residue. Other evidence suggested that this is a particular serine and is in the active site (Section 4.7.6). The reaction between DIFP and the enzyme is shown schematically in Figure 4.20.

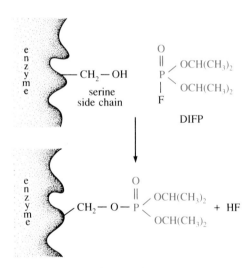

Figure 4.20 Reaction of DIFP with the side chain of a particular serine residue present in the enzyme.

However, though it proved possible to show that in the case of these particular proteases and esterases DIFP mimics a normal substrate, DIFP is a bit of a curiosity. It does not closely resemble the normal substrates of these enzymes in any sense of overall shape. It just happens to have the appropriate chemical groups and reacts with the active site, a somewhat lucky accident and, incidentally, one of some historical importance.

In general we would feel on safer ground if the substance chosen to label the active site actually resembled the normal substrate closely. But it must react with the active site to form a complex, which unlike a normal ES complex, is stable enough to be isolated. Substances of this type, called *active site-directed irreversible inhibitors* or **affinity labels**, have been recently specifically designed.

◇ Affinity labels must have the property of resembling a substrate well enough to bind at the active site. Can you recall such a class of substances, other than actual substrates?

◆ Competitive inhibitors (Section 4.4.3).

However *unlike* competitive inhibitors, which are reversible inhibitors, an affinity label is specially designed to be an *irreversible* inhibitor and so form a stable, covalently-bonded, complex with the active site. This then allows isolation and identification of the amino acids in the active site, by partial hydrolysis of the inhibited enzyme, as was done for DIFP.

For example, a substance called tosyl-L-phenylalanine chloromethyl ketone (mercifully, usually abbreviated to TPCK) resembles the substrate of the enzyme chymotrypsin. On reaction with the enzyme it irreversibly binds to the active site, i.e. it is an affinity label for chymotrypsin. Following partial hydrolysis of the enzyme and analysis of the resulting peptides, it can be seen that TPCK has reacted with a histidine (His) residue, specifically His-57 (i.e. the 57th amino acid in the polypeptide chain, counting from the N-terminus; Chapter 3). This provides powerful evidence that His-57 is in the active site of chymotrypsin and is involved in catalysis with normal substrates (which TPCK has, of course, been designed to resemble). Incidentally, chymotrypsin is also one of the proteases that reacts with DIFP, as shown above; of which more later (Section 4.7.6).

4.5.3 pH optima

pH has a profound effect on enzyme activity. For each enzyme there is a particular pH, the optimum pH, at which its activity is greatest. Thus while lysozyme has an optimum pH of about 5, that of pepsin, a protease, is about 2, suited to the acid environment of the stomach where pepsin operates. The existence of an optimum pH is readily explicable when one considers enzymes as proteins. Proteins contain a large number of ionizable groups, in the side chains of amino acids; that is groups that can carry a charge (e.g. glutamic acid $\rightleftharpoons$ glutamate; Table 3.2 in Chapter 3). The ionization of a chemical group, and hence the charge it carries, depends on its **pK_a**. The pK_a of an acidic group is the pH at which it is 50% ionized, and so at any particular pH, some groups will be ionized and some will not. (A fuller treatment of pK_a is given in Box 4.2.) Altering the pH of a solvent therefore alters the degree of ionization of these groups. If one or more groups at the active site must be ionized for catalysis to occur, the pH of the system will affect the activity of the enzyme. If pH is plotted against activity, we get a **pH–activity curve**; an example is shown in Figure 4.21.

From such curves, pK_a values for the pH-sensitive step or steps in the enzyme-catalysed reaction can be estimated. So when such a curve is known for a particular enzyme, it would seem to be a relatively simple matter to identify the amino acids involved in catalysis (at least, those that ionize); presumably those whose pK_a values (obtained by independent studies on *free* amino acids) correspond to the pK_a values determined from the pH–activity curve. Simple it may seem, but there are possible complications:

□ Several different ionizable groups with somewhat similar pK_a values might be ionized at the optimum pH.

□ The pH might also affect the state of the substrates or coenzymes in some significant way.

□ The pH might affect the overall shape of the protein, affecting the active site only as a secondary consequence.

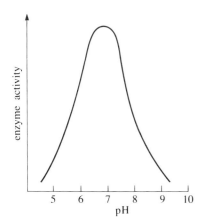

Figure 4.21 The pH–enzyme activity curve for amylase.

Box 4.2 The derivation and significance of pK_a

An acid is a substance that dissociates to give a proton (H^+) and a 'conjugate base'; that is, in the dissociation of a weak acid, HA,

$$HA \rightleftharpoons H^+ + A^-$$

A^- is the conjugate base.

The equilibrium constant for the dissociation, K_a, is defined by

$$K_a = \frac{[H^+][A^-]}{[HA]} \qquad (4.4)$$

Therefore, dividing both sides of Equation 4.4 by K_a and $[H^+]$, we get

$$\frac{1}{[H^+]} = \frac{1}{K_a} \frac{[A^-]}{[HA]}$$

Taking logarithms we get

$$\log_{10} \frac{1}{[H^+]} = \log_{10} \frac{1}{K_a} + \log_{10} \frac{[A^-]}{[HA]}$$

As $\log_{10} 1/[H^+] = -\log_{10}[H^+]$ and this is the definition of pH,

$$pH = \log_{10} \frac{1}{K_a} + \log_{10} \frac{[A^-]}{[HA]}$$

That is

$$pH = -\log_{10} K_a + \log_{10} \frac{[A^-]}{[HA]}$$

pK_a is defined as $-\log_{10} K_a$, so

$$pH = pK_a + \log_{10} \frac{[A^-]}{[HA]} \qquad (4.5)$$

Where HA is 50% ionized, $[A^-] = [HA]$.

Thus by substitution in Equation 4.5

$$pH = pK_a + \log_{10} 1 = pK_a \quad (\text{as } \log_{10} 1 = 0)$$

Therefore the pK_a for an acidic group is the pH at which that group is 50% ionized.

pK_a is a useful way of comparing the relative strengths of acidic groups; the stronger the acidic group, the lower its pK_a.

Thus trichloracetic acid ($pK_a = 0.65$) is a stronger acid than acetic acid ($pK_a = 4.75$), which is in turn a stronger acid than the ammonium ion ($NH_4^+ \rightleftharpoons NH_3 + H^+$; $pK_a = 9.25$); because the lower the pK_a, the lower the pH at which the acidic group is a proton donor. Some compounds, including many of biological importance, have more than one ionizable group in the molecule and hence more than one pK_a. For instance, all free amino acids have at least two ionizable groups, the carboxyl group and the amino group:

$$
\begin{array}{cc}
\overset{\displaystyle NH_2}{\underset{\displaystyle COOH}{R-C-H}} &
\overset{\displaystyle NH_2}{\underset{\displaystyle COO^-}{R-C-H}} \\[2em]
\overset{\displaystyle \overset{+}{N}H_3}{\underset{\displaystyle COO^-}{R-C-H}} &
\overset{\displaystyle \overset{+}{N}H_3}{\underset{\displaystyle COOH}{R-C-H}}
\end{array}
$$

Some amino acids also have ionizable side chains; these contribute to the three-dimensional structure of proteins (Chapter 3) and are often important in enzyme catalysis (Section 4.7).

◇ Why might a change in pH affect the overall shape of a protein?

◆ The three-dimensional structure of a protein depends in part on inter-actions between charged groups (+ and −) on amino acid side chains (Chapter 3). The existence of these charges depends on the pH of the surrounding medium and the pK_a values (Box 4.2) of the amino acid side chains in question (how is more complicated, as Section 4.7.6 will show). Thus alterations in pH can cause alterations in charges on amino acid side chains other than those involved in catalysis. If these charges are involved in maintaining the three-dimensional structure of the protein, altering them might lead to a change in the three-dimensional structure overall. (Indeed, large-scale changes in pH can cause denaturation of proteins; but this is not an easily reversible event, unlike that shown in Figure 4.21.)

□ The pK_a of a group in a free amino acid (used as the reference value to identify the group ionized) may differ from that of the same group when the amino acid is in a protein, because of the influence of neighbouring groups. This last point is very significant and will be elaborated later (Section 4.7.6).

Because of the various possible complicating factors much caution must be exercised in interpreting pH–activity curves.

4.5.4 X-ray diffraction

If you want to know the amino acid side chains present in, and the shape of, the active site, a logical approach is to 'take a picture'. 'Pictures' can be taken of proteins and three-dimensional models built from them by the technique of X-ray diffraction (Chapter 3, Box 3.1). If this is done for an enzyme, the problem of identifying where precisely the active site is, in what is a highly complex structure, is considerable; the active site rarely 'leaps out of the page or model'. Unlike our drawing of lysozyme in Figure 4.22, it does not come with a handy arrow!

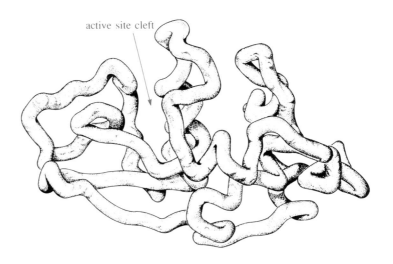

active site cleft

Figure 4.22 Simple model of egg white lysozyme showing its active site (arrowed), as derived from X-ray diffraction.

Ideally we would like to crystallize the ES complex and subject *it* to X-ray diffraction. As we stated in Section 4.4.1, we might expect this to be very difficult, because once a molecule of enzyme has interacted with its substrate(s) to form an ES complex, this can then break down to give products. *In solution*, by keeping the enzyme well supplied with its substrate(s) (as at v_{max}; Section 4.4.1), we can ensure that virtually all of the enzyme present at any one time is in the form of ES complex. But though it may be possible to diffuse substrates into enzyme crystals, they break down there too. So to get substrate into crystals quickly enough to keep the enzyme virtually saturated with it might be impossible under normal circumstances. Sometimes, however, there are circumstances or tricks whereby this problem of the instability of ES complexes can be circumvented. We shall consider some here.

So far we have largely considered the ES complex of an enzyme to be a single entity formed from enzyme plus substrate. But for enzymes that catalyse reactions with more than one substrate (e.g. $L + M \rightleftharpoons Z$), we might expect there to be more than one ES complex (e.g. see Figure 4.9). The full assembly of E plus L plus M, the *complete* ES complex, would proceed to give products. So this ES complex (ELM) would likely be much too unstable to examine by X-ray diffraction. But it might also be possible to obtain crystals of ES complexes where the enzyme is combined with *just one* of the two substrates, say EL or EM. These complexes lacking one or other of the two substrates, what we can call 'partial' ES complexes, might be stable enough to examine by X-ray diffraction. Indeed this has proven the case for a number of enzymes with two or more substrates. Studying the structure of one of these 'partial' ES complexes, can allow us to visualize what the complete ES complex (ELM) might be like.

However, where forming stable 'partial' ES complexes proves impossible, notably where the enzyme catalyses a single-substrate reaction (S→P), there may be still other means available to gain some knowledge of an ES complex.

For example, sometimes it is possible to form relatively stable ES complexes, even in the crystalline state, by using a substance that is a *poor* substrate of the enzyme, i.e. one that reacts very slowly. Or an equivalent procedure is to use an appropriate *inhibitor* of the enzyme, instead of a substrate, and subject the enzyme–inhibitor complex to X-ray diffraction.

◇ What sort of (appropriate) inhibitor would you use?

◆ A competitive inhibitor, as this is known to bind at the active site because of structural similarity to the substrate (Section 4.4.3) and it is the active site that we wish to locate. This is one obvious reason why it is useful to be able to distinguish whether a particular inhibitor is competitive or not (as in Question 4, for example).

Lysozyme was the first enzyme whose three-dimensional structure and location of its active site were solved by this combination of competitive inhibitor and X-ray diffraction (Figure 4.22). Lysozyme has been a key example in the study of enzymes and has had an interesting scientific history.

Lysozyme has antibacterial activity and was first discovered in 1922 by Alexander Fleming (later to discover penicillin) during his search for just such agents. Fleming was running a cold, and, as an experiment, he put some of his nasal mucus onto a bacterial culture plate. Some time later, he was excited to find that bacteria near the mucus had apparently been dissolved away—had he discovered the antibacterial agent he was looking for?

He was able to show that the active ingredient was an enzyme, and it was given the name *lysozyme*. For, despite any impression we may have given to the contrary (Section 4.2), naming enzymes in those pre-systematic days did have some logic. In the name 'lysozyme', *lyso* denotes its capacity to lyse (that is, dissolve) bacteria. And *zyme* was a word commonly associated in those days with enzymes (from the Greek for yeast, on which the earliest studies of enzymes, some 90 years ago, were made).

◇ What is the systematic ending now given to enzyme names?

◆ -*ase* (Section 4.2).

A good source of lysozyme proved to be tears, and soon a number of volunteers were suffering 'ordeal by lemon' to provide them. Fleming, the later popular hero through penicillin, may on this occasion be cast as the 'comic book' villain—'victims' were paid for their tears. Apparently, the laboratory technicians had it down to a fine art, getting threepence a time and keeping regular accounts paid at the end of the month. In 1922, threepence was not be sniffed at.

Though lysozyme proved of little use therapeutically, it has been very useful in the study of enzyme catalysis, and, fortunately less 'cruel' sources are available—egg white, for one.

Over the past 25 years or so, X-ray diffraction has yielded a great deal of useful information on enzymes, particularly where 'pictures' with and without substrate (or inhibitors) have been built up. As with any technique, however,

X-ray diffraction does have some problems. The technique is not that simple and is time-consuming. There is sometimes also the nagging question of whether what is observed from a protein crystal really reflects what happens in solution. Of course, in some cases, evidence supporting the similarity of crystal and dissolved enzyme comes from the fact that the crystals catalyse the breakdown of substrates diffused into them.

Generally what is needed is some corroborating evidence, from other techniques, that what is learnt from X-ray diffraction is consistent with the interpretation put on the functioning of the observed structure (see Section 4.6 for an example).

4.5.5 Protein engineering

One recent addition to the study of enzymes is, in essence, a particularly sophisticated form of chemical modification (Section 4.5.2); one that potentially allows us to change amino acids, one for another, at specific places in a protein, as we wish. The technique goes by the name of **protein engineering**, which, although aptly describing the *result* of the technique, is a bit of a misnomer for how it is actually carried out. What is in fact engineered is, not the protein, but the *gene* coding for it. We will not go into how the gene coding for a specific protein is obtained nor how it can be altered (engineered); this broadly falls within the province of the rapidly growing field of *genetic engineering*, itself a branch of the increasingly fashionable *biotechnology* (Section 4.9.2). Essentially, one can obtain the gene for the protein of interest and alter specifically one or more of its nucleotide bases (Chapter 3). This can be done so that one or more amino acids in the protein for which this gene normally codes will themselves be altered, and altered in a predictable manner under the control of the experimenter. On inserting the altered gene into a suitable host organism, such as a bacterium or a yeast, the altered protein is produced. A very sketchy summary of some of the principles underlying protein engineering is given in Figure 4.23.

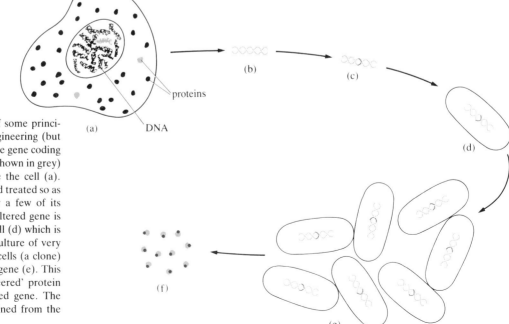

Figure 4.23 A summary of some principles underlying protein engineering (but not the actual methods). The gene coding for the protein of interest (shown in grey) is one of very many inside the cell (a). The gene is obtained (b) and treated so as to alter specifically one or a few of its nucleotide bases (c). The altered gene is inserted into a microbial cell (d) which is then grown to produce a culture of very many millions of identical cells (a clone) that all contain the altered gene (e). This clone produces the 'engineered' protein corresponding to the altered gene. The engineered protein is obtained from the cells (f). (Not to scale!)

This technique has great potential for our understanding of enzymes. Let us take an example.

The enzyme carboxypeptidase A is a protease that catalyses the hydrolysis of certain specific amino acids from the carboxyl end of peptide chains (Chapter 3). The enzyme has been extensively studied by X-ray diffraction. This has revealed that, on binding substrate, movement occurs in a number of amino acid side chains in the active site. The largest movement is by the side chain of a tyrosine (Tyr-248); this movement is 1.2 nm, a dramatically large distance in active site terms; in fact equivalent to about a quarter of the diameter of the whole enzyme. This and other evidence has implicated Tyr-248 as being involved in catalysis by carboxypeptidase A; the hydroxyl group of Tyr-248 (see Figure 4.24a), in particular, has been suggested to be of importance in catalysis. In 1985, this putative role for Tyr-248 was further tested by means of protein engineering. Tyr-248 was replaced by a phenylalanine.

◇ Do you recall the difference between tyrosine and phenylanine?

◆ As shown in Table 3.2 (Chapter 3), and in Figure 4.24, they differ by a *single hydroxyl group* on the benzene ring.

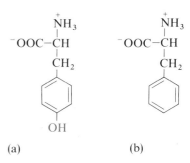

Figure 4.24 The formulae of (a) tyrosine and (b) phenylalanine, with the tyrosine hydroxyl ($-OH$), by which they differ, shown in blue.

When the engineered enzyme was then compared to the natural one it was found that the rate of catalysis with the engineered enzyme was virtually identical to that with the natural enzyme.

◇ What does this suggest about the catalytic role of Tyr-248; its hydroxyl group in particular?

◆ Replacing Tyr-248 by a phenylalanine is tantamount to removing the hydroxyl from Tyr-248. Yet this does not affect the catalytic activity of the enzyme. This suggests that the putative role in catalysis for the hydroxyl of Tyr-248 is in fact *incorrect*.

However the engineered enzyme did have an increased K_M value for some peptide substrates.

◇ What does this suggest about the role of the hydroxyl of Tyr-248?

◆ K_M values can sometimes give an indication of the relative affinities of enzymes for their substrates (Section 4.4.2; Box 4.1). A higher K_M indicates a lower affinity, a weaker binding of enzyme to substrate. Thus the raised K_M of the engineered enzyme might indicate a weaker binding of enzyme to substrate than in the natural enzyme. This in turn implies that the role of the hydroxyl of Tyr-248 (in the natural enzyme) is not in catalysis *per se* (i.e. not in the actual chemical change to substrate) but is in *binding of peptide substrates*.

Thus protein engineering has allowed us to 'dissect' this enzyme-catalysed reaction into two distinct phases—the binding of substrate to form an ES complex ($E + S \rightleftharpoons ES$) and catalysis to produce product ($ES \rightarrow E + P$)—and to identify towards which particular phase a single hydroxyl, that of Tyr-248, contributes. Protein engineering is as yet in its infancy, but, as this example shows, it is certainly a technique that offers a level of precision in studying the way enzymes work that has hitherto only been dreamed of. It also has promising commercial applications, of which more in Section 4.9.2.

4.5.6 Interpretation of data on enzyme mechanisms

You will have realized that no one technique can give total information on how any enzyme works. Each technique has its particular strengths and weaknesses, and all data must be carefully scrutinized in that light. X-ray diffraction, for example, can certainly help locate the active site and identify the amino acid side chains projecting into it. But it does not so readily identify which of the side chains are actually involved in catalysis rather than say helping to bind the substrate (Figure 4.19). Chemical modification—by affinity labels, for example—can help to identify catalytically important side chains but it does not so easily tell us their three-dimensional relationship to the substrate nor to the rest of the active site, as might X-ray diffraction. In theory, protein engineering could be used to alter every amino acid in an enzyme, one at a time, thereby yielding a vast number of variants of that enzyme for study—thus nailing down those amino acids directly involved in catalysis—but in practice this is virtually impossible. With some indication, from other techniques, of which amino acids are likely to be involved in catalysis or binding and are therefore worth altering (as was Tyr-248 above), the precision of protein engineering need not be used in such an indiscriminate manner. So what is always needed to build up an accurate picture of how any particular enzyme works is a combination of techniques.

Summary of Section 4.5

1 To unravel the mechanism of catalysis of an enzyme we need to: find out the nature of the chemical changes to the substrate(s); identify the chemical groups comprising the active site; elucidate how those groups effect catalysis.

2 Some enzymes possess one or more non-protein components in their active sites that are important for catalysis; these are often known as coenzymes.

3 Some coenzymes can carry out catalysis on their own, to some extent at least (an extent, however, that is insignificant compared to the catalysis by the whole enzyme that occurs *in vivo*). By studying these simpler coenzyme model systems, we can learn something about the enzymic reactions too.

4 Coenzymes can often be seen to be related to certain vitamins, from which the coenzymes are synthesized by the organism.

5 Inhibition of enzymes with irreversible inhibitors can be useful in identifying the amino acid side chains present in the active site. Affinity labels, irreversible inhibitors specifically designed to resemble substrates, are particularly useful in that regard.

6 Enzymes exhibit distinct pH optima. Determining these can sometimes give clues as to which groups in the active site have to be ionized in order for catalysis to occur. However, interpretation is difficult.

7 X-ray diffraction is a technique that can give detailed information on the three-dimensional structure of enzymes. Analysis of an ES complex, or an enzyme with competitive inhibitor bound, can help identify the active site and the amino acid side chains that contribute to binding and catalysis.

8 Protein engineering is a new technique that allows us to change selected amino acids in an enzyme as we wish. This can enable us to examine the consequences to catalysis of such changes and thereby confirm or refute the purported roles of the amino acids concerned.

Question 6 (*Objectives 4.1 and 4.5*) Which of the following statements are true and which are false?

(a) Vitamins are coenzymes.

(b) All living organisms need the same vitamins.

(c) Affinity labels are irreversible inhibitors.

(d) NAD^+ is chemically altered by some enzymes.

(e) Vitamins are never metabolized by animals.

(f) All enzymes have the same optimum pH, that typical of cells (i.e. around pH 7.4).

(g) Enzymes are very sensitive to pH because acid hydrolyses peptide bonds.

(h) X-ray diffraction of enzymes produces some photographs of the enzymes performing catalysis.

(*Note*: Questions 7 and 8 at the end of Section 4.7 also relate in part to Section 4.5.)

4.6 RIGIDITY OR FLEXIBILITY: LOCK AND KEY VERSUS INDUCED FIT

Though many techniques may be brought to bear on the problem of how an enzyme works (Section 4.5.6), much of our discussion has centred on *structure*: how to determine the detailed structure of an active site, or, better still, of an ES complex. This emphasis on structure owes much to what is undoubtedly one of the central tenets of biochemistry and of its offspring, molecular biology, in particular—that understanding structure is the key to understanding *function*. For enzymes, this concept is brought to mind most readily by the phenomenom of specificity; something that conjures up an image of a precise fit between substrate and active site, a complementarity of size and shape, in Emil Fischer's memorable metaphor, a *lock and key*. This lock and key view of enzymes can be used to explain not just specificity, but catalysis too. Just as a key must have the right size and shape to fit *and* turn a lock, so a substrate must have the right size and shape to bind to an active site *and* lead on to catalysis (Section 4.4.3). (Presumably a competitive inhibitor has the size and shape to fit an active site but, for some reason, no catalysis occurs—the 'lock' does not turn with the wrong 'key' in it.) There are, however, some problems with the lock and key hypothesis.

Firstly, it ignores something that we know now about protein structure that was unknown in Fischer's day—that proteins, enzymes included, show a considerable degree of flexibility in their conformation (Chapter 3, Box 3.3). So active sites need not be the rigid structures that the word 'lock' implies. And indeed, as X-ray diffraction studies have revealed, in some enzymes, such as carboxypeptidase A (Section 4.5.5), there are marked changes in the conformation of the active site that occur on binding of substrate.

Secondly, on the lock and key hypothesis, what is to stop substances that are smaller than the substrate, and perhaps have some chemical groups in common with it, entering the active site and disrupting catalysis? This they could perhaps do by simply getting in the way ('jamming the lock'). Or, if they possess some of the same reactive chemical groups as the substrate, why do they not compete with it for catalysis; why are they not substrates too? Water, for example, is a small molecule that might be expected to compete in many enzyme-catalysed reactions where −OH groups are involved. So why does water not disrupt many such reactions? There is certainly plenty of it in cells. This is a problem of specificity—lock and key readily accounts for specificity among substances of roughly the same size but does not so readily exclude much smaller ones that have the right chemical groups for reaction.

A hypothesis proposed by Koshland in 1958 might make up for these apparent deficiencies in Fischer's. Koshland suggested that the shape of substrate and active site are not precisely complementary, but on binding the substrate induces a change in conformation in the active site such that the two (substrate and altered active site) are then complementary. Figure 4.25 gives a highly stylized comparison between the classical lock and key hypothesis and Koshland's **induced fit hypothesis**.

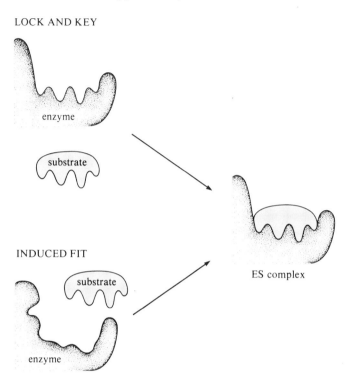

LOCK AND KEY

enzyme

substrate

ES complex

INDUCED FIT

substrate

enzyme

Figure 4.25 The lock and key and induced fit mechanisms of enzyme–substrate interaction. In the lock and key model the active site (lock) is precisely complementary in shape to (and the right size for) the substrate (key). In the induced fit model the fit is *not* precise *before* binding. But on binding, the substrate induces a change in conformation in the active site so that it is now precisely complementary in shape to the substrate.

So on the induced fit hypothesis, specificity is visualized *not* as a rigid 'pre-fit' of substrate and active site (as with lock and key), but as a more flexible response of active site to substrate. In Koshland's hypothesis, Fischer's 'lock and key' has become 'hand and glove'—a floppy glove, like a woollen one, of the right general form and size (the active site) only attains a shape that truly fits the hand (the substrate) when the hand is in it. Interestingly, Koshland originally advanced his hypothesis to explain specificity and before the inherently flexible nature of proteins was fully appreciated. However the induced fit hypothesis, like the lock and key one, can be advanced to account for catalysis too. Only the substrate is capable of *inducing* the correct change in the active site; 'correct', that is, for subsequent catalysis to occur. Substrate induces and thereby 'activates' the active site. Smaller substances might enter the active site but either they cannot induce the required change in its conformation, or, if they do (say by possessing appropriate chemical groups), they then do not fit the site as a whole well enough for catalysis (i.e. specificity discriminates against them).

So much for an induced fit explanation of how enzymes might work, but how well does it fit the facts?

Though there is now no doubt that all proteins are flexible to some degree or other, what significance, if any, this has for enzymes in general is far less clear. Certainly a flexible active site means that *potentially* it can undergo substantial change in conformation when substrate binds (as it does in carboxypeptidase A). But not all enzymes seem to undergo such change, indeed on current counts most do not—so flexibility and induced fit would seem to be of little relevance to how those enzymes work. Even in those enzymes where substantial change in the conformation of the active site does occur on binding substrate, it is another matter altogether to determine the relevance of such change, if any, for how the enzyme works; to whether the induced fit hypothesis holds true for them. Obviously it is *tempting* to believe that such changes in conformation are significant—after all, if we subscribe to the view that structure and function are related, then a change in three-dimensional structure (conformation) would seem to have functional significance. Tempting, but is it always true? And, if true, is the significance of the change in conformation for specificity and/or catalysis (induced fit) or for some other function? Is an observed change in the conformation of an active site *necessary* for specificity and/or catalysis, or not? This is the key question when trying to test whether or not an observed change in conformation supports the induced fit hypothesis. Finding the answer is less easy; however, for a few enzymes there is reasonable evidence that the active site does undergo a change in conformation when substrate binds *and* that this is connected to specificity and/or catalysis.

One such enzyme is hexokinase isolated from yeast (Section 4.3.3). Hexokinase catalyses the phosphorylation of glucose by ATP producing glucose 6-phosphate and ADP:

$$\text{glucose} + \text{ATP} \overset{\text{hexokinase}}{\rightleftharpoons} \text{glucose 6-phosphate} + \text{ADP}$$

glucose ATP → glucose 6-phosphate ADP

If you examine the above equation, you will appreciate that another way of viewing the reaction is to say that the hydroxyl ($-\text{OH}$) of glucose removes the terminal phosphate of ATP. This removal of the phosphate can also be achieved by the $-\text{OH}$ of water. But hexokinase, water and ATP alone (no glucose) only remove the phosphate from ATP at $1/40\,000$ the rate of when glucose is present. So, it seems that though water is small enough to enter the active site of hexokinase and has an $-\text{OH}$ group required for reaction, apparently it cannot 'activate' the enzyme as well as glucose can. The enzyme can discriminate (by specificity) between the $-\text{OH}$ of glucose and that of water. X-ray diffraction shows that both ATP and glucose cause substantial change in the three-dimensional structure of the enzyme (Figure 4.26).

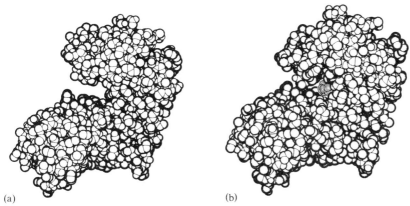

(a) (b)

Figure 4.26 Three-dimensional structures of hexokinase derived from X-ray diffraction data: (a) free enzyme; (b) enzyme with glucose (in blue) bound.

143

It thus appears that in the case of hexokinase a change in the shape of the active site does occur on binding substrate *and* that this is necessary for specificity and rapid catalysis (discriminating between water and glucose). Hexokinase would seem to conform to an induced fit mechanism. This has the important effect on preventing a potential alternative substrate which is in plentiful supply (water) from disrupting the reaction with the actual substrate, glucose. You should note how both structural data (Figure 4.26) and kinetic data (comparing the rates of phosphate removal) are needed to show a correlation between structural change and enzyme activity (induced fit).

There are very few enzymes where such a correlation between structural change in the active site and specificity and/or catalysis can be demonstrated. And we also know that many enzymes do not undergo substantial change in the shape of their active site on binding substrate. In these cases the active site may be flexible but for all intents and purposes of specificity and catalysis it behaves essentially like a relatively rigid 'lock'. So, as is often the case with enzymes, no one hypothesis—lock and key or induced fit—seems to describe the behaviour of all enzymes equally well. 'Lock and key', 'hand and glove', 'rigid bowler or floppy hat "induced" to sit neatly on the head'—all have been invoked to describe active sites and their interactions with substrate. There is evidently no lack of imagery. This may be the 'poetic soul of science' peeping through or, as you might suspect, it may reflect merely that there is still a lot about enzymes that we do not know.

It is easy enough, in principle, to visualize how the close juxtaposition of an array of amino acid side chains next to a substrate in an active site can ensure specificity, a precise complementarity of shape between active site and substrate (Figure 4.19). Ultimately this is true irrespective of whether that complementarity comes from a relatively rigid 'pre-shaped' active site or from one 'fine-tuned' by induced fit (Figure 4.25). It is also easy to see that some side chains closely aligned to the substrate might be involved in the making and breaking of bonds (type A in Figure 4.19), in catalysis. This, however, does not of itself tell us where the increase in reaction rate, the catalytic power, comes from. To take an analogy: switching on the ignition, turning the starter motor, engaging the gears and releasing the handbrake are all necessary *before* a car can move off. But this does not tell us where the power to drive the car comes from (the engine, plus fuel). Likewise, having the right amino acid side chains 'engaged' (whether already in place or following induced fit) alongside the substrate is necessary *before* catalysis can occur, but does not identify the source of catalytic power. It is to this topic that we now turn.

4.7 CATALYTIC POWER

Though the ability of enzymes to produce increases in reaction rate of the order of 10^9–10^{12} times has been appreciated for many years, *how* they do so remains one of the '$64\,000$ questions' of biology. There is, as yet, no one totally convincing answer. There are, however, a number of serious candidate hypotheses.

It seems that taking enzymes as a whole, there are a number of quite different mechanisms by which they might effect catalysis. For example, they may bring substrates close together and in orientations favouring reaction (Section 4.7.1); provide catalytic groups close to particular bonds in the substrates (Section 4.7.2); place substrates under stress (Section 4.7.3); involve substrates in alternative, covalent reactions that are more energetically favourable (Section 4.7.4). The extent to which any one mechanism operates in a particular enzyme, if indeed at all, and how much it contributes to catalytic power, can vary from enzyme to enzyme. Thus, if there is to be a prize for answering *how* enzymes in general work so fast, it may well have to be shared out among several hypotheses.

The hypotheses themselves are quite complex, involving detailed structural and kinetic arguments, well beyond the scope of our treatment here. We shall, therefore, introduce briefly some of the major current ideas concerning enzyme catalytic power and try to indicate some of the pros and cons.

4.7.1 The proximity effect

Consider a reaction involving two or more reactant molecules (a multi-molecular one) e.g. $A + B \rightleftharpoons C$. The molecules must collide before they can react. The rate of reaction will be related to the rate of collision; the more collisions, the more reaction. Various things can affect the rate of collision. These include, of course, the concentration of the molecules involved in the reaction; the greater the concentration, the greater the rate of collision.

It has long been believed that a general feature of *many* catalysts, biological or not, is that they provide a physical surface to which reactants can bind and thus be brought closer to each other. This increases the **effective concentration** of the reactants.

In living cells, the reactants (i.e. substrates) for reactions are usually in low concentration. So we might reasonably suppose that an important contribution of enzymes to catalysis of reactions involving more than one molecule is to raise their effective concentrations. This an enzyme can do by *specifically* holding each of the molecules in its active site. By holding the reactant molecules close together, the enzyme increases the probability of reaction between them.

But in an uncatalysed reaction, not all collisions of reactant molecules will necessarily lead to formation of products. As well as colliding with the required energy, the colliding molecules must be in the correct orientations to bring the potentially reactive groups in them together, but the orientations will be random. Beyond simply holding reactants close to each other, an enzyme active site can also ensure that the molecules are held in a relative orientation to each other that favours reaction. This combination of holding substrate molecules close together and in the correct orientation for reaction is sometimes referred to as the **proximity effect**.

The possible contribution of the proximity effect to catalytic power can be demonstrated by *non-enzymic* model systems. For example, in the reaction shown in Figure 4.27a the two molecules, which we shall term simply A and B, must collide and in the correct orientation for reaction to occur between the carboxyl group of A and the carbonyl group of B. In an analogous reaction, shown in Figure 4.27b, the two reactive groups (carboxyl and carbonyl) are part of the same molecule and in an orientation appropriate for reaction. This relative proximity of the reactive groups results in a reaction 200 000 times faster than that shown in Figure 4.27a.

Recent calculations have shown that in *some* enzymes the contribution to catalytic power of the proximity effect *could* be very considerable; in others it might be very much more modest. Thus though all enzymes exert a proximity effect, its precise degree and its relative contribution to catalytic power will vary from enzyme to enzyme. Other mechanisms contributing to catalytic power, that will operate in some enzymes and not in others, are detailed below (Sections 4.7.2 to 4.7.4).

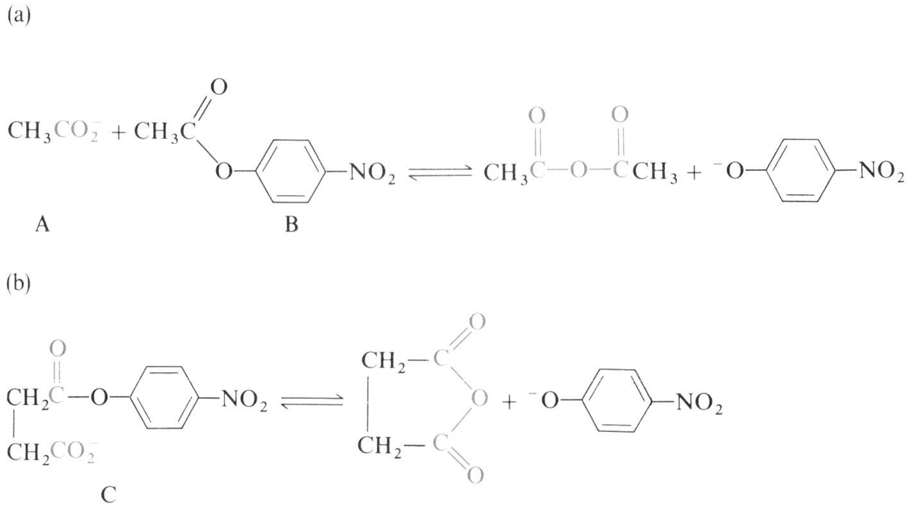

Figure 4.27 The proximity effect as demonstrated in a non-enzymic model system. For reaction to occur in (a) the two molecules, A and B, must collide in an orientation such that the reactive carboxyl group in A is in contact with the reactive carbonyl group in B. In (b) the the reactive groups (carboxyl and carbonyl) are part of the same molecule (C). The relative rate of reaction is 2×10^5 times faster in (b) than (a). The reactive carboxyl and carbonyl groups that react to form the anhydride product are shown in blue.

(a)

$CH_3CO_2^- + CH_3C(=O)O-\langle\rangle-NO_2 \rightleftharpoons CH_3C(=O)-O-C(=O)CH_3 + {}^-O-\langle\rangle-NO_2$

A   B

(b)

$CH_2C(=O)-O-\langle\rangle-NO_2$ with $CH_2CO_2^-$ $\rightleftharpoons$ anhydride $+ {}^-O-\langle\rangle-NO_2$

C

4.7.2 Acid–base catalysis

Many chemical reactions go faster when in the presence of both acidic and basic groups (i.e. they are *catalysed* by acids and bases). We can term this **acid–base catalysis**, though it is more correctly called *general acid–general base catalysis*; where any chemical group with an ionizable proton is considered to be *general acid*, and any group that can accept a proton is a *general base*.

It is known that many enzyme-catalysed reactions involve the movement of protons; that is, they involve acid–base equilibria.

We also know that pH affects enzyme activity (Section 4.5.3), suggesting a *possible* catalytic role for ionizable groups (perhaps acids and bases) in the active site.

Putting these three facts together it seems reasonable to propose that some enzymes operate by acid–base catalysis. That is, the active site of such enzymes contains both acidic and basic groups that participate in catalysing the reaction.

◇ Can you suggest some possible acidic and basic groups that might be present in enzyme active sites?

There are a number of such groups in the side chains of amino acids. The most obvious ones are the carboxylic acid group of glutamic and aspartic acids $(-COOH \rightleftharpoons -COO^- + H^+)$ and the amino group of lysine $(-NH_2 + H^+ \rightleftharpoons -NH_3^+)$. Perhaps less obvious are the hydroxyl group of tyrosine:

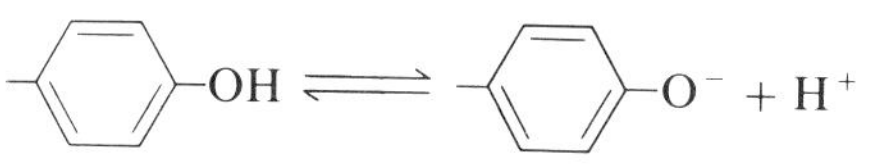

$-\langle\rangle-OH \rightleftharpoons -\langle\rangle-O^- + H^+$

the sulphydryl group of cysteine $(-SH \rightleftharpoons -S^- + H^+)$ and the imidazole group of histidine:

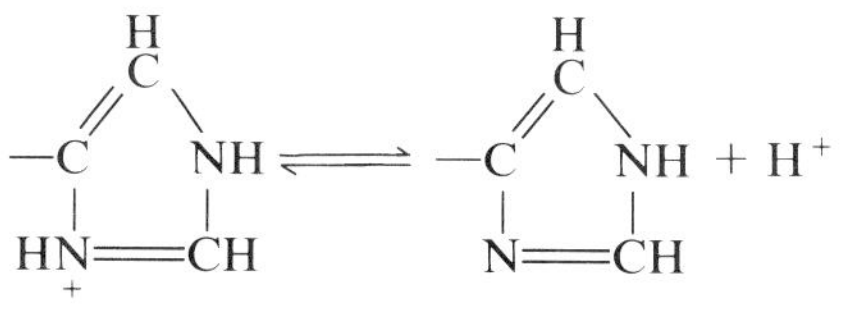

As well as providing catalytic acidic and basic groups, there is another important aspect of acid–base catalysis, where operating in enzymes. The catalytic acidic and basic groups are not in free solution but are present as part of the same entity, the active site of the enzyme. Thus, like bound substrate molecules, they too can be close and in the correct orientation for reaction (with the substrate). A kind of proximity effect (Section 4.7.1) operates for the catalytic groups (acid and base) and for the substrate molecules too. Having all such participants in the reaction (reactants and catalytic groups) held together can greatly enhance the probability of reaction, hence an increased rate i.e. catalysis.

This aspect is not exclusive to acid–base catalysis, as all active sites provide the catalytic groups (acid, base or others) close to the substrate and in appropriate orientations. Providing catalytic groups in appropriate arrays is also of course equally true for reactions where only one substrate molecule is involved (i.e. $S \rightarrow P$).

4.7.3 Stress and strain; electronic strain

On the binding of a substrate to an enzyme, the substrate may be distorted into an energetically unfavourable (and hence unstable) conformation to fit the active site more perfectly. This distortion of a substrate may well introduce **strain** into some of its chemical bonds (Figure 4.28). Such strain may be part of catalysis, the bonds now being weaker and more susceptible to reaction. That, in a nutshell, is the *strain hypothesis* explanation of catalytic power. But does it in fact happen?

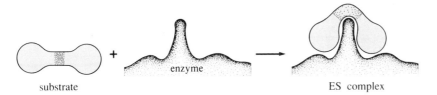

substrate   enzyme   ES complex

Figure 4.28 Strain catalysis. The blue stippled area indicates the reactive bond, in the substrate, which is being strained.

When lysozyme was analysed by X-ray diffraction (Figure 4.22) and three-dimensional models built, it seemed that, on binding to form the ES complex, part of the substrate was distorted. Thus lysozyme might derive at least some of its catalytic power from the strain introduced into the substrate following the distortion this suffered on forming the ES complex. However more recent evidence suggests that appearances may be deceptive and considerable doubt has been cast on this 'strain interpretation' for lysozyme and on the strain hypothesis in general.

One problem with the strain hypothesis is that to stretch or twist chemical bonds severely (Figure 4.28) considerable energy must be applied *over a short distance*; in other words a *strong* force is needed. Though an enzyme may have plenty of energy available (e.g. derived from the *binding energy* of substrate–enzyme interaction, or from the initial folding of the polypeptide chain to form the three-dimensional enzyme protein) it is doubtful whether this could be concentrated over the short distance required for a strong force. There is also the point that enzymes, as proteins, tend to be *more flexible* than their substrates, and so, if anything were to be distorted, the enzyme would be a better candidate than the substrate! As a whole, explanations invoking a strained distorted substrate as a major reason for catalysis are much less favoured than they were, say, 10 years ago. Even where X-ray diffraction studies suggest that the substrate is distorted (as for lysozyme), other explanations, not invoking strain, can be advanced (Section 4.7.5).

If you think of trying to break some rigid object, say a piece of wood, you will appreciate that strain need not necessarily produce actual distortion. Strain is undoubtedly being applied to the wood, but it may not budge. In engineering terms, it is more accurate to say that the object being strained, but showing no distortion, is under **stress**; it will have a *tendency* to distortion, but will not undergo any actual change in shape. As enzymes tend to be the more flexible element in substrate–enzyme interactions, it may well be more common for active sites to show strain (i.e. change in shape) and substrates to be stressed. Nevertheless there may be instances where actual strain of the substrate, and not just stress, is involved importantly in catalysis.

Of course one question common to all such stress and strain hypotheses is: where would the required energy come from? Two possible sources are:

☐ Energy from binding the substrate(s)

☐ Energy stored in the enzyme originating from its specific folding into its characteristic three-dimensional shape

Another form of strain in which there would be no 'visible' distortion is **electronic strain**, and this is probably sometimes important in catalysis. In electronic strain, interaction between adjacent chemical groups (either groups in amino acid side chains forming part of the active site itself or in bound substrates or coenzymes) results in a weakening of certain bonds in those groups. This can have the affect of endowing some groups with unusual chemical reactivity; that is, activity not normally associated with those groups when they are not involved in such interactions. For example, the abnormal reactivity of certain amino acid side chains in active sites may be explicable in such terms; that is, chemical activity possessed by these side chains that is exhibited only when the amino acids are part of a protein, not when they are free (Section 4.7.6).

4.7.4 Covalent intermediate catalysis

In all enzyme-catalysed reactions the substrate undergoes steps that it does not in the equivalent uncatalysed reaction, the formation of an ES complex being one obvious such step. However such steps do not necessarily involve the formation of any strong, covalent, bonds other than those that would be formed in the uncatalysed reaction anyhow. But many enzyme-catalysed reactions *do* involve the formation of *new* covalent complexes (i.e. ones not formed in the uncatalysed reaction on the path from reactant to product). Such formation of a transient covalent complex between enzyme and the substrate (or reaction intermediates formed from the substrate) can contribute to catalysis. For this to hold true, the covalent complex must 'take' the substrate down a different, energetically more favourable, reaction pathway. In other words, the alternative pathway in this **covalent intermediate catalysis** proceeds more quickly; as shown in Figure 4.29, for catalysis to occur, reactions 2 plus 3 must be faster than reaction 1 (that of the uncatalysed pathway).

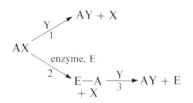

Figure 4.29 Covalent intermediate catalysis. E−A is the supposed covalent intermediate.

4.7.5 Transition state theory—a unifying hypothesis?

We have now briefly reviewed some of the hypotheses currently advanced to explain the catalytic power of enzymes. What we have said is undoubtedly a gross oversimplification of the situation. It should also be evident that more than one of the mechanisms suggested could operate together, each making a contribution to catalytic power, the relative contributions varying from enzyme to enzyme. There is, however, another way of considering catalysis that may offer a more coherent view of catalytic power; a view that can encompass many, perhaps all, of the hypotheses that we have discussed, and others too. This view depends on **transition state theory**.

Put simply, the transition state (also sometimes called the *activated complex*) is the chemical entity or state at the stage in a reaction where bonds are in the process of being made or broken. (It is also the term applied to the stage itself.) The transition state, with some of its bonds part-formed, must be very unstable. There is no possibility of isolating such an entity but, from knowledge of the reaction, we can put down on paper what its formula probably looks like. The transition state is also defined as being at the peak of the energy curve in a **reaction-coordinate diagram** (Figure 4.30).

Catalysts, enzymes included, can be said to operate by reducing the activation energy (lowering the hump in the curve) for reaction; e.g. from A down to B in Figure 4.30. From transition state theory, anything that stabilizes the transition state will reduce the activation energy. Thus according to transition state theory, *enzymes can be said to possess catalytic power by virtue of their ability to stabilize the transition state*. The challenge to transition state theory is to recast all of the various suggested catalytic mechanisms (as in Sections 4.7.1 to 4.7.4, for example) in terms of stabilization of the transition state. *We shall examine just a couple of cases.*

In many chemical reactions it appears likely that the transition state is very unstable because of a number of highly unstable positive and negative charges thought to exist on it.

◇ How might such charges be stabilized?

◆ By interaction with charges of the opposite sign (positive with negative, and vice versa).

So acid–base catalysis by an enzyme can be regarded as the active site providing appropriately charged side chains (Section 4.7.2) strategically placed to 'neutralize' the destabilizing charges on the transition state, and thereby stabilizing it.

We have seen that in some enzymes the substrate appears to be distorted when in the ES complex (e.g. lysozyme). This has been taken as evidence that strain occurs on binding of substrate and that this contributes to catalysis (Figure 4.28). Transition state theory can provide an alternative explanation. Namely, as the transition state makes a better fit with the active site than does unaltered substrate it is the transition state that is stabilized. So on binding, the substrate undergoes a change in conformation to realize the transition state conformation. There is no implication of strain leading to distortion of the substrate.

◇ The change in substrate conformation to that of the transition state, to fit the active site better, seems to presuppose a flexible substrate and a relatively rigid active site. Can you see a possible problem with this idea?

◆ We now know that proteins, enzymes included, often have considerable degrees of flexibility in their structures (Section 4.6 and Chapter 3), frequently more than do substrates. Flexibility has actually been observed in some active sites themselves on binding of substrate (Section 4.5.5). The problem is how much any such flexibility might affect transition state stabilization as an explanation for why the substrate changes conformation.

Irrespective of the precise reasons, the transition state stabilization explanation for the observed conformational change in some substrates is not

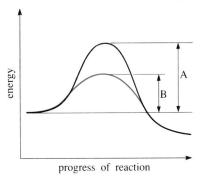

Figure 4.30 Simplified reaction–coordinate diagrams for an uncatalysed reaction (black) and its enzyme-catalysed counterpart (blue), showing the difference in activation energies. The respective activation energies are A and B.

universally accepted, and indeed it is likely to be difficult *operationally* to distinguish always between strain-induced changes and transition state stabilized ones. And, of course, in some enzymes one may be the more correct explanation for any observed change in substrate conformation, while in other enzymes the other may be.

4.7.6 The micro-environment

As we stated at the outset, the verdict on the relative merits of the different hypotheses to explain catalytic power of enzymes has not yet been fully returned. It is likely that a number of the hypotheses that we have discussed, and others, provide partial explanations of the catalytic power of various enzymes and that any particular enzyme may derive its catalytic power via a number of mechanisms. Irrespective of which catalytic mechanisms are actually operating, what any enzyme in effect does is to provide a **micro-environment** in which reaction can occur efficiently. In such a micro-environment substrates can be stressed or strained, effective concentrations raised, or whatever—all unimpeded by the surrounding medium. This last point could be critical.

There are many reactions in the cell for which water is one of the substrates (see Table 4.1: hydrolases), but for many others it is not. Indeed, some reactions proceed best in a hydrophobic environment; yet the cell is an aqueous environment. Such reactions can, and do, occur in cells. They occur in the active sites of certain specific enzymes, sites that are hydrophobic. (Induced fit may also have a role in limiting disruption by water; Section 4.6.) The non-polar environment of such hydrophobic sites can facilitate a variety of reactions that would be virtually impossible in aqueous solution. For example, the charge on a group would be effectively much more powerful in a non-polar environment than in a polar aqueous one, perhaps rendering that group a more powerful catalyst.

Different types of active site provide different types of specialized micro-environment—hydrophobic (non-polar) sites, more hydrophilic (i.e. more polar) ones, sites with charged groups needed for acid–base catalysis or with metal ion coenzymes important in certain types of electronic interactions, and so on. In these various micro-environments the range of properties of chemical groups can be 'extended', they can act in ways unexpected (by us) from their known 'normal' properties in free aqueous solution, thus contributing to catalysis. One final example may make this clear.

You may recall that chymotrypsin is a member of that class of proteases that reacts with DIFP (Section 4.5.2) via the side chain of a serine residue; Ser-195 in the case of chymotrypsin. There are 28 serine residues in chymotrypsin, yet *only* Ser-195 reacts with DIFP. This and other evidence suggests that Ser-195 is 'unusual' and is important in catalysis. It also seems that to be active, the Ser-195 side chain must act *as if* it carries a negative charge, that is *as if* the −OH is ionized or at least partially ionized. This is highly surprising to a chemist. Chymotrypsin operates effectively at pH 7, and, at this pH, the −OH of *free* serine would be virtually un-ionized. How is it that the −OH of Ser-195, present in the active site of chymotrypsin, apparently can act *as if* appreciably ionized at this pH?

From X-ray diffraction it is known that Ser-195 is very near His-57, and it is therefore possible that they interact (Figure 4.31).

◇ This might suggest that His-57 (as well as Ser-195) is involved in catalysis in chymotrypsin. What evidence supports such a role for His-57?

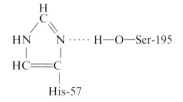

Figure 4.31 Interaction between the −OH group of Ser-195 and the side chain of His-57; the interaction may be via a hydrogen bond.

◆ The reaction of chymotrypsin with an affinity label, TPCK, followed by partial hydrolysis of the enzyme protein, helped reveal the label bound to His-57 (Section 4.5.2).

It appears that His-57 can accept a proton from the hydroxyl of Ser-195, this occurring during the vital *catalytic* step where the serine oxygen 'attacks' the substrate. His-57 itself may well be affected by interaction with a nearby acidic group, the ionized carboxyl group ($-COO^-$) of an aspartic acid side chain, Asp-102 (Figure 4.32a). This helps orient His-57 so that it can accept the proton from Ser-195. It probably also has the effect of stabilizing the positively charged form of histidine produced after binding of substrate and the transfer of the proton from Ser-195 to the histidine (Figure 4.32b).

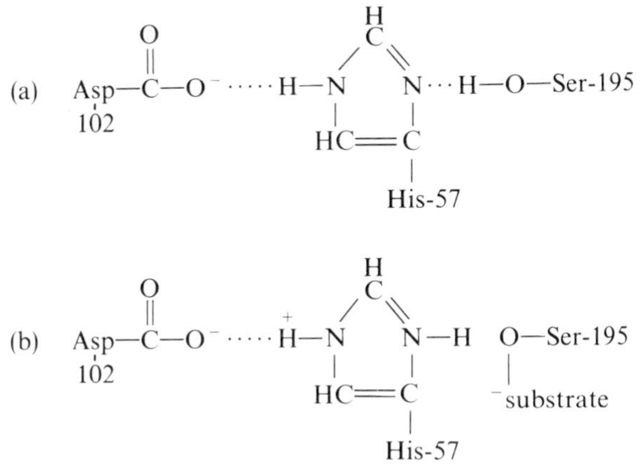

(a)
$$\text{Asp} - \overset{\overset{\displaystyle O}{\|}}{C} - O^- \cdots\cdots H - N \qquad N \cdots H - O - \text{Ser-195}$$

His-57

(b)
$$\text{Asp} - \overset{\overset{\displaystyle O}{\|}}{C} - O^- \cdots\cdots \overset{+}{H - N} \qquad N - H \quad O - \text{Ser-195}$$

His-57

Figure 4.32 The postulated interaction between Ser-195, His-57 and Asp-102 in the active site of chymotrypsin, before (a) and after (b) substrate binds. A proton is transferred from Ser-195 to His-57 during reaction between Ser-195 and the substrate; hence facilitating this reaction.

The story is yet more complex and, it should be said, not totally uncontroversial. For our present purposes what is important to appreciate is how Asp-102, His-57 and Ser-195 act *in concert*. A vital *net* effect of this *catalytic triad*, as it is sometimes known, is to increase markedly the reactivity of the serine oxygen towards the substrate—something, that, as we characterized above, makes Ser-195 appear *as if* its $-OH$ was ionized; not a property which free serine would possess at this pH. This may also be an example of electronic strain (Section 4.7.3), the interaction between His-57 and Ser-195 inducing strain in the bond between the H and O of the serine $-OH$ (Figures 4.31 and 4.32a), thereby enhancing the reactivity of the serine oxygen towards the substrate.

◇ Three amino acids act together thereby modifying the properties that each would have if alone in free aqueous solution; in particular the reactivity of the serine oxygen is massively enhanced. These amino acids are at positions 57 (Ser), 102 (Asp) and 195 (His) in the polypeptide chain. Yet all are involved in catalysis and are hence in the active site. Does this imply that the active site occupies a large proportion of the total enzyme; from 57 to 195 at least, out of the total of 241 amino acids in the enzyme? Or is there another explanation?

◆ Remember that polypeptide chains are virtually never extended linear structures, but that secondary and tertiary levels of structure result in a highly folded and twisted chain (Chapter 3) where various, often distant, parts of the linear sequence are brought into contact with each other. It is quite probable, therefore, that residues that are distant in the linear sequence are adjacent when the polypeptide is in its final folded three-dimensional shape. A look back at Figure 4.22 should make this clear.

As the example of chymotrypsin shows, the bringing together of various amino acid side chains to form a three-dimensional array, in the active site, can endow an enzyme with almost 'magical' chemical properties—in fact, chemical groups in substrate, coenzyme or enzyme amino acid side chains themselves (such as the −OH of serine) can take on 'unusual' reactivities. The more we learn about actual enzyme-catalysed reactions, the more we can understand how it is that the various micro-environments can alter the properties of chemical groups from those familiar to us from their behaviour in free aqueous solution. In such alterations, of course, lies the key to catalytic power. From the cell's 'point of view', by possessing a large variety of micro-environments, a range of minute 'alcoves'—the active sites of enzymes—a single cell can encompass a wide range of different reactions, all at efficient rates; the many reactions in a few 'pots' (Section 4.1).

From *our* point of view, not only will increased understanding of catalytic power help answer one of the most fascinating questions in biology—how do enzymes work?—but it should also enable us to better exploit enzyme power for industrial purposes (Section 4.9.2).

Summary of Sections 4.6 and 4.7

These sections have dealt with difficult topics. You might find it useful to read the following summary and then try the questions that follow. These questions also relate in part to Section 4.5; they help tie together the interpretation of how enzymes work, based on data gained from examining the structure of active sites and enzyme mechanisms (Sections 4.5 and 4.6) and theories about catalytic power (Section 4.7).

1 The specific fit between enzyme active site and substrate can be visualized as being like that between lock and key.

2 An alternative view of the relationship between substrate and enzyme is that of induced fit, where binding of the substrate causes a change in conformation of the active site. Induced fit is said to be necessary for specificity and/or catalysis. However, there is limited evidence from just a few enzymes that for them the induced fit hypothesis may be correct.

3 Where enzymes derive their catalytic power from is still a topic of considerable uncertainty. Many hypotheses have been advanced: acid–base catalysis, the proximity effect, stress and strain, electronic strain, covalent intermediate catalysis, and others. An increasingly favoured idea to be applied to enzyme catalysis is transition state theory.

4 A general feature of all the hypotheses concerning catalytic power is that enzyme active sites provide catalytic groups in orientations favourable for catalysis.

5 The proximity effect of bringing substrates in multimolecular reactions close together in the active site leads to an increase in the effective concentration of substrates.

6 In enzymes, acid–base catalysis operates by the strategic positioning of acidic and basic groups in the active site. On transition state theory this can be viewed as a way of stabilizing positive and negative charges that are present in the transition state and that tend to destabilize it.

7 Strain may occur on binding of substrate to enzyme causing distortion of the substrate and a weakening of some of its bonds. This may contribute to catalysis. An alternative view, based on transition state theory, is that the

apparent distortion is not due to strain but merely reflects the substrate adopting a conformation appropriate to the transition state; the transition state being postulated to fit the active site better than does the substrate. Strain need not produce an alteration in shape, it may just produce stress in the substrate.

8 Covalent intermediate catalysis supposes that an alternative reaction pathway, to that of the uncatalysed reaction, is followed. It also involves formation of a transient covalent intermediate.

9 Energy for catalysis can derive from binding energy (of substrate to enzyme) or stored energy from folding of the enzyme protein into its three-dimensional shape.

10 All-in-all, enzymes provide the cell with a series of different micro-environments, allowing it to carry out a wide range of different reactions at high rates.

Question 7 (*Objectives 4.1, 4.5, 4.9 and 4.10*) The enzyme carboxypeptidase A catalyses the hydrolysis of peptide bonds. The enzyme contains one atom of zinc in each molecule of enzyme. On removal of this zinc, the enzyme is inactivated. Activity is restored by addition of zinc *or* cobalt. X-ray analysis shows that addition of a substrate to the enzyme causes a tyrosine (Tyr-248) side chain to move by 1.2 nm to a position over a peptide bond in the substrate, a bond that is to be split. Acetylation of carboxypeptidase A (i.e. treating it with a reagent that adds acetyl, CH_3CO-, groups) inactivates it.

In the light of these data, which you can take to be true, classify each of the following statements as 'true', 'false', or 'possibly true but data insufficient to allow a conclusion'.

(a) Koshland's induced fit hypothesis is wrong.

(b) Zinc is involved in the structure of the protein, so that on its removal, the protein is destroyed.

(c) A tyrosine (Tyr-248) side chain is involved in catalysis.

(d) Acetylation removes zinc from the enzyme.

(e) The transition state involves stabilization by the zinc ion, Zn^{2+}.

(f) Zinc is a coenzyme of carboxypeptidase A.

Question 8 (*Objectives 4.1, 4.5, 4.9 and 4.10*) Consider the data given in Question 7 plus the following (which again you can assume are true). On acetylation of carboxypeptidase A, a modified tyrosine residue can be detected. X-ray diffraction shows that removal of zinc does not change significantly the conformation of the enzyme.

Classify the following statements as 'true', 'false', or 'possibly true but data insufficient to allow a conclusion'. (Use the data from Questions 7 *and* 8.)

(a) The functional role of Tyr-248 is to bind the zinc.

(b) Zinc is involved in catalysis and not in maintaining the overall structure of the enzyme.

(c) The tyrosine (Tyr-248) side chain is involved in catalysis.

(d) The movement of the tyrosine (Tyr-248) side chain produces strain in the bound substrate.

4.8 ISOENZYMES

When biologists speak of a particular enzyme such as urease or chymotrypsin or lactate dehydrogenase (LDH), they are sometimes referring to a purified, chemically analysed entity, but more often they are describing a preparation, of varying degrees of purity, that has a particular catalytic activity. So urease is operationally a biological preparation that catalyses the breakdown of urea to carbon dioxide and ammonia; chymotrypsin catalyses the hydrolysis of certain peptide bonds; LDH catalyses the oxidation of lactate to pyruvate.

Presumably all living organisms have many such reactions in common and therefore each must possess the enzymes necessary to catalyse these common reactions. But are the enzymes that catalyse the same reactions in different organisms the same chemically? Is urease from rats the same as that from mice? Is the lysozyme from children's tears the same as that from egg white or that from crocodiles? If the techniques developed for purifying proteins and analysing their structures (Chapter 3) are applied to enzymes with the same catalytic role but isolated from different organisms, we generally find that they are chemically different. The differences increase as do other more obvious differences between the organisms—thus chymotrypsin from humans would be expected to be chemically more like that from the chimpanzee than that from the haddock.

However, when we speak of a particular enzyme from human or ape or haddock, are we in each case speaking of a single chemically unique type of enzyme? Is there sometimes more than one type of enzyme for a particular catalytic activity, in the same organism? To carry this further, do different organs or tissues in a complex organism have different enzymes with the same catalytic activity? Why should they? It would seem a waste of effort.

To answer these questions, we have designed the following exercise. You are presented with three blocks of evidence. You should examine each block in turn. After examining a block of evidence, you should examine the interpretations offered immediately after it, and from these select that which you consider the most justified *up to that point*. Note that at certain points, you may not necessarily be able to decide firmly between the alternatives given, more than one may be reasonable, and the process of evaluation *should be progressive*. That is, in examining later evidence and choosing between interpretations, *bear in mind evidence given earlier*. When you have finished, read the comments given afterwards.

Evidence I

(a) Look at Figure 4.33 and read its caption.

(b) The five bands in Figure 4.33 are labelled 1–5 in decreasing order of negativity (i.e. 1 is the most negatively charged and therefore nearest to the anode).

(c) Analogous patterns to that shown in Figure 4.33 can be obtained on examining the tissues of some other species of animals.

Interpretations of Evidence I

(i) There are five different types of LDH in humans, differently distributed between the different organs.

(ii) Several different proteins have been stained, not all of which are LDH.

(iii) Only one type of LDH is present, but different impurities in the different tissues produce the different patterns.

(iv) Only two basic 'types' of LDH are present, 1 and 5. The others are mixtures of 1 and 5.

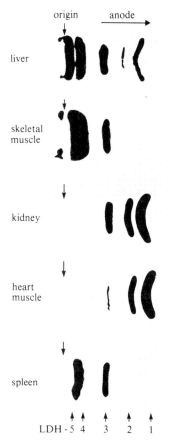

Figure 4.33 Starch gel electrophoresis pattern of five separate human organ extracts. The extracts applied to the starch gel in buffer at the 'origin' were subjected to an electric current so that all the charged substances in the extracts moved in accordance with their net electric charges. After the current was switched off, the gel was stained with a substrate of lactate dehydrogenase (LDH) plus dyes specific for the product (so that only where LDH was present would dye be trapped).

Evidence II

The various types of LDH (as seen in Figure 4.33) were separated, purified and examined.

(a) Both type 1 and type 5 have relative molecular masses (M_r) of 134 000.

(b) Treatment of type 1 or type 5 with a chemical that separates protein subunits gives proteins of M_r 34 000 having no enzymic activity.

(c) Figure 4.34 shows the results of mixing pure type 1 and pure type 5 in sodium chloride, then freezing and thawing. (If no freezing–thawing step is done, just two bands, corresponding to types 1 and 5, are observed.)

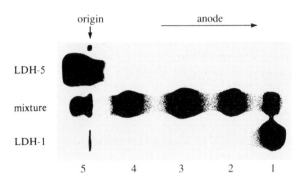

Figure 4.34 The effect of freezing and thawing on a mixture of type 1 and type 5 LDH. (The actual electrophoresis and staining were according to the same procedure as in Figure 4.33.)

Interpretations of Evidence II

(i) As types 1 and 5 have the same M_r, they are the same enzyme, LDH. Impurities of low M_r give them different electrophoretic properties.

(ii) Types 1 and 5 are different kinds of LDH; types 2, 3 and 4 are mixtures of 1 and 5 in differing proportions.

(iii) All are types of LDH. Type 1 is composed of four identical subunits, say HHHH; 5 is composed of four identical (but differing from those of 1) subunits, say MMMM. Types 2, 3 and 4 are mixtures of M and H subunits in differing proportions.

Evidence III

(a) All the five types are LDH. Antibodies produced against pure type 5 inhibit the enzymic activities of types 5, 4, 3, 2 and 1 by 86, 68, 41, 23 and 0% respectively. (Remember that antibodies are proteins produced by injecting a 'foreign' protein from an animal of one species into an animal of another species. The antibody proteins produced by the recipient animal can combine *specifically* with protein of the kind that was injected; Chapter 3.)

(b) Genetic studies on humans suggest that two separate genes exist for LDH.

(c) All the five types are LDH. They all have the same M_r.

Interpretations of Evidence III

(i) Only two different types of LDH are present—1 and 5. Types 2, 3, and 4 are mixtures.

(ii) Five different types of LDH are present. Type 1 consists of 4 subunits, HHHH (H_4, for short); type 5 consists of 4 subunits, MMMM (M_4). Type 2 is H_3M, type 3 is H_2M_2, and type 4 is HM_3. (*Note:* The order in our HM notation is irrelevant; e.g. HM_3 and M_3H are the same.)

Comments on interpretations of Evidence I

Interpretation (i) is best, because it merely reflects what can be seen from Figure 4.33 (however, you will be aware from Chapter 3, that there could be more than five types present; some bands could contain more than one type of protein with the same charge). Interpretation (ii) is unlikely, because the staining technique depends on the reaction catalysed by LDH specifically. Interpretation (iii) is unlikely, because the pattern is similar in some other organisms, and it is unlikely that 'impurities' would be the same in different organisms or lead to a similar pattern. Interpretation (iv) is possible but assumes too much; 'mixtures' is vague.

Comments on interpretations of Evidence II

Interpretation (iii) is best. The idea of a four subunit LDH is compatible with the dissociation from an M_r of 134 000 to one of 34 000 (as $4 \times 34\,000$ is as close to 134 000 as experimental error might likely allow). You would not, however, be justified in concluding that the four subunits are identical because of the identity of relative molecular mass. Their identity is, however, supported by (c), as mixing of H and M to give tetrameric structures (i.e. ones containing four subunits) could give three 'new' types M_3H, M_2H_2 and MH_3. Interpretation (ii) is also possible *at this stage*, but if 2, 3 and 4 are mixtures of 1 and 5, we would expect them to have a proportionately higher M_r than 1 and 5, something we do not know about, as yet. Interpretation (i) is poorly reasoned, because similarity in M_r is an insufficient criterion of identity.

Comments on interpretations of Evidence III

Interpretation (ii) is almost certainly correct. Evidence (c) rules out Interpretation (i) (and, incidentally, Interpretation (ii) of Evidence II), because the M_r of 'mixtures' should be greater than that of pure type 1 or pure type 5. We know from genetics that each specific gene codes for one specific polypeptide (one gene–one polypeptide chain). So two genes, each coding for one type of subunit (H or M), could in principle give rise to five possible tetramers; i.e. one gene codes for H type subunits, one for M. (If H and M subunits can combine freely in the cell, five possible tetramers can be formed: H_4, H_3M, H_2M_2, HM_3 and M_4.) The suggested composition (Interpretation (ii)) of types 1 to 5 is compatible with the inhibition by antibodies produced against pure type 5 (Evidence (a)). These antibodies are presumably specific against subunit M and when bound to it interfere with its functioning. Therefore they inhibit the different types of LDH in proportion to the proportion of M subunits that each contains.

Conclusions

You probably concluded correctly that the structures of types 1 to 5 are those suggested in Interpretation III (ii). Therefore it seems that different tissues have different types of LDH.

So far we have referred to the different forms of LDH as 'types'. The name usually applied where several forms (types) of an enzyme exist in a single species of organism is **isoenzymes** (or isozymes). For the phenomenon is by no means unique to LDH. Many other enzymes, from a wide variety of organisms, have been found to consist of several isoenzymes. The number of component isoenzymes varies from enzyme to enzyme and from species to species of organism.

The existence of isoenzymes raises some intriguing questions. As, by definition, each isoenzyme of a particular enzyme catalyses the same reaction (e.g. LDH 1, 2, 3, 4 and 5 all catalyse the interconversion of lactate and pyruvate), why do several isoenzymes exist? Why is more than one form of an enzyme needed? Why, for example, should different tissues in the same organism have different proportions of the five LDH isoenzymes? What are the biological functions, the roles, of isoenzymes?

Answering 'why' things exist, assigning function, is never easy in biology. Nevertheless, there is good evidence for many isoenzymes that they have important roles in the regulation and control of metabolism, the topic of the next chapter. This is indeed the usual function assigned to the LDH isoenzymes that we have considered above. It has been proposed that they have a role in the regulation and control of the metabolism of glucose, a vital source of energy in very many cells. But what that role is precisely is a topic of considerable controversy—LDH isoenzymes provide perhaps the best example of how difficult it can be to reach a single view about the biological function of something. As the regulation and control of glucose metabolism is a central topic of Chapter 6, we consider there some suggestions for the function of LDH isoenzymes and a little of the controversy surrounding them.

In addition to the biological functions of isoenzymes, whatever they may be, the study of isoenzymes is in itself important in a number of fields. Two areas where studies on isoenzymes are important are evolutionary biology and medicine. Within the same species of organism, variation in the pattern of isoenzymes between individual organisms is often found. Such variation, for example, in fruit-flies, humans and some plants, has been a source of much study and speculation in evolutionary biology. Medical interest in isoenzymes is of a more practical kind, a topic we discuss briefly in the next section.

4.9   ENZYMES IN MEDICINE AND INDUSTRY

Though our main concern in this chapter is the biological function of enzymes, it is interesting to make some mention of their application outside their normal biological role. In fact, and not surprisingly, their application depends heavily on their biological role—that is, their great catalytic power and specificity make them attractive agents for medical and industrial exploitation—and so you should note how this section serves to reinforce some of what we have discussed earlier. First we shall consider briefly some medical uses of enzymes; then some industrial ones.

4.9.1   Enzymes in medicine

So far, the major use for enzymes in medicine has been in diagnosis. There are two broad ways in which they are used: firstly, enzymes can be used to assay substances which are *molecular markers* for particular clinical conditions; secondly, enzymes themselves can be used as molecular markers.

There are a number of situations where particular substances present in body tissues are indicators (**molecular markers**) of clinical conditions and it is necessary both to detect these substances and measure (assay) their levels. A classical case is diabetes mellitus, where an abnormally elevated blood sugar (glucose) level is both a molecular marker of the disease and a useful guide as to appropriate treatment (e.g. whether to inject insulin or not). There are many analogous situations.

In all cases two basic things are needed:
□ The assay should be specific and accurate. As a diagnosis of the patient's condition and prescribed treatment may depend heavily on the assay, it must obviously detect *only* the marker sought after and measure its amount accurately.

☐ The assay should be capable of measuring small amounts of the molecular marker; because the patient may have to provide blood (or other tissues), an assay that needs 'drops' rather than 'bucketfuls' is desirable.

Enzymes certainly fit the specificity bill: name any biological substance and you could probably find an enzyme for which it is a specific substrate.

◇ Do enzymes fit the accuracy bill? Consider what an enzyme assay, as discussed in Section 4.3, actually measures.

In Section 4.3 we were concerned with assaying *enzymes*; that is measuring how much *enzyme* was present, by measuring the rate of reaction that a preparation catalysed, given a certain amount of substrate. However, if a *known amount of enzyme* is used, then the rate of reaction can be just as easily used to determine the amount of a *substrate* (such as a molecular marker) present. Where we use the assay to measure substrate (rather than to measure enzyme, as in Section 4.3) we could for convenience call it an *enzyme-based assay*; where used to measure enzyme (Section 4.3), simply an *enzyme assay*. Generally this distinction is not made and both are called 'enzyme assay'.

Finally, there is the question of convenience. In many instances the taking of samples (blood or whatever) can be done by trained staff, as can the assays using whatever sophisticated apparatus (pH meter, spectrophotometer, etc.) is needed. This is obviously alright within the confines of a hospital or perhaps a doctor's surgery, something one can resort to for occasional testing. But where routine testing is desirable, such as, say, for diabetics, this is highly inconvenient. Considerable effort has gone into designing assay systems (enzymic or not) that can be used safely, accurately and simply by a GP without recourse to hospital laboratories (so-called 'bedside kits'), or increasingly by patients themselves.

Enzyme-based assays are not the easiest things to do without training and laboratory equipment. However, a recent marriage between enzymology and electronics promises to revolutionize 'bedside' or self-administered testing. The idea is to couple an enzyme-based assay to an electronic device (such as an electrode or a transistor). The formation of specific *product* by the enzyme, that depends on the presence of its specific substrate (the *molecular marker*), is in turn translated into an electronic signal. This can be registered readily on a meter. This general idea—the construction of specific **biosensors**—has already come to fruition with a biosensor that can detect blood glucose levels and which is suitable for home use by diabetics. The enzyme bit of the biosensor is glucose oxidase, an enzyme that catalyses the conversion of glucose to gluconolactone. This enzyme has been used by diabetics for some years in Clinistix. In Clinistix the glucose level is monitored by a blue colour, this being the result of a secondary colour reaction (Section 4.3.1) involving gluconolactone. In the biosensor the output is an electronic one. Though we have concentrated on one molecular marker, blood glucose, there are many others that enzymes are used to assay. The other use of enzymes that we wish to mention is where the *enzymes themselves* are used as molecular markers.

Blood plasma contains numerous enzymes, many of which have no role there. They merely reflect a certain degree of 'leakage' from various tissues. This leakage of enzymes into the plasma and their removal from, or destruction in, the plasma sets up a sort of balance. Thus for each specific enzyme found in plasma there is a normal concentration range. Departure from this *can* indicate a clinical condition. Diseases can damage tissue cells and this can cause increased leakage of cellular enzymes into the plasma. (So, of course, can other things—perhaps a 'couple of double Scotches' the night before a blood test might push the plasma level of certain liver enzymes over the top of the normal range.)

For example, the enzyme aspartate aminotransferase (AST) is found widely in the body and some is normally present in the plasma. Large increases in the plasma level (say, 10–100 times normal) often indicate severe damage either to cardiac cells (as might result from a heart attack) or to liver cells (as, say, from viral hepatitis). More moderate increases in plasma AST may indicate a variety of other diseases. Likewise, increased plasma levels of trypsin or α-amylase can indicate problems with the pancreas, the organ in which they (or their precursors) are produced.

The plasma level of lactate dehydrogenase (LDH), an enzyme that we discussed in Sections 4.1, 4.3 and 4.8, can also be a useful molecular marker of disease.

◇ Will an elevated plasma level of LDH of itself indicate which organ is damaged?

◈ No. LDH occurs in many types of cell and so damage to one or more different tissues/organs could cause an elevated plasma level (Section 4.8).

◇ Can you think of a way of making an elevated plasma level of LDH into a more specific marker of which tissues are damaged?

◈ Consider Section 4.8 on isoenzymes. Different tissues have different patterns of the five forms of LDH. Determining the isoenzyme pattern of LDH in the plasma, rather than just the elevated plasma level of LDH as a whole, might provide additional clues as to the tissue(s) damaged.

The tissue specificity of LDH isoenzymes is far from precise (see Figure 4.33), but LDH isoenzymes could be of clinical use, specifically in distinguishing between cardiac or liver damage. This might be important, as, for example, following a heart attack some patients may show a *second* rise in plasma LDH. This could indicate either a second heart attack or congestive liver damage.

◇ Look at Figure 4.35. Which of the two electrophoresis patterns of LDH, (b) or (c), is consistent with cardiac damage and which with liver? (Reference to Figure 4.33 will probably be necessary.)

◈ (b) shows marked elevation in isoenzymes 1 and 2, and is thus consistent with LDH containing predominantly H subunits characteristic of *heart* LDH. (c) has a marked elevation in type 5 LDH, that comprising M subunits more characteristic of *liver*. Therefore pattern (b) is more consistent with cardiac damage and pattern (c) with liver damage.

However, there are much better enzyme indicators of which organ, heart or liver, is involved. Creatine kinase is released into plasma following a (second) heart attack but not from liver damage. Reciprocally, liver damage releases isocitrate dehydrogenase into plasma, but a heart attack does not.

A whole range of enzymes, some with diagnostically useful isoenzymes, are now used as molecular markers of a whole range of clinical conditions.

Of course in reaching a diagnosis no doctor uses molecular markers, be they plasma enzyme levels or not, as the sole evidence for a particular clinical condition. For one thing, the patient presumably exhibits other, grosser symptoms; patients are unlikely to present themselves at the surgery because

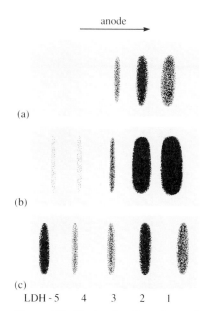

Figure 4.35 Electrophoretic patterns of human LDH from specimens of blood plasma. (a) is from a normal healthy individual. (b) and (c) are from two different patients.

they noticed an elevated enzyme level! The doctor will consider and interpret a whole range of symptoms before ordering a search for molecular markers. Molecular markers may provide the last, but sometimes decisive, piece in the diagnostic jigsaw; but they too require interpretation. Thus enzymes, used either as molecular markers themselves or to assay other molecular markers, are useful weapons in the diagnostic armoury; no more, no less.

4.9.2 Enzymes in industry

Though enzymes are unique to living organisms, catalysts as a whole are not. Ever since the discovery of catalysis in the 19th century, chemists have employed a variety of catalysts in both research laboratories and industrial processes. The modern petrochemical industry, from which we derive both fuel and a huge range of plastics, dyes, fertilizers, adhesives, food additives, pharmaceuticals, detergents, herbicides and so on, is heavily dependent on synthetic catalysts. Given our present-day appreciation of the catalytic power and, most impressive, the specificity of enzymes, it is not surprising that we should seek to exploit this for industrial, or even domestic, use.

At one end of a spectrum are enzymes with relatively mundane uses. Most notable among these are certain proteases. In particular, protease from the bacterium *Bacillus licheniformis* has found a considerable market niche in so-called 'enzyme' or 'biological' washing powders. Its ability to digest protein (such as ones associated with blood, egg or whatever) at the relatively high temperature, for an enzyme, of 65 °C, and at the alkaline pHs typical or detergents, has rendered it an ideal source of domestic enzyme power. Protease also finds some use as a meat tenderizer. Some 500 tonnes per annum (!) of protease are used commercially, worth tens of millions of pounds. Many other enzymes are fairly commonly used, particularly in the food processing industry: glucoamylase, α-amylase and glucose isomerase, three enzymes prepared from bacteria and moulds, are used in producing so-called 'high fructose syrups', used in soft drinks; rennin, from calf stomachs, and some microbial substitutes for rennin, are used in cheese making; pectinase is used to prepare fruit and vegetable purees or remove tomato skins; anthocyanase is used in colouring of jams and jellies; and so on.

At the other end of the spectrum are enzymes needed to catalyse extremely precise reactions, for highly specialized products, which are difficult to achieve otherwise. Among these is penicillin acylase, used for one step in the conversion of natural penicillin (itself a product of whole cell, mould, metabolism) to so-called semisynthetic penicillins. To do this step without penicillin acylase actually involves three tricky chemical steps.

More and more enzymes are finding commercial applications. Yet there are often some drawbacks to their use. Enzymes are sometimes expensive to produce, especially where they must be prepared to very high levels of purity, and the techniques involved are often suitable for the small scale typical of research laboratories and not for an industrial scale. Many enzymes are unstable, necessitating their frequent replacement; something that can be tedious and expensive. Coenzymes (where needed) can also be unstable and expensive, and of course some are actually consumed during the reaction (e.g. NAD^+; Sections 4.3.1 and 4.5.1). Recently there has been considerable research and development applied to these problems. Better techniques of bulk purification of enzymes have been developed; some ways of stabilizing some enzymes are known; ways of more cheaply recycling coenzymes, or replacing them with substitutes, are gradually being developed. These developments fall under the broad blanket of the field of biotechnology.

Ideally the day will come when we can find a suitable enzyme for any chemical process that we wish to perform—with the necessary specificity and catalytic power, cheap to obtain and stable and cheap to use. The wealth of living organisms, and hence vast variety of different enzymes, makes this possible. And should nature have 'failed to provide', protein engineering (Section 4.5.5) may perhaps do so, with time. For example, suppose that a natural enzyme with a particular required activity cannot be found or, if found, will not operate under convenient industrial or domestic conditions. Then protein engineering might be used to modify an existing enzyme, tailoring it to our precise requirements; say, a protease that is very active at 15 °C, suitable for cold washing of clothes.

Obviously the more we understand about features such as the specificity and catalytic power of enzymes (Sections 4.5 to 4.7) the more we shall be able to employ protein engineering on a rational basis. That is, detailed knowledge of how a particular enzyme works might allow us to change its active site, by protein engineering, to produce a new enzyme with predictable properties. Some such work is already underway. As for commercial use, protein engineering of enzymes has yet to take off on any real scale, but the potential is certainly there.

Summary of Sections 4.8 and 4.9

1 Some enzymes in a species of organism exist in several different related forms, isoenzymes. An example is human lactate dehydrogenase (LDH). Each isoenzyme of an enzyme probably has a particular metabolic role.

2 Enzymes are used quite widely in medical diagnosis. Some are used to assay molecular markers of disease; others themselves serve as molecular markers. Isoenzymes are among molecular markers for certain clinical conditions.

3 A number of enzymes have found medical/commercial applications. For example, enzymes are used in making semisynthetic penicillins, in 'biological' washing powders, in food processing and for measuring blood sugar levels in diabetics.

Question 9 (*Objectives 4.1 and 4.11*) Which of the following statements are true and which are false?

Isoenzymes are important because they:

(a) Support the theory of evolution.

(b) Help in clinical diagnosis.

(c) Are useful currently in chemotherapy.

(d) Complicate enzyme purification.

Question 10 (*Objectives 4.1, 4.11 and 4.12*) Classify each of the statements below as true or false.

(a) An increased plasma LDH level indicates a heart attack.

(b) Acute pancreatitis (inflammation of the pancreas) could lead to an elevated level of α-amylase in the plasma.

(c) An increased plasma level of LDH was suspected to derive from a heart attack; this diagnosis would be corroborated by an increase in the plasma level of both type 1 LDH isoenzyme and isocitrate dehydrogenase.

Question 11 (*Objectives 4.1 and 4.13*) Classify each of the statements below about enzymes as being either advantageous or disadvantageous to their use medically or industrially/commercially.

(a) They are unstable.

(b) Some of them need coenzymes.

(c) They are specific.

(d) They can often be isolated from micro-organisms.

4.10 SPECIFICITY—ITS WIDER SIGNIFICANCE

There is little doubt that the single most important distinction between enzymes and non-biological catalysts is *specificity*. Synthetic catalysts often have considerable catalytic power but have rarely been produced with specificity even approaching that of enzymes. Specificity is a major attraction for the use of enzymes in industrial processes and medicine (Section 4.9), and, of course, in living cells it is a key to the existence of specific metabolic pathways (e.g. the directive effect; Section 4.1). As we have seen, specificity depends on a close fit between substrate and active site, whether totally 'preformed' or induced (Section 4.6).

The concept of specificity is of more general importance in biological systems than for enzyme–substrate interactions alone. For example, enzymes are often specific not just for their substrates at the active site, but for other substances too, at other distinct sites on the enzyme. This forms the basis of much of the enzymic role in regulation and control of metabolism, the subject of Chapter 5 (see Chapter 6, too). Other non-enzymic proteins also exhibit specificity and these too have important biological roles.

For example, *cell surface receptors* are sites at which so-called *signal molecules* (hormones, neurotransmitters and the like) are received. Obviously any signalling system must be specific; the correct message must get to the correct place. Having a receptor specific for the signal is one way to ensure this—only a cell with that receptor will interact with the corresponding signal molecule and respond accordingly.

Likewise, other proteins embedded in the various membranes present in the cell (Chapters 2 and 3) help discriminate between those substances to be allowed to pass through the membranes and those to be denied passage (Chapter 7).

The fundamental requirement of any *immune system* is to be able to distinguish 'friend' from 'foe', to distinguish between the organism's own normal chemical components and unusual, or 'foreign' ones (such as from invading organisms like bacteria or viruses; or from abnormal materials produced by the organism itself, such as perhaps occurs in cancerous states). Thus an immune system exhibits specificity. And this specificity is very sophisticated; for not only can it distinguish 'friend' from 'foe' but it can distinguish between a huge range of 'foes'. To each foreign component it is able to mount a specific response. Once again this specificity depends on molecule to molecule interaction. For example, *antibodies*, proteins that form a vital part of the immune system in higher organisms, are highly specific for foreign *antigens* (Chapter 3, Section 3.4.4). Each antigen (foreign substance) provokes the formation, by the host organism's immune system, of a particular antibody specific to *that* antigen and capable of binding to it and thereby helping to take it out of action.

In developing organisms, the assembly of cells to form tissues and organs evidently requires some degree of cell–cell recognition; so that the correct cells are in the correct numbers at the correct place at the correct stage in development. This is once again a case of specificity. It is probably dependent, in part at least, on molecules on the surface of cells interacting *specifically* with molecules on the surface of other cells (Chapter 7). Thus a prospective liver cell in a developing organism can specifically associate with another prospective liver cell, a prospective kidney cell with a prospective kidney cell, and so on. The surface molecules postulated to be involved are probably glycoproteins (Chapter 3), and the sugar portions of these molecules probably have an important role in determining the specificity. (As the 'probablys' and 'postulated' indicate, this study of the molecules involved in development is itself a still developing field; little is certain, as yet.)

You will doubtless come across many other examples of specificity as you study biology.

◇ Can you think of a rather different example of specificity from Chapter 3—one that does not involve proteins?

◈ The base complementarity that helps determine the double-stranded nature of DNA is one example. Each member of a complementary base pair (adenine—thymine; cytosine—guanine) is specific for the other.

The specificity evident in various biological systems varies in detail and at the levels it operates—ranging from the molecular (as discussed here) to the specific interactions between whole organisms and their environments, so important in ecosystems and a determining force in evolution. But as far as the molecular level is concerned, there is little doubt that much of our current understanding of specificity (be it for enzymes, antibodies, cell surface receptors, or whatever) emanates from the study of the exquisitely precise interaction between enzyme and substrate; a relationship well worth understanding.

SUMMARY OF CHAPTER 4

1 The principles underlying intermediary metabolism can be understood by a careful study of enzymes.

2 Catalytic power, specificity and responsiveness to regulation and control are all important features of how enzymes help to effect intermediary metabolism.

3 Enzymes can be assayed by measuring any feature in which the products differ from the substrates. Techniques exist that exploit changes in colour or in the ultraviolet spectrum, changes in pH, or coupling one assay to another.

4 From a plot of reaction versus time, the initial rate of reaction, v, can be calculated. This initial rate is used as the basic measure of reaction rate.

5 The specific activity of an enzyme preparation is a useful indicator of the relative purity of that preparation.

6 The first step in an enzyme-catalysed reaction involves the substrate binding to the active site of the enzyme to form an enzyme–substrate complex (ES complex).

7 When, during a reaction, the enzyme is saturated with substrate, as ES complex, v_{max} is attained. The Michaelis–Menten equation describes the

relationship between v and the substrate concentration, [S]. K_M is [S] when $v = \frac{1}{2}v_{max}$.

8 Lineweaver–Burk and Hofstee–Eadie plots are convenient transformations of the Michaelis–Menten equation from which to estimate v_{max} and K_M.

9 Enzyme inhibitors can be either reversible or irreversible. Reversible inhibitors can be subdivided into various classes including competitive and non-competitive.

10 Some enzymes have non-protein parts which are involved in the reaction. These are often called coenzymes.

11 Many coenzymes seem to be related to, and synthesized from, vitamins.

12 Chemical modification of an active site can help to identify those amino acid side chains in the active site that are directly involved in the catalysed reaction. Affinity labels are particularly useful.

13 Each enzyme has an optimum pH. This can give some clue as to the amino acid side chains involved in catalysis; but interpretation is difficult.

14 X-ray diffraction can permit the construction of an accurate three-dimensional model of an enzyme and indicate the location and structure of the active site.

15 Protein engineering provides a way of changing, specifically, individual amino acids in an enzyme and thereby testing their role, if any, in catalysis.

16 The relationship between enzyme and substrate can be visualized to be like that between a lock and key. An alternative hypothesis is that involving induced fit.

17 Where enzyme catalytic power comes from is still a difficult question. Several contributing sources may exist. These include: acid–base catalysis; the proximity effect; stress and strain; electronic strain; covalent intermediate catalysis. Transition state theory provides another useful way of considering catalytic power.

18 Some enzymes in an organism exist in several different related forms, isoenzymes. Each isoenzyme of an enzyme probably has some particular, and possibly subtly different, metabolic role.

19 Enzymes are used quite widely in medical diagnosis. Some are used to assay the levels of molecular markers of disease. Some enzymes, themselves, can be used as molecular markers.

20 Many enzymes have found a medical or industrial/commercial use.

21 The concept of specificity, so important in studying enzymes, can be extended to other biological systems. It is particularly useful when considering other molecular interactions such as, for example, those between antibody and antigen, or between signal molecule and cell surface receptor.

OBJECTIVES FOR CHAPTER 4

Now that you have completed this chapter you should be able to:

4.1 Define and use, or recognize definitions and applications of each of the terms printed in **bold** in the text.

4.2 Name major ways in which enzymes resemble or differ from synthetic catalysts. (*Question 1*)

4.3 Define or recognize the main features of the systematic naming of enzymes. (*Question 1*)

4.4 From suitable data on enzyme-catalysed reactions, choose possible assay techniques. (*Question 2*)

4.5 Given suitable data, predict the effect of factors such as pH, temperature and inhibitors on the structure and function of enzymes. (*Questions 1, 6, 7 and 8*)

4.6 Distinguish between and recognize the effects of the different classes of enzyme inhibitors. (*Question 4*)

4.7 Provide or select evidence for the existence of enzyme–substrate complexes. (*Question 3*)

4.8 From suitable data, interpret or derive parameters such as K_M and v_{max} for enzymes. (*Questions 4 and 5*)

4.9 Decide whether given data support the lock and key or induced fit mechanisms for enzyme specificity and/or catalysis. (*Questions 7 and 8*)

4.10 From suitable data, choose between possible mechanisms for enzyme specificity or catalytic power. (*Questions 7 and 8*)

4.11 Recognize the essential features of isoenzymes. (*Questions 9 and 10*)

4.12 Given suitable clinical and enzymological data, diagnose, or comment on diagnoses of, certain clinical states. (*Question 10*)

4.13 Distinguish between features of enzymes that are advantageous, or disadvantageous, or irrelevant to, their medical or industrial/commercial use. (*Question 11*)

ENZYMES: REGULATION AND CONTROL

5.1 STUDYING THE REGULATION AND CONTROL OF INTERMEDIARY METABOLISM

In Chapter 4 you saw how two key properties of enzymes, *specificity* and *catalytic power*, are largely responsible for the speed and accuracy of the hundreds of reactions that constitute intermediary metabolism. In turn, these properties of enzymes can be understood in terms of their detailed structure; *structure and function* are linked. Adopting this theme in the present chapter, we shall show how the other remarkable aspect of intermediary metabolism— its controlled response to changes in conditions (i.e. it can be regulated)—can also be understood, at least in part, in terms of enzyme structure. But before getting down to the nitty gritty of enzyme structure, it is worth reminding ourselves of what precisely it is about intermediary metabolism that we seek to explain.

Within certain limits, all cells can regulate their metabolism in response to environmental changes. Take, for example, *Escherichia coli*, a common bacterium, resident in the intestines of mammals, ourselves included. Outside intestines, and inside laboratory glassware, *E. coli* can grow and reproduce at temperatures ranging from 15 to 42 °C and in dilute aqueous media or highly viscous glycerol-containing ones. It can use a wide variety of 'foods': any one of over 20 different organic compounds can serve as its sole source of energy and carbon from which *E. coli* ultimately must synthesize all its other carbon-containing (organic) compounds. Other bacteria exhibit yet other types of flexibility; for example, *Alcaligenes faecalis* can grow and reproduce in media ranging in pH from 5 to 9. Presumably differences in a cell's external environment result in some differences in its *internal* one. And indeed, in such cases as those above, this is reflected in variation in the *rate* of growth and reproduction. *But growth and reproduction do still occur.* To appreciate how truly remarkable this apparently simple statement is, let us pause for a moment and think about intermediary metabolism, on which growth and reproduction so much depend.

Intermediary metabolism involves hundreds of different chemical reactions which must be enzyme-catalysed to occur at any appreciable rate (Chapter 4, Section 4.1). But enzymes are very sensitive to factors such as pH, tempera-ture, concentrations of ions and so on, and each type of enzyme has its own particular set of requirements, its own optimal conditions for activity. So if in response to changes in the external environment, conditions inside a cell alter, we might anticipate that different enzymes are affected to different extents. Yet the overall network of enzymes, that helps constitute the system that we call intermediary metabolism, apparently still operates, as witnessed by continued growth and reproduction, albeit at a different rate. Somehow the system as a whole can compensate, adjust to altered conditions, regulate its activity. *How?*

Even if external conditions are relatively constant, the problem of how a cell manages to control the rates of many hundreds or thousands of different chemical reactions still remains. Intuition suggests that some degree of control must exist, a point readily brought home by the highly simplified (!) chart of intermediary metabolism seen in Figure 5.1.

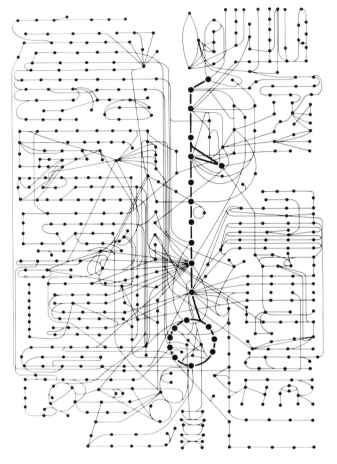

Figure 5.1 A neat scheme devised by Bruce Alberts and colleagues to show *some* of the major pathways of intermediary metabolism. No names of substrates, products or enzymes are shown but each substrate/product is represented by a dot and each reaction by a line joining dots. There are 520 dots.

◇ Look at Figure 5.1. One route is picked out by printing the dots and lines in bolder type. Can you identify the pathway(s) involved?

◈ The route traces the fate of glucose (at the top), via glycolysis, down to pyruvate and thence, via the link reaction, to acetyl CoA and into the TCA cycle (readily seen as the cyclic pathway towards the bottom of the figure). This key metabolic route (glycolysis through TCA cycle), crucial for energy metabolism in many cells, forms the basis of much of the discussion in Chapter 6.

It would seem that the most likely way to regulate a complex system is to control *components* of the system. To take an analogy, consider the design of a road network in a city, where the prime objective is to keep traffic moving as freely and fast as is compatible with safety. If the system is to cope well with varying conditions—say, a rainstorm slowing down the average vehicle rate or a breakdown severely restricting flow on one road—we need to build in ways of controlling segments of the system and thereby regulate traffic flow in the system as a whole, as conditions demand. Roundabouts, traffic lights and suchlike placed at major intersections would be examples of how to effect such control. Likewise, if a cell is to regulate its overall output, as manifest, say, by its rate of growth and reproduction, it must be able to control specific components of the system. The next, obvious questions are what constitute the cell's 'intersections and traffic lights'—*what cellular components are subject to control, and how?*

5.1.1 Mechanisms and timescales

From one viewpoint, *individual metabolic pathways* are the components of the whole system of intermediary metabolism that are subject to control. By controlling a metabolic pathway its net output (of its products) can be adjusted as conditions 'demand'; that is, its net output can be *regulated*. Control and regulate individual metabolic pathways and intermediary metabolism as a whole can be more readily controlled and regulated. But an individual metabolic pathway is composed of steps—a sequence of individual enzyme-catalysed reactions. So, from another viewpoint, the regulation and control of a metabolic pathway rests on the ability to regulate and control the output of *one or more enzyme-catalysed steps* in that pathway. As you will see, these two viewpoints are fully compatible. However, having rather taken the stance that enzymes are the key to understanding 'life, the universe and everything' (Chapter 4, Section 4.1), it on the whole behoves us to take an 'enzyme's eye view' and ask how the rate of enzyme-catalysed reactions can be regulated and controlled.

Very broadly speaking, the rate of an enzyme-catalysed reaction inside a cell depends on two things: the amount present of the particular enzyme, and its catalytic activity, which depends on the conditions (including the level of substrate) in which it is operating. In principle, it therefore follows that each of these factors can form the basis of mechanisms of regulation and control: that is, vary the *amount* of enzyme in a controlled manner (i.e. regulate the amount), or regulate the *catalytic activity* of a fixed (i.e. unchanging) amount of the enzyme. In fact both types of mechanism exist.

Mechanisms of the first type depend on increasing or decreasing the amount of enzyme per cell. In energetic terms this can often be an expensive process. For example, if the response to an environmental change is to increase the amount of an enzyme, this generally involves making it from scratch, synthesizing it from its component amino acids. If, on the other hand, a decrease in the amount of an enzyme is the appropriate response, this may mean destroying an enzyme that previously took a lot of energy to synthesize. In addition, these responses are *slow*, though 'slow' is a highly relative term. For example, in some bacteria an increased amount of enzyme can be detected within a very few minutes of an environmental change and the level of enzyme continues to rise rapidly. In plant and animal cells, such responses tend to be much slower and many minutes or even hours can be needed before an appreciable increase in the amount of enzyme, and hence in reaction rate, occurs. However, to a bacterium whose whole cell cycle may be less than an hour, even a few minutes to achieve the start of an increase in enzyme level may be relatively sedate! In general, regulation of metabolism that involves altering (regulating) the amount of one or more particular enzymes in a cell is reserved for (relatively) long-term processes—adaptations to environmental changes or development. For example, some of the changes in metabolism that occur during development of multicellular organisms seem to depend on such mechanisms. In this chapter we are concerned only with rapid responses to environmental changes that depend, not on altering the amounts of enzymes, but on the second type of mechanism—regulating and controlling the activity of fixed amounts of enzymes.

Minutes or even hours might elapse before a change in the *amount* of an enzyme is large enough to cause an appreciable change in the rate of an enzyme-catalysed reaction. In contrast, regulating the *catalytic activity* of a fixed amount of enzyme can produce a substantial change in reaction rate within seconds. Here the key questions are: *how is catalytic activity regulated and controlled* and *how does this help produce a rapid and 'appropriate' response to changes in conditions?* Answering these questions forms the bulk of this chapter.

In detail, there is more than one answer—hence our reference above to 'types' of mechanisms; there are several ways of regulating and controlling the catalytic activity of enzymes. In this chapter we shall concentrate chiefly on two major mechanisms: *allosteric interaction* and *reversible covalent modification*. First we must make a few very general points about the experimental approach to studying the regulation and control of intermediary metabolism.

5.1.2 *coli* the 'king'

To some extent or other, all cells can regulate and control their intermediary metabolism. But much of the fundamental work in this field has been performed on bacteria. This is true for both broad types of mechanism: those affecting the amount of enzyme, and, the topic of this chapter, those affecting the catalytic activity of a fixed amount of enzyme. The reasons for this pre-eminence of bacteria are of an essentially practical and pragmatic nature.

As single-celled organisms, bacteria can be exposed rapidly and readily to a wide variety of environmental conditions; the experimenter has simply to change the growth medium, as desired. To change the surroundings of animal and plant cells can be much more difficult—imagine trying to change, in a specific manner, the surroundings of, say, a group of liver cells inside an intact rat or some cortical cells inside a root. Admittedly this problem of access can be overcome by isolating cells from animal and plant tissues and getting them to grow as single cells in appropriate laboratory media. But this technique, *tissue culture*, is far from easy; not surprisingly, the animal or plant cells rarely take readily to such foreign surroundings. Compared to bacteria, animal and plant cells grown in tissue culture tend to be very finicky about their nutritional requirements, are fragile, and grow and reproduce very slowly. And cells in tissue culture may not represent their 'normal' counterparts in intact tissues. It is unlikely, for example, that isolated liver cells grown in culture as single cells are identical to those in their more normal environment of the intact liver. Quantity can also be a problem; if needed, it may well be possible to grow tens of kilograms of bacteria—very useful if you wish to isolate some component that occurs only in very small amounts per cell. Even where possible to grow such quantities of cells in tissue culture, for most purposes it is prohibitively expensive; for, unlike bacteria, the nutritional requirements of tissue culture cells are usually far from frugal. For these, and other, reasons bacteria have often been the 'organisms of choice'. And among bacteria, *E. coli* is 'king'; it is likely that more experiments have been performed on *E. coli* and thus more is known about its biochemistry than for any other species of organism.

The problem then is to decide how relevant findings from *E. coli* (or other bacteria) are to other organisms, such as multicellular ones like ourselves. This question of *universality* applies to any findings (and indeed to some extent to comparisons between any species), but it is particularly pertinent in the study of regulation and control where bacteria, and *E. coli* in particular, have been so much the favoured experimental organisms. Fortunately, it appears that many of the basic principles underlying the mechanisms of regulation and control of catalytic activity are universal, though in detail these will vary from case to case and species to species. We are concerned here with just such basic principles. So, in this chapter we adopt the strategy of using whatever examples most readily illustrate these principles, generally bacterial examples, in the knowledge that the essentials usually hold true for other organisms too. If this leaves us open to the jibe oft laid at the door of those supreme exploiters of bacteria—the molecular biologists—that they believe that what is true for *E. coli* must also be true for '*E. lephant*', so be it. We will, however, take the precaution of also discussing briefly some mechanisms of regulation and control that occur in cells in multicellular organisms for which there is no exact equivalent in bacteria. These relate to compartmentation (Chapter 4, Section 4.1) and form the penultimate section of this chapter (Section 5.6).

Finally, before we start our discussion proper of regulation and control of catalytic activity, a word of warning. We have already pointed out that regulation and control of intermediary metabolism can be viewed at different levels—the 'enzyme's eye view' or that of an entire metabolic pathway, for example—and that these can be quite compatible. As will become evident, so too is the view that enzyme structure is a vital component—the *'structure–function'* theme. But all of these viewpoints are just that, and they are human ones, interpretations based on 'how we think it must be'; almost as if we put ourselves in the place of the cell and ask 'how would we cope with this, given such and such . . . ?' The temptation to do this is virtually irresistible, for, after all, we are trying to find out what *we* consider the 'logic' of the cell; how the cell 'copes' with a change in environment, how the mechanisms help in cellular 'economy' and so on. Obviously this is not what the cell does in any literal sense; it does not 'design' a system nor 'think' how to 'cope' with changes or how to 'balance its economy'. But for the investigator of how regulation and control operates, the experimenter, assuming the mantle of the cell (considering the cell's 'problems' and 'solutions') is a very fruitful way of thinking. So we make no apologies for adopting a *teleological* approach in this chapter and indeed we would encourage you to do so too. However, as we shall point out, there are some dangers inherent in this approach. Above all else you should be able to recognize which arguments depend on our adopting this 'designer' role and appreciate that we are certainly not suggesting that the cell 'knows what it wants or needs'!

5.2 FEEDBACK INHIBITION IN LINEAR PATHWAYS

In *E. coli* the amino acid isoleucine is synthesized from another amino acid, threonine. This, in turn, is synthesized from yet another amino acid, aspartic acid (aspartate), which itself derives *ultimately* from glucose (or some other carbon sources), via the TCA cycle, and ammonium ions. In *E. coli* the conversion of threonine into isoleucine occurs via a linear pathway comprising five enzyme-catalysed steps, which we can represent simply as:

$$\text{threonine} \xrightarrow{\ 1\ } M \xrightarrow{\ 2\ } N \xrightarrow{\ 3\ } O \xrightarrow{\ 4\ } P \xrightarrow{\ 5\ } \text{isoleucine}$$

During investigations of this pathway some interesting features came to light. It became apparent that if an amount of isoleucine was *added* to the growth medium far in excess of what the cells needed (to maintain their normal rate of protein synthesis), *the cells stopped converting threonine into isoleucine*. It appears as if the cells could detect the amount of isoleucine present and, finding it more than they needed, stopped making any more from threonine. Presumably such a system, revealed by the experimenter adding isoleucine to the cells, normally operates in the cells to control their rate of synthesis of isoleucine.

If you look at the pathway from threonine to isoleucine, you will see that the *end-product* of the pathway, isoleucine, appears to be inhibiting the functioning of the pathway, and hence is stopping the further production of any isoleucine. The amount of isoleucine in the cell is *controlling* the functioning of the pathway that produces it. In fact it does this by inhibiting one of the enzyme-catalysed steps in the pathway.

◇ In principle, isoleucine could inhibit the pathway, and thus its own synthesis, by inhibiting *any one* of the five enzyme-catalysed steps. Assuming that the only substances in the pathway that the cell ultimately needs for other processes (protein synthesis) are threonine and isoleucine, can you suggest which step or steps might be inhibited?

◈ Step 1, the conversion of threonine to M. This would ensure that in the presence of excess isoleucine no wasteful conversion occurs of threonine to the (now) not needed intermediates on the path to isoleucine, M, N, O and P.

Indeed this is the case, as shown in 1956 by the American biochemist Edwin Umbarger. He made the important observation that step 1, threonine → α-oxobutyrate (M in our scheme), is catalysed by an enzyme, threonine deaminase, which is inhibited *in vitro* by isoleucine. The inhibition was specifically by isoleucine, other amino acids having a much smaller or no effect. In a solution containing 10^{-2} mol l^{-1} isoleucine, the inhibition of the enzyme was 100%; at 10^{-4} mol l^{-1} isoleucine, it was 52%. Thus, these *in vitro* observations indicate that a high level of isoleucine inside the cell inhibits threonine deaminase and hence less isoleucine is synthesized. This inhibition is *readily reversible*, so as isoleucine is used up (for example, in synthesizing proteins) and its level falls, the inhibition is released; threonine deaminase operates again and more isoleucine is synthesized. Thus the level of isoleucine inside the cell controls its own synthesis: supply (by synthesis) is geared to demand; synthesis is regulated. Such a system is called a *feedback system*, as the amount of an end-product (here isoleucine) controls a step in its own production (here threonine deaminase) and the amount of the end-product produced by the pathway is regulated. In this particular instance it is a **negative feedback system**, as it involves a negative (inhibitory) effect of the end-product on an earlier step. Here this ensures that the cell does not produce too much isoleucine and thus waste threonine.

There are many other examples known, in both bacteria and higher organisms, where the end-products of metabolic pathways inhibit the operation of the pathways in this manner. The phenomenon as a whole is called **feedback inhibition**, or, sometimes, *end-product inhibition*. The substance causing the inhibition (e.g. isoleucine) is called the **feedback inhibitor** or *end-product inhibitor*. Enzymes thus inhibited (e.g. threonine deaminase) are said to be *sensitive* to feedback (end-product) inhibition and are examples of **regulatory enzymes**. In effect, such regulatory enzymes *both detect the level of end-product and produce an 'appropriate' response* (i.e. if the level is high they slow down their rate of catalysis). *Feedback inhibition ensures no overproduction when an end-product is already in adequate supply.*

An analogous, and in effect opposite, phenomenon is **enzyme activation**. In this, high levels of certain substances, often ones that are either substrates of early steps in a pathway (what we can call *starting substances*) or related to substrates, *activate* (i.e. increase the catalytic activity of) an enzyme catalysing an early step in the pathway. *Enzyme activation ensures a rapid utilization of a starting substance when it is present in large amounts.*

Look at the two sentences in italics. Note how feedback inhibition and enzyme activation both act to achieve steady levels of substances within a cell, avoiding overproduction or under-utilization, respectively. (Note also our 'designer' role in concluding this!) In some metabolic pathways, such as glycolysis (Chapter 6), where both types of control operate, they must be finely tuned so that there is a harmonious balance between activation and inhibition so as to allow the rate of operation of the pathway as a whole to be well adjusted to the cell's needs.

In many instances both feedback inhibition and enzyme activation operate on the same pathway and in such cases these are, of course, opposing forces. As you will see shortly (Section 5.4), in some cases even a single regulatory enzyme can be sensitive to both activation and feedback inhibition.

It is convenient to have a number of conventions for graphical representation of feedback inhibition and enzyme activation. We adopt the following:

☐ Enzyme-catalysed reactions (steps) are indicated by *straight black arrows*, as elsewhere in this book.

☐ Feedback inhibitors or **enzyme activators** (i.e. substances that activate enzymes) are shown in *blue*.

☐ The line joining the activator or inhibitor to the enzyme reaction it affects is also shown in *blue*.

☐ Where activation is occurring, the blue line ends with an *asterisk* at the reaction affected, and it indicates that activation is of the enzyme catalysing the reaction indicated by the black arrow.

☐ Where inhibition is occurring, the blue line ends in a *broad blue bar* at the reaction affected, and it indicates that inhibition is of the enzyme catalysing the reaction indicated by the black arrow.

◇ Using the above conventions, sketch out the pathway from threonine to isoleucine via M, N, O and P, indicating that isoleucine (when in high amounts) inhibits the enzyme catalysing step 1, that from threonine to M (i.e. α-oxobutyrate).

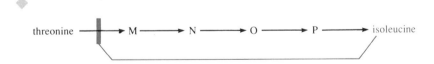

We shall now discuss features of the regulation of the threonine to isoleucine pathway that have points in common with many other systems showing feedback inhibition.

The end-product, isoleucine, inhibits step 1. This prevents conversion of threonine to useless intermediates, that is to substances needed *only* as intermediates on the pathway to production of isoleucine (M, N, etc.). Thus the *control point* of this pathway, as for many others, is the *first enzyme-catalysed step unique to that pathway*. The end-product 'feeds back' onto this first step and controls it. This it does by controlling the catalytic activity of the enzyme catalysing that step (threonine deaminase). The system by which an end-product 'feeds back' onto an earlier control point (as does isoleucine onto threonine deaminase) is sometimes called a **feedback loop.**

In the case of threonine deaminase, the end-product inhibitor (isoleucine) has a substantial degree of structural similarity to the substrate (threonine) of the enzyme that it inhibits:

threonine   α−oxobutyrate ammonium   isoleucine

◇ What might this suggest about the nature of the inhibition?

◈ That inhibition is *competitive* (Chapter 4, Section 4.4.3); that is, isoleucine, being somewhat similar in structure to threonine, competes with the substrate for the active site of the enzyme.

However, *in fact*, the degree of structural similarity between isoleucine and threonine is coincidental and irrelevant as far as the mechanism of feedback inhibition is concerned. Feedback inhibition is *not* of the competitive type and the feedback inhibitor (in this example, isoleucine) does *not* bind at the active site. In many examples of feedback inhibition the irrelevance of structural similarity between inhibitor and substrate is brought home forcefully by the fact that there is little or no such similarity! One such example concerns glutamine synthetase, an enzyme that catalyses the formation of glutamine from glutamate, ammonium ions and ATP:

glutamate   ammonium   ATP (adenosine triphosphate)

glutamine   ADP (adenosine diphosphate) P_i (phosphate)

The glutamine thus produced can feed into several different metabolic pathways, for example, into one leading to the formation of glucosamine 6-phosphate. This particular end-product of glutamine metabolism is an inhibitor of glutamine synthetase, the first enzyme of the pathway. Yet, as can be seen readily, there is very little structural resemblance between end-product inhibitor (glucosamine 6-phosphate) and the substrates or products of the enzyme that it inhibits (glutamine synthetase).

Such feedback inhibitors bind at separate sites from active sites, at so-called *allosteric sites*, and do not depend on structural similarity to the substrates of the enzymes that they inhibit for their mode of action. This feedback inhibition, typical of regulatory enzymes such as threonine deaminase and glutamine synthetase, is known as *allosteric inhibition*. Thus such regulatory enzymes, inhibited by end-products or activated by starting substances, are sometimes also called **allosteric enzymes**. *How* such enzymes are activated or feedback-inhibited is discussed at some length in Section 5.4.

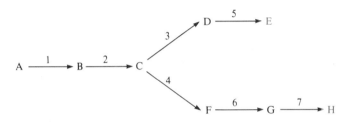

glucosamine 6-phosphate

5.3 FEEDBACK INHIBITION IN BRANCHED PATHWAYS

The pathway from threonine to isoleucine is a linear one; there are no branches. Many metabolic pathways are branched: a simple hypothetical branched pathway leading from starting substance A to end-products E and H is shown in Figure 5.2.

$$A \xrightarrow{1} B \xrightarrow{2} C \xrightarrow{3} D \xrightarrow{5} E$$
$$C \xrightarrow{4} F \xrightarrow{6} G \xrightarrow{7} H$$

Figure 5.2 The pathway from A to E and H.

◇ What would happen if, when one of the end-products, say E, was at a high level inside the cell, it inhibited step 1 in Figure 5.2?

◈ Inhibition of step 1 would lead to a reduced production of E. But it would also lead to a reduced production of the *other* end-product, H. This, of course, might prove 'undesirable' to the cell, for though it may be able to tolerate a reduced production of E (indeed with a high level of E present this is apparently 'desirable'), it may still require unabated production of H. The converse argument applies if step 1 is inhibited by H rather than by E, when H is at a high level.

There are several *theoretical* ways (note our 'designer' role again) of overcoming this problem. In practice, many of these ways are actually used in living organisms. In different pathways and different organisms, different mechanisms may operate. The overall result is, however, much the same—a more balanced production of *all* end-products. In essence, the problem is to ensure that reduced production of one end-product does not reduce production of other end-products of the same pathway. We shall consider just two of the ways in which this is achieved: *enzyme multiplicity* and *multiproduct inhibition*.

5.3.1 Enzyme multiplicity

In Chapter 4 we discussed the occurrence of *isoenzymes* (Section 4.8)—enzymes with similar catalytic activities but differing structures coexisting in the same organism or cell. Though they share a similar catalytic activity, by differing in structure isoenzymes may be subject to different controls. Isoenzymes affected differently by different end-products of branched pathways can form the basis of regulation and control of those pathways. The control of such a system is said to depend on **enzyme multiplicity**.

For example, consider the hypothetical pathway shown in Figure 5.3, where in a single cell two isoenzymes for catalysing step 1 (A→B) coexist. One (isoenzyme 1e) is inhibited by a high level of end-product E, the other (isoenzyme 1h) by a high level of end-product H.

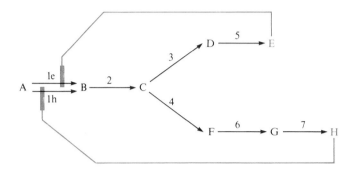

Figure 5.3 The pathway from A to E and H, involving two isoenzymes (1e and 1h) for step 1.

The consequences of such a scheme are very interesting. Suppose that the level of E is high. Isoenzyme 1e will be inhibited. Isoenzyme 1h will still operate, so B will be produced, but at a reduced rate (as only one isoenzyme, 1b, is working). Reduced B will mean reduced C, *but* C can still be converted to H via F and G. So H will still be produced despite the high level of E.

However, what is to stop some of C being converted to D (and D to E)? If this occurs, some of the output of the remaining active isoenzyme, 1h, will lead to production of unneeded E. Thus the production of E cannot be cut off completely, despite its high level in the cell, *and* in consequence, the production of H (not at a high level) may well be reduced. Likewise an excess of H would not prevent some of the output of isoenzyme 1e being used in producing H. Such production would defeat the apparent purpose of the feedback inhibition—prevention of wasteful production. So other additional control points are needed as well as those operating on step 1. Look at the branches from C to E (steps 3 and 5) and from C to H (steps 4, 6 and 7). They form two linear pathways. So one obvious way of preventing a wasteful flow into one or other branch is to ensure that the *first step in each branch* (3 or 4) is also inhibited by feedback from the end-product of that branch (Figure 5.4).

As can be seen from Figure 5.4, if, for example, E is at a high level (i.e. in excess of the cell's current needs), it inhibits 1e. 1h still operates to produce B and hence C. But D is not produced from C, as step 3 is also inhibited by excess E. So all the remaining C is diverted towards F, G and H. Production

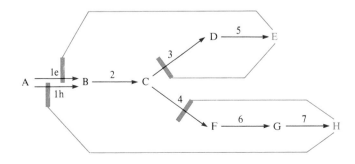

Figure 5.4 The three control points in the branched pathway, A to E and H, where step 1 involves two isoenzymes (1e and 1h).

of H continues, though production of E is inhibited. The converse is true where H is in excess. Ideally, each isoenzyme is capable of converting enough A to B just to accommodate the necessary rate of production of the end-product to which it is sensitive. So, when one isoenzyme is inhibited, the normal rate of output of the other end-product continues.

You may well wonder why, if steps 3 and 4 are inhibited by E and H respectively, the organism needs to inhibit step 1 also. After all, if high E inhibits step 3, then all of C is diverted to produce H; a similar argument applies for high H inhibiting step 4. However, consider what would happen when *both* E and H are in excess. Steps 3 and 4 will be inhibited, but if step 1 is *un*inhibited by E or H, then A will be converted to C, a useless intermediate. Inhibition of step 1 by E and H prevents this.

One example of a branched pathway in which a strategy somewhat similar to that shown in Figure 5.4 seems to operate is that leading from aspartate to lysine and threonine in *E. coli*. An isoenzyme catalysing the first step common to both branches of the pathway (called step 1, above) is inhibited by lysine, and another isoenzyme by threonine. Later steps immediately *after* the branch point are also inhibited by the respective end-products, as shown in Figure 5.5.

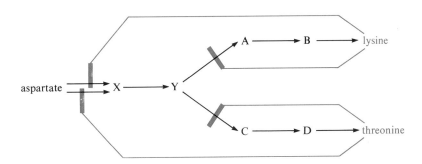

Figure 5.5 A simplified version of the pathway from aspartate to lysine and threonine in *E. coli*.

5.3.2 Multiproduct inhibition

It is perfectly possible to control branched pathways and hence regulate their outputs *without* isoenzymes at step 1. For example, one could have a *single* enzyme catalysing step 1 that is sensitive to more than one of the end-products such that no one end-product alone, even in great excess, completely inhibits the enzyme. Such pathways can be said to be controlled via **multiproduct inhibition**.

In some actual pathways, though each end-product can *partially* inhibit enzyme 1 (i.e. that catalysing step 1), independently of the others, *maximum* inhibition is only achieved by *all* end-products being simultaneously in excess (Figure 5.6). In other pathways, high levels of all end-products are needed for inhibition of enzyme 1 to occur at all—no one end-product *on its own* inhibits it (Figure 5.7). In either case the enzymes present after the branch point are also sensitive to inhibition by their own particular end-products (as was also the case for the type of system shown in Figures 5.4 and 5.5). Figures 5.6 and 5.7 show these two alternative schemes in operation.

(*Note*: The basic schemes we have described—Figures 5.4, 5.6 and 5.7—can, in principle, operate for pathways with more than two end-products.)

As will be seen in Section 5.5.2, glutamine synthetase from *E. coli* (an enzyme introduced in Section 5.2), provides an elegant example of the type of control shown in Figure 5.6.

Figure 5.6 A hypothetical branched pathway where each end-product when at a high level can independently inhibit enzyme 1 up to the maximum value shown. If all three end-products, X, Y and Z, are at the high levels giving their individual maximum inhibitory effects, and we assume that the effects are cumulative (multiplied), then enzyme 1 will be inhibited by 79%, i.e. it will operate at 21% of its uninhibited rate. (Since 50%, 40% and 30% inhibition leave 0.5, 0.6 and 0.7 residual activity respectively, the net effect leaves $0.5 \times 0.6 \times 0.7 = 0.21$ residual activity; i.e. 79% inhibition.)

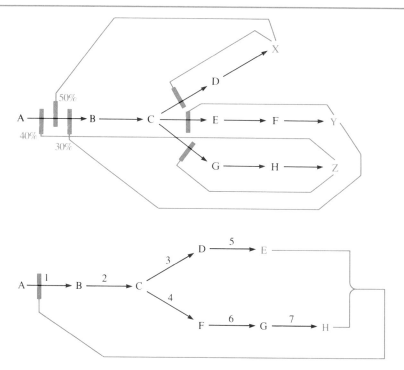

Figure 5.7 A hypothetical branched pathway where all end-products (E and H) must be simultaneously at high levels for feedback inhibition of enzyme 1 to occur.

The synthesis of lysine and threonine from aspartate in *Rhodopseudomonas capsulatus* provides an example of the type of system shown in Figure 5.7, as *both* lysine and threonine must be present at high levels for any inhibition of the first enzyme common to the branched pathway (enzyme 1).

But the pathway from aspartate to threonine and lysine in *R. capsulatus* is the same as that in *E. coli*, where, as we have already seen, control involves a different strategy: isoenzymes at step 1 (Figure 5.5). Thus in two species of bacteria, let alone a bacterium and an elephant (!), different mechanisms have evolved apparently for effecting the same process—the regulation and control of the metabolic pathway producing lysine and threonine from aspartate. This should warn us not to overdo the 'designer' role; there is evidently frequently more than one 'reasonable' way of regulating and controlling a pathway and, where this so, the diversity of possible solutions is often to be found displayed among actual living organisms. The *general* mechanism is however the same—one involving *feedback inhibition*—and whatever the variations (isoenzymes, multiproduct inhibition or yet other mechanisms that we will not discuss), the evolutionary selective pressure was presumably the same—to maintain a balanced production of end-products, without either over- or under-production.

SUMMARY OF SECTIONS 5.1 TO 5.3

1 Cells can regulate their metabolism in response to changes in their environment.

2 Regulation of metabolism overall can be achieved by regulation and control of individual metabolic pathways. This involves regulating and controlling particular enzyme-catalysed steps in those pathways.

3 The rate of operation of an enzyme-catalysed step can be controlled by either altering the amount of enzyme present, or by altering the catalytic

activity of a fixed amount of enzyme (i.e that already present). The latter (the main subject of this chapter) involves the more rapid and energetically economical mechanisms.

4 The catalytic activity of some enzymes, called regulatory enzymes, can be either increased (activation) or decreased (inhibition). A regulatory enzyme is often one catalysing the first step in a metabolic pathway.

5 Activation is controlled by the level of a starting substance of the pathway or a substance related to it. Inhibition is controlled by the level of an end-product of the pathway, a so-called end-product or feedback inhibitor; the phenomenon as a whole is known as feedback inhibition or end-product inhibition or negative feedback. A high level of the end-product results in feedback inhibition of the regulatory enzyme; this results in a reduction in the output of the pathway as a whole and, hence, reduced production of the end-product.

6 In branched metabolic pathways, the rate of the first step may be controlled by more than one end-product. Such control may involve isoenzymes catalysing the first step (enzyme multiplicity). Alternatively, it may involve a single regulatory enzyme that is not fully inhibited by a high level of any one end-product alone (multiproduct inhibition), wherein either a high level of each end-product partially inhibits the enzyme or all end-products must be at a high level for any inhibition to occur.

7 In branched pathways, in addition to the enzyme catalysing the first step, there are generally other regulatory enzymes involved. The first steps after the branch point are commonly also control points (they are catalysed by regulatory enzymes).

Question 1 (*Objectives 5.1 and 5.2*) Which of the following statements reveal a 'designer' approach in our understanding of the subject matter?

(a) Threonine deaminase is inhibited by a high level of isoleucine.

(b) Wings enable birds to fly.

(c) The active site of an enzyme is shaped so as to allow only its substrate to fit it.

(d) The role of the inhibition of threonine deaminase by isoleucine is to ensure that the cell does not synthesize too much isoleucine.

Question 2 (*Objectives 5.1 and 5.3*) On adding substance X to some fungal cells, there is a marked increase within 20 minutes in the production of substance Y by those cells. Y is known to be produced by a metabolic pathway leading from X. Which of the following statements are possibly true and which probably false?

(a) X activates an enzyme in the pathway from X to Y.

(b) X is a feedback inhibitor of an enzyme in the pathway from X to Y.

(c) X causes an increase in the amount of one or more enzymes in the pathway from X to Y.

Question 3 (*Objectives 5.1 and 5.3*) Read again the information given in Question 2. In addition, assume that if X is added to the cells at the same time as a substance that inhibits specifically the synthesis of proteins, no rise is seen subsequently in the production of Y (which is not itself a protein). Now which of the statements (a)–(c) given in Question 2 are likely to be true and which false?

Question 4 (*Objectives 5.1 and 5.4*) Look at the hypothetical metabolic pathway shown in Figure 5.8 whereby two end-products X and Y are generated from A. Assuming overproduction of X and Y is to be avoided, what possible feedback loops are likely to exist? Use the conventions developed in Section 5.3 to sketch these loops.

Question 5 (*Objectives 5.1 and 5.5*) Assume the pathway shown in Figure 5.8 is controlled as shown in the answer to Question 4 (Figure 5.19). Now answer the following:

(a) If excess Y is added to the cells, which of the nine substances shown (A, B, C, D, E, F, G, X and Y) will decrease in amount? (Assume 1y produces enough B to satisfy normal requirement of Y.)

(b) If a large amount of D is added to the cells, what might the effects be on the synthesis of A, B, C, D, E, F, G, X and Y in the cells?

Figure 5.8 Hypothetical pathway from A to X and Y. 1x and 1y are two isoenzymes (inhibited by high levels of X and Y, respectively) each catalysing step 1, the conversion of A to B.

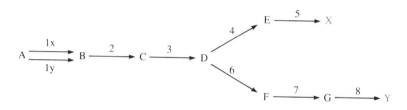

5.4 ALLOSTERIC INHIBITION AND ACTIVATION

We have seen the 'logic' of how feedback loops allow the cell to regulate the output of a metabolic pathway by controlling the catalytic activity of one or more enzymes in the pathway. Though there are a number of variants on this theme (Sections 5.3.1 and 5.3.2), one feature common to all of them is the central role of the first step in each pathway as a *control point*; the first enzyme is a *regulatory enzyme*, the common target for feedback inhibition. What is it about such enzymes that makes them *sensitive* to feedback inhibition (and/or to activation)?

As we noted earlier (Section 5.2), feedback inhibitors are reversible inhibitors that do not necessarily resemble chemically the substrates or products of the enzymes that they inhibit. Given a lack of structural similarity between a feedback inhibitor and the enzyme it inhibits (the sensitive regulatory enzyme), it is reasonable to postulate that such an inhibitor binds not at the active site of the sensitive enzyme but at a distinct separate site. As we shall see shortly there is often evidence that this is indeed the case. Such a separate site where a feedback inhibitor binds is called an **allosteric site** (from the Greek *allos*, other and *steros*, shape). This raises the fascinating question of *how binding of a substance (inhibitor/activator) to one part of an enzyme (the allosteric site) can affect the functioning of another part of the enzyme (the active site)*. The answer is not yet fully known. Nevertheless we do have considerable insight into how a few such allosteric enzymes operate, and these provide some of the most impressive examples of the subtlety of the link between enzyme structure and function (Chapter 4). We shall concentrate on just one example, arguably the most extensively studied and one of the best understood of all allosteric enzymes, *aspartate transcarbamoylase (ATCase)* from *E. coli*.

ATCase is the enzyme catalysing the first step in the metabolic pathway from aspartate (and carbamoyl phosphate) to end-product cytidine triphosphate (CTP). ATCase is inhibited by high levels of CTP, something that ensures that overproduction of CTP is avoided. This inhibition is *reversible* (i.e. high levels of CTP inhibit it, and at low levels of CTP this inhibition is released). Much more obviously than for threonine deaminase (Section 5.2), the (end-product) feedback inhibitor, CTP, has little structural resemblance to either the substrates or products of the reaction (Figure 5.9).

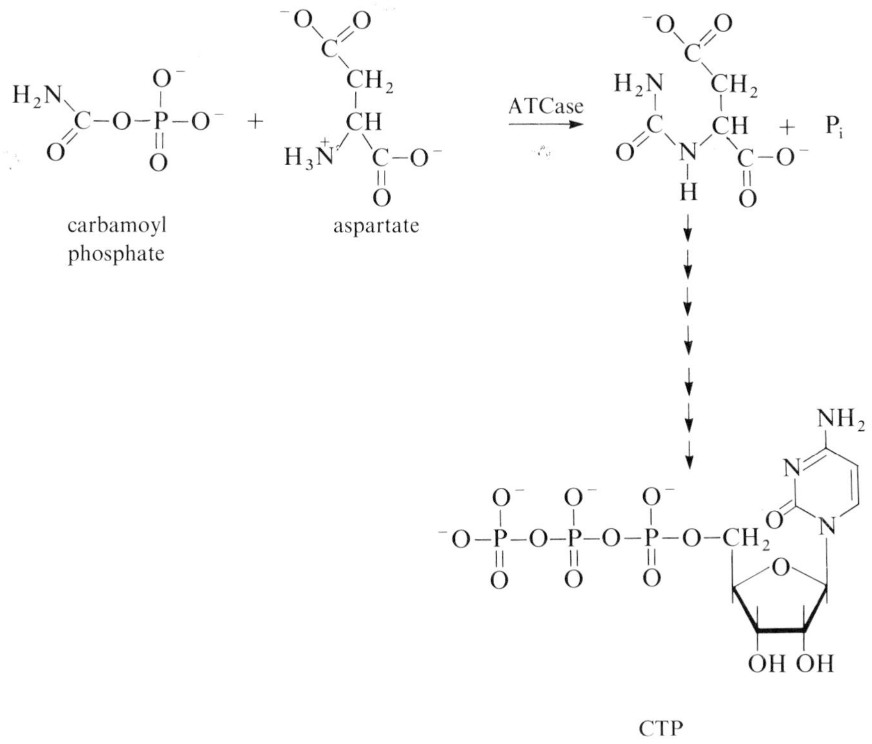

Figure 5.9 ATCase, the first enzyme of the pathway from aspartate and carbamoyl phosphate to CTP, is feedback-inhibited by high levels of CTP.

But ATCase can be also *activated* by ATP. This activation is also *reversible* (i.e. high levels of ATP activate the enzyme, and at low levels of ATP the activation ceases). Furthermore, high levels of ATP can overcome inhibition by CTP. What is the 'logic' behind this dual control of ATCase by CTP and ATP?

When energy is plentiful in a cell, the time is ripe for growth, and, among other things, rapid growth means plenty of RNA synthesis. Such an abundance of energy is evident as a high level of ATP (Chapter 6). And when ATP is high, ATCase activity is stimulated and thus plenty of CTP is made. Thus when energy is abundant, the cell will have plenty of ATP *and* CTP. *Both* are needed for RNA synthesis, so the cell is able to support a high level of RNA synthesis.

On the other hand, if energy (ATP) is low in a cell, neither rapid growth nor enhanced RNA synthesis 'make good sense' (*our* logic again). In such circumstances there is little point in the cell synthesizing large amounts of CTP (as *both* ATP and CTP are needed for RNA synthesis it 'makes sense' to maintain an appropriate balance between them). In this absence of a high level of ATP, the activity of ATCase will be controlled by the level of CTP; as CTP is a feedback inhibitor of ATCase, it will in these circumstances ensure no overproduction of CTP.

In essence then, ATP and CTP compete for effect on the catalytic activity of ATCase, and they do so by *competing* directly for the same *allosteric site*. When CTP binds to the allosteric site, this *decreases* (inhibits) the activity of

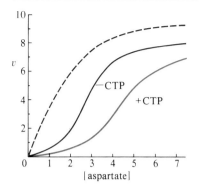

Figure 5.10 Plot of v against [S] (i.e. [aspartate]) for ATCase in the presence (blue line) and absence (solid black line) of CTP. The dashed black line is a classical Michaelis–Menten curve (see the text, later on, for how this is obtained).

ATCase. When ATP binds to the site (instead of CTP), it *increases* (activates) the activity of the enzyme. Actually CTP binding to the allosteric site decreases the affinity of ATCase for its substrates; ATP binding to the allosteric site increases the affinity of ATCase for its substrates. (*Note*: Do *not* confuse this with competitive inhibition described in Chapter 4, Section 4.4.3. CTP is neither a competitive nor a non-competitive inhibitor as described there. The competition described here is between the feedback inhibitor, CTP, and the enzyme activator, ATP, *not* competition between an inhibitor and a substrate. Competition between CTP and ATP is for an allosteric site, *not* for an active site.)

For ATCase the key question then becomes: *how does the binding of either CTP or ATP at the allosteric site affect the binding of substrates at the active site?* As we have already said, this type of question—how binding at one site (the allosteric site) can affect activity at another site (the active site)—is central to all allosteric enzymes. The problem is made even more interesting by an examination of the kinetics of allosteric enzymes such as ATCase. In Figure 5.10 we show the velocity against substrate (aspartate) concentration plot (v against [aspartate]) for ATCase with CTP (blue line) and without CTP (solid black line).

You can see at once that even in the *absence* of CTP there is something odd about the curve (Figure 5.10).

◇ What is odd about the curve as compared with the 'usual' curve one gets from plotting v against [S]?

◈ It is *not* a rectangular hyperbola, the shape associated classically with enzymes displaying Michaelis–Menten kinetics (Chapter 4, Section 4.4.1).

The classical rectangular hyperbola (shown for comparison as the dashed black line in Figure 5.10) has been replaced by an **S-shaped** or **sigmoid curve**. (Note that despite the non-hyperbolic curve, ATCase and other allosteric enzymes *do* function via the initial formation of an enzyme–substrate (ES) complex.) In the presence of CTP, the curve is even more sigmoid.

Think what a sigmoid curve might mean. Look at Figure 5.10 and, for the moment, concentrate on the rectangular hyperbola, typical of *non*-allosteric enzymes, and the curve for ATCase in the *absence* of CTP. At low substrate levels (low [aspartate]), v for ATCase is very low, and, as these levels increase, the increase in v is initially very gradual relative to an enzyme with Michaelis–Menten kinetics. However, as the concentration of substrate is increased further, *the rate of increase* in v becomes very dramatic; for each increment in the substrate concentration, the rise in v is now much greater than for the same increment at lower substrate levels. For example, for the solid black line (ATCase without CTP), compare the increase in v where [aspartate] is increased from 2 to 3 units with that for a similar increase in [aspartate] from 1 to 2 units. It seems that at very low levels of substrate, the substrate molecules have little effect on the rate of the ATCase reaction (v), but somehow, as it were, 'prepare' the ATCase so that subsequent molecules of substrate (as [aspartate] is increased) produce a larger relative increase in v (till eventually the rate approaches v_{max}).

One way of explaining this phenomenon is to *assume* that *each molecule of ATCase has more than one active site*. And, though all these active sites are identical, binding of a molecule of substrate to any one of them facilitates subsequent binding of substrate at other site(s). This phenomenon is known as **cooperativity**, as the active sites appear to cooperate with each other. As these active sites are presumably separate entities on the ATCase molecule, the binding of substrate to one active site somehow affects substrate binding at other active site(s): *how?*

If you think back for a moment, you will appreciate how similar this question is to the one we posed earlier: how can binding of an *inhibitor* or *activator* (here CTP or ATP) to an allosteric site affect activity at a distant (active) site? Perhaps an understanding of the cooperativity between active sites of ATCase (a phenomenon common to other allosteric enzymes) can also help explain precisely how *allosteric effectors* work. **Allosteric effector** is the collective name for inhibitors or activators that, respectively, inhibit or activate allosteric enzymes.

This hope seems justified as the hyperbolic curve shown in Figure 5.10 (dashed black line), where *no* cooperative events seem to occur, is also the curve obtained experimentally if the ATCase has first been treated with mercurials (mercury-containing compounds) like *para*-chloromercuri-benzoate. As this treatment *also* abolishes the sensitivity of ATCase to CTP and ATP, it seems that loss of cooperativity between active sites and loss of sensitivity to allosteric effectors go hand in hand. Furthermore, the accentuation of the sigmoid nature of the v against [aspartate] curve by the presence of CTP also suggests some sort of connection between sensitivity to allosteric effectors and the mechanism of active site cooperation.

In line with our faith in the connection between structure and function (Section 5.1.2), further elucidation of the apparently related phenomena of active site cooperativity and allosteric inhibition/activation comes from a detailed consideration of enzyme structure.

ATCase is a large enzyme with an M_r of around 306 000 and roughly equivalent in size to a sphere of 13 nm diameter. We have seen that treating ATCase with mercurials abolishes sensitivity to ATP and CTP as well as cooperativity between active sites. Mercurials react with $-SH$ groups and they disrupt the quaternary structure of ATCase so that it dissociates into two types of subunit (Chapter 3, Section 3.4.4), which can be separated one from the other in the laboratory. One is a large subunit (M_r 102 000), the other a smaller one (M_r 34 000). The large subunit is capable of catalysing the reaction between aspartate and carbamoyl phosphate (Figure 5.9) and is hence called the **catalytic subunit**. Catalysis by this subunit is, however, *unaffected* by ATP or CTP. On the other hand, the small subunit has no catalytic activity but *can* bind ATP and CTP. Thus it seems that this small subunit carries the allosteric site, and hence it is called the **regulatory subunit**. If catalytic and regulatory subunits are mixed together, a molecule with the same size as the original ATCase is produced, and this then shows both catalytic and regulatory responses; in other words, ATCase can be reconstituted.

These distinct functions of the two types of subunit can be readily demonstrated by separating catalytic and regulatory subunits on a sucrose density gradient (Box 5.1) and testing each fraction of the gradient for catalytic and regulatory activity (Figure 5.11). Each fraction taken from the gradient is split into two portions: one portion is tested for catalytic activity (dashed black line in Figure 5.11), the other for whether it can respond to CTP (what we have termed 'inhibition'—blue line in Figure 5.11).

To test a sample for 'inhibition', catalytic subunit is added to it, and then the catalytic activity is measured in the presence and absence of CTP.

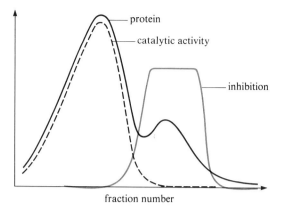

Figure 5.11 Catalytic activity and its inhibition by CTP in catalytic and regulatory subunits of ATCase as separated on a sucrose density gradient. The solid black line shows the amount of total protein present in each fraction. The dashed black line shows catalytic activity. The blue line shows the degree of 'inhibition' by CTP.

◇ Why add catalytic subunit when testing for inhibition by CTP?

◈ Fractions containing only regulatory subunit (higher fraction numbers in Figure 5.11) would have *no* catalytic activity anyhow, and so obviously their inhibition by CTP cannot be tested. On adding catalytic subunit to them, they then have both catalytic and regulatory subunit (equivalent to whole enzyme) and hence show catalytic activity in the absence of added CTP and inhibition of this activity in its presence. Adding catalytic subunit to portions of lower number fractions from the gradient (Figure 5.11), which already contain catalytic subunit, results, of course, in their still having catalytic activity, but as they have no regulatory subunit this is not inhibited by added CTP (blue line in Figure 5.11).

Thus in ATCase, not only do effectors (inhibitor, CTP; or activator, ATP) bind at a separate site from the active site, but this site (the allosteric site) is actually located on a separate subunit (regulatory subunit) from that bearing the active site (catalytic subunit).

In fact each molecule of ATCase consists of *two* catalytic subunits and *three* regulatory subunits. And each catalytic subunit comprises three identical polypeptide chains (each of M_r 34 000); so we can represent a catalytic subunit as c_3. Each regulatory subunit comprises two identical polypeptide chains (each of M_r 17 000) and can thus be represented as r_2.

◇ Using the c_3/r_2 notation, write down the structure of the intact enzyme, ATCase. What will be the M_r of this intact enzyme?

◈ As ATCase has *two* catalytic subunits (each being c_3) and *three* regulatory subunits (each being r_2), the intact ATCase is c_6r_6. The M_r of ATCase should therefore be $(6 \times 34\,000) + (6 \times 17\,000) = 306\,000$; and this is indeed the measured M_r of ATCase given earlier.

William Lipscomb and his colleagues have performed some elegant X-ray diffraction studies of the intact ATCase molecule with and without a substrate analogue, *N*-phosphonacetyl-L-aspartate (PALA). PALA is in fact an inhibitor of the enzyme by virtue of its chemical resemblance to a combination of the *two* substrates of the enzyme, aspartate and carbamoyl phosphate (Figure 5.9). PALA is thus a *bi*substrate analogue which may well resemble closely the *transition state* that forms during the normal catalytic process (Chapter 4, Section 4.7.5). It thus inhibits the enzyme by fitting the *active site*, and, crucially, its use in X-ray diffraction allows the location of the active site to be identified (Chapter 4, Section 4.5.4). Similarly, Lipscomb's group have studied ATCase with bound allosteric inhibitor, CTP, and this has allowed them to locate the allosteric site.

Box 5.1 Sucrose density gradient centrifugation

Separating components on a sucrose density gradient involves spinning components of different size (such as catalytic and regulatory subunits), by means of a centrifuge, through a graded solution of sucrose (i.e. a solution where the concentration, hence density, of sucrose increases from top to bottom of the centrifuge tube). After some time the spinning is stopped, the tube is removed from the centrifuge and fractions collected from different levels in the gradient (i.e. distances down the tube) by puncturing the bottom of the tube and successively collecting drops into a series of test-tubes. The sucrose gradient helps stabilize the positions reached by the various components against turbulence that, say, might occur on taking the tube out of the centrifuge. Components will be in fractions according to how far they have travelled through the sucrose during centrifugation. This depends on a quantity termed their *sedimentation coefficient*. In essence, large components will be further down the tube (and hence in the early fractions to be collected from the bottom of the tube, i.e. those with low fraction numbers) than will small ones (Figure 5.12).

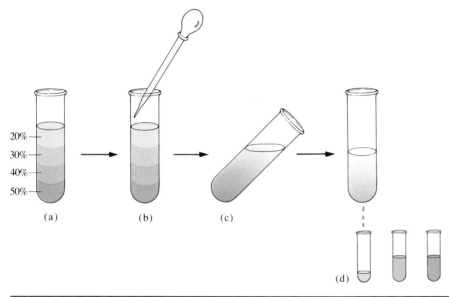

Figure 5.12 Sucrose density gradient centrifugation. A gradient of dissolved sucrose is produced by layering solutions of different concentrations (here, 20% to 50%) on top of each other in a plastic centrifuge tube (a). The sample containing the components to be separated is then layered on top of the gradient (b). The tube is then spun in a centrifuge (c) for some time, typically a few hours. It is then removed, punctured at its base and equal measured volumes (or the same number of drops) are collected into each of a successive series of test-tubes, thus separating different fractions from the gradient (d).

The two views of free ATCase (i.e. without PALA or CTP) seen in Figure 5.13 show that the enzyme comprises a 'sandwich' with one catalytic subunit (comprising three polypeptide chains; i.e. c_3) lying above and another (c_3) lying below a 'filling' of three regulatory subunits (each one being r_2). As the view in Figure 5.13b shows, the 'filling' of regulatory subunits in the 'sandwich' is a bit on the mean side, all towards the edges, there being a large cavity in the centre of the enzyme. This cavity appears to be accessible from

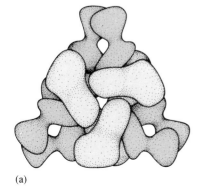

(a)

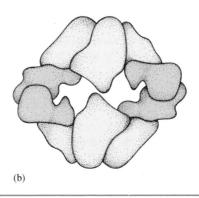

(b)

Figure 5.13 Two views of a model showing the arrangement of the catalytic and regulatory subunits in ATCase, as derived from X-ray diffraction. The six polypeptide chains (c) that comprise the two catalytic subunits (i.e. $2 \times c_3$) are shown in grey, the six (r) that comprise the three regulatory subunits (i.e. $3 \times r_2$) are in blue. (a) is looking down the three-fold axis of symmetry (from the top of the 'sandwich'); (b) is a side-on view, i.e. perpendicular to that of (a).

various directions (see, for example, the triangular hole between the three c polypeptide chains seen at the centre of Figure 5.13a). The enzyme has a three-fold axis of symmetry; in Figure 5.13a we are looking down this axis.

The studies with PALA have revealed that ATCase has six identical active sites, thus satisfying the need for more than one active site in order to explain cooperativity theoretically (see above). Each active site is actually formed by the interaction of amino acid side chains from two separate, adjacent, c polypeptide chains.

◇ As there are in total six c polypeptide chains and it takes contributions from *two* to form an active site, how can the enzyme have six active sites?

The key is that each c polypeptide chain is identical but each of a pair of c polypeptide chains contributes *different* amino acid side chains to form any particular active site. Thus, for example, say one polypeptide chain (call it A) contributes amino acid side chains 80 and 84 (i.e. those of amino acids at positions 80 and 84) and the adjacent polypeptide chain (call it B) contributes amino acid side chains 134, 229 and 231 to help form an active site. Then B can also contribute *its* amino acid side chains 80 and 84 and A contribute *its* 134, 229 and 231 to help form another identical active site.

Likewise, X-ray diffraction studies of ATCase with CTP bound have revealed that there are six identical allosteric sites, but in this case each site is wholly in each one of the six r polypeptide chains.

The structure shown in Figure 5.14 of *half* an ATCase molecule (i.e. c_3r_3) reveals the locations of the allosteric sites and the active sites (or at least that portion of each active site contributed by one c polypeptide chain).

Take a look at the model in Figure 5.14. In the actual (three-dimensional) molecule, the distance between active sites in the same catalytic subunit (i.e. the same c_3) is 2.2 nm; and that between an allosteric site and the *nearest* active site is 6 nm. These are very large distances in molecular terms (to give some sort of scale, a carbon–carbon single bond is only about 0.154 nm in length). Yet recall why we embarked on this discussion of the *structure* of ATCase: to try to elucidate what we *assumed* must be happening—*how does binding something at one site on an enzyme affect what happens at another site?* So far our discussion of structure has done little to answer this key question, but it has certainly emphasized the magnitude of the problem, quite literally. In both cases—cooperation between active sites, and inhibition/activation between allosteric sites and active sites—we must explain *long-range interactions*. In one case (cooperativity between active sites) the interaction must be over a distance of 2.2 nm at least, in the other (allosteric inhibition/activation) over at least 6 nm; both much too far to be due to direct molecular interaction between the sites involved (recall again that a typical bond length is of the order of 0.1–0.2 nm). Indeed, as shown by some cleverly designed experiments performed in Howard Schachman's laboratory, these interactions may occur over even greater distances.

What Lahue and Schachman did was to take a preparation of separated catalytic subunit (i.e. c_3) and react it with a substance that adds trinitrophenyl groups to two lysine side chains (Lys-83 and Lys-84) which lie in the active site region in each of the three c polypeptide chains. This has two effects. Firstly, and not surprisingly, it almost entirely abolishes the catalytic activity of the catalytic subunits. Secondly, trinitrophenyl groups are coloured and their spectrum when bound to the protein can be readily and accurately measured. Moreover, the precise spectrum of the bound trinitrophenyl groups depends on their surrounding environment, how much they are 'shielded' by adjacent amino acid side chains, for example. A change in their surroundings can lead to a measurable change in their spectrum; thus these coloured groups in the active site region act as 'reporters' of such changes. The modified catalytic subunits (which we can call c_3^{tri}) were then combined with normal regulatory subunits to form a hybrid enzyme, $c_6^{tri}r_6$ (remember each molecule of ATCase has two catalytic subunits, each themselves c_3, and three regulatory subunits, each r_2).

$c_6^{tri}r_6$ had low catalytic activity but did show the typical sigmoid v against [S] curve and inhibition by CTP and activation by ATP (Figure 5.10). Thus the

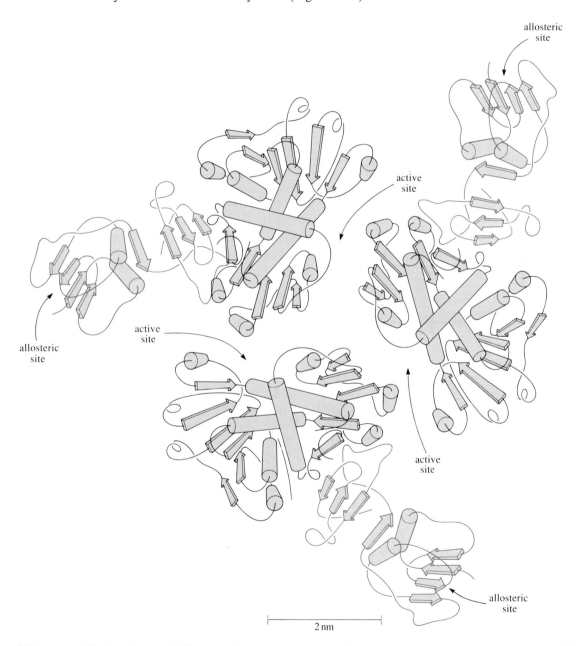

Figure 5.14 A view of half a molecule of ATCase (c_3r_3) looking down the three-fold axis of symmetry (i.e. the angle in Figure 5.13a). Some of the detail of the folded polypeptide chains is shown (β-structures as ribbons, α-helical regions as cylinders), as are the locations of the allosteric sites and the active sites (portion from one c polypeptide chain). The c polypeptide chains are shown in black and grey, the r polypeptide chains in blue.

modified enzyme still showed cooperativity and allosteric inhibition/ activation. And when either CTP or ATP were bound to it, there was a *significant change in the spectrum of the trinitrophenyl groups*. The trinitrophenyl groups at the active site were 'reporting' a change there, when ATP or CTP bound at the (distant) allosteric sites. (In fact ATP and CTP cause *opposite* changes in the spectrum of the reporter groups.) The most likely type of change to account for the change in spectrum is a *change in conformation* in the region of the reporter groups, i.e. in the active site region.

◇ What significance could such a change in conformation have?

⬧ As you recall from Chapter 4, the shape of an enzyme active site is critical to its functioning. A change in its conformation might either enhance (as perhaps in *induced fit*; Section 4.6) or inhibit this function.

Thus it seems likely that CTP inhibits enzyme activity by *causing a change in conformation of the active site* that is deleterious to binding of substrate (as stated earlier, it decreases the affinity of enzyme for substrate). ATP seems to have the opposite effect. But in each case (CTP and ATP) their action is *long-range*: their binding at allosteric sites on regulatory subunits (six such sites, one on each r polypeptide) causes changes in distant active sites (Figures 5.13 and 5.14). There is what has been called 'cross-talk' between regulatory and catalytic subunits, or, more specifically here, binding at an allosteric site being 'communicated' to an active site.

A similar experiment, again performed by Lahue and Schachman, showed that there is also *cross-talk between active sites* and this could account for cooperation between those sites in terms of substrate binding, as witnessed by the sigmoid curve (Figure 5.10). Here a hybrid enzyme was produced that contained one normal catalytic subunit, one modified one and normal regulatory subunits, that is, $c_3c_3^{tri}r_6$. On adding the bisubstrate analogue PALA to this enzyme, there was a change in the spectrum of the trinitrophenyl reporter groups. PALA binds virtually exclusively to the active sites on the *normal* catalytic subunit (c_3) but the reporter groups are located in the active sites of the *modified* catalytic subunit (c_3^{tri}). Thus binding of substrate to active sites on one catalytic subunit (c_3) can be communicated to active sites *as far away as on the other catalytic subunit* (Figure 5.13). Once again this 'communication received' (as reported, in the experiments, by the trinitrophenyl groups) presumably leads to a change in conformation of those active sites receiving the message; here a change that favours further binding/catalysis of substrate, hence cooperativity. Yet further experiments demonstrate that there is also somewhat more local cross-talk between active sites within the *same* catalytic subunit (the same c_3). Indeed, as other experiments from Schachman's laboratory show, binding of substrate to just any one of the six active sites is communicated to all the other five (i.e. those on the same catalytic subunit and those on the other one).

Finally, X-ray diffraction studies from Lipscomb's group also show that binding of the bisubstrate analogue PALA causes large changes in ATCase both at the tertiary and quaternary levels (Chapter 3, Section 3.4). At the tertiary level, there are substantial changes in the conformation of each c polypeptide. This seems to help bring together those amino acid side chains from the two adjacent c polypeptides that contribute to the same active site. At the quaternary level, binding of PALA causes the two complete catalytic subunits (the two c_3) to move apart and rotate, and, also, each of the regulatory subunits (i.e. each r_2) to rotate.

Thus we can begin to tie together what probably happens in ATCase. When binding of substrate to an active site, or effector (ATP or CTP) to an allosteric site, occurs, this event is communicated to (other) active sites; the communication over long distances being via changes in the conformation of the molecule. Once received, this communication results in more local changes in the conformation, and hence functioning, of those active sites. The current challenge presented by ATCase is to work out the precise route by which binding of substrates, and ATP and CTP cause *different specific* changes in quaternary and/or tertiary structure and how these effect *different specific* changes in the affinity of active sites for their substrates.

By concentrating, at some length, on just one allosteric enzyme, ATCase, we trust we have been able to explain some of the intricacies of the problem of how allosteric enzymes work and some of the methods used to tackle it. Obviously other allosteric enzymes will not operate in precisely the same way as ATCase; for example, there may be very few that actually have active and allosteric sites on quite separate subunits as does ATCase. Nevertheless ATCase does have some very instructive take-home messages no doubt relevant to many other allosteric enzymes and in some cases to protein function in the wider sense. It is worth noting those here:

□ Once again *specificity* is the key to much of the way in which proteins operate. Not only does ATCase have a specific active site (for aspartate and carbamoyl phosphate), but it also has an allosteric site specific for an inhibitor (CTP) and an activator (ATP); i.e. for allosteric effectors. This dual specificity, for catalysis and regulation, endows ATCase with its *specific role* as a regulatory enzyme in the pathway leading to CTP (and hence to RNA synthesis).

□ *Protein flexibility* (Chapter 3, Box 3.3; Chapter 4, Section 4.6), which permits change in shape, is important both for mediating the long-range communication between different sites (active and active, and allosteric and active) and for translating that communication into modification of the active site, thus altering its capacity for substrate binding (and/or catalysis, in some enzymes).

□ **Allosteric regulation**, as the phenomenon exemplified by the behaviour of ATCase is known, bears powerful witness to the elegant and subtle relationship between *structure and function*.

Summary of Section 5.4

1 An allosteric effector (i.e. feedback (end-product) inhibitor, or an activator) of a sensitive regulatory enzyme acts by binding at a site(s) on the enzyme that is separate from the active site(s). Such inhibitor/activator sites are called allosteric sites. The sensitive enzyme is said to be an allosteric enzyme and the phenomenon as a whole is known as allosteric regulation.

2 As the level (i.e. concentration) of a feedback inhibitor in the cell determines ultimately how much of it is bound to an allosteric enzyme, inhibition is readily reversible. At high levels of inhibitor, binding to the enzyme will also be high, and so inhibition will occur; if the level of the inhibitor drops, so does the amount of binding to the enzyme and inhibition is reduced. A similar argument applies to activation.

3 Aspartate transcarbamoylase (ATCase), the first enzyme on the pathway from aspartate and carbamoyl phosphate to CTP, is a much studied case of an allosteric enzyme. It is inhibited by CTP and activated by ATP. CTP and ATP can compete for binding at the same allosteric site.

4 ATCase gives a sigmoid curve when v is plotted against [S]. This indicates that cooperativity is occurring between the multiple active sites on ATCase;

i.e. binding of substrate to one active site facilitates subsequent binding of substrate to other active sites.

5 Mercurials abolish the sensitivity of ATCase to CTP and ATP and destroy cooperativity between active sites. They do so by separating the enzyme into two types of subunit: catalytic subunits, that bear the active sites; and regulatory subunits, that bear the allosteric sites. Intact ATCase has two catalytic subunits and three regulatory subunits. Each catalytic subunit has three identical c polypeptide chains; each regulatory subunit has two identical r polypeptide chains. ATCase is thus c_6r_6.

6 X-ray diffraction studies reveal that individual active sites are large distances apart as are active sites and allosteric sites.

7 Binding of substrate to one active site or ATP/CTP to an allosteric site is communicated to (other) active sites some distance away. These communications ('cross-talk') cause changes to the active sites receiving them, resulting in enhanced or diminished affinity of those active sites for substrate.

8 There can be shown to be 'cross-talk' between catalytic and regulatory subunits and between c polypeptide chains in the same catalytic subunit. Cross-talk is achieved by changes in the enzyme at the tertiary and/or quaternary levels.

9 Allosteric regulation, as in ATCase, demonstrates the importance of specificity, protein flexibility and the relationship between structure and function in enzymes.

Question 6 (*Objectives 5.1 and 5.6*)   Which of the following statements are true, and which are false?

(a) Feedback inhibitors of allosteric enzymes are always closely related structurally to substrates of the enzymes they inhibit.

(b) Enzymes showing cooperative effects have more than one active site.

(c) Allosteric enzymes do not form an enzyme–substrate (ES) complex.

(d) If a metabolic pathway is subject to allosteric inhibition *and* activation, there must be at least two enzymes in the pathway involved (i.e. one enzyme inhibited, the other activated).

Question 7 (*Objectives 5.1, 5.6 and 5.7*)   An enzyme catalysing the first step in a six-step linear pathway (enzyme 1) is inhibited by the end-product of the pathway, substance X. Which of the following would you expect *might* be true and which false?

(a) Treatment of enzyme 1 with a substance that destroys some amino acid side chains in the enzyme renders the enzyme catalytically unaltered but insensitive to X.

(b) Enzyme 1 is irreversibly inhibited by X.

(c) Enzyme 1 is activated by substance Y.

(d) Z is a competitive inhibitor of enzyme 1. Binding of Z to enzyme 1 is inhibited by X.

5.5 REVERSIBLE COVALENT MODIFICATION

Though allosteric regulation is undoubtedly a widespread and major force in the regulation of metabolic pathways, it is not the only mechanism. Another important mechanism is **reversible covalent modification**. This too operates at the level of individual regulatory enzymes and, as the name implies, involves a modification of enzyme structure that depends on the making or breaking of *covalent* bonds; it is this modification that determines (i.e. controls) the activity of the enzyme.

◇ In what way does this differ from allosteric control, as say typified by ATCase?

◈ Allosteric control of ATCase involves a substantial change in the tertiary and quaternary structure of the enzyme but this does not appear to involve any change in *covalent* bonds in the enzyme protein.

In reversible covalent modification, some small chemical group can be covalently attached to, or detached from, an enzyme and in so doing the activity of the enzyme is affected. As the group can be either attached or detached, the modification of enzyme structure, and hence of activity, is *reversible*. Thus, though reversible covalent modification of enzyme activity occurs by quite a different mechanism from allosteric regulation, it does share the property of reversibility.

◇ Why is reversibility important?

◈ Remember the function of enzymes involved in regulation and control of metabolism (Section 5.1). As levels of certain substances (e.g. feedback inhibitors) rise and fall, so must the activity of the enzymes that are sensitive to those substances. Thus, for example, where an enzyme is capable of reducing its activity in the presence of an inhibitor, it must also be capable of returning to a high (uninhibited) level of activity when the level of inhibitor is low; i.e. for it to function as a regulatory enzyme the effect of inhibitor must be reversible.

There are many examples of reversible covalent modification, involving different enzymes and different chemical groups effecting the modification.

5.5.1 Reversible covalent modification involving a phosphate group

The best-known example of reversible covalent modification, one that involves the addition and removal of a phosphate group, comes from studying the regulation of glycogen metabolism in mammals.

Glycogen metabolism is complex, the synthesis and degradation of glycogen occurring by separate metabolic pathways. Each pathway is regulated by a number of highly sophisticated and equally complex controls, and, furthermore, there is coordination between the regulation of the two pathways.

◇ Putting on your 'designer' hat; why must there be coordination between the two regulatory systems?

◈ There would seem to be little point in a cell expending energy synthesizing a substance (here, glycogen) at the same time that it was actively engaged in degrading it. Thus if systems are in place to regulate synthesis of glycogen, and other systems to regulate degradation, it would seem 'logical' to coordinate both systems such that when one is serving to step up one process (say, synthesis) the other is acting to step down the opposite process (say, degradation).

Fortunately, for our current purposes—a consideration of reversible covalent modification—we can ignore most of the complexity of the regulation of glycogen metabolism overall (see, however, Section 5.7) and concentrate on just one regulatory enzyme, *glycogen phosphorylase* (*phosphorylase*, for short). This is involved in the first step in glycogen degradation where it catalyses the 'release' of glucose residues (as glucose 1-phosphate), one at a time from the ends of the chains of the glycogen molecule (Chapter 3, Section 3.6); as shown in Figure 5.15.

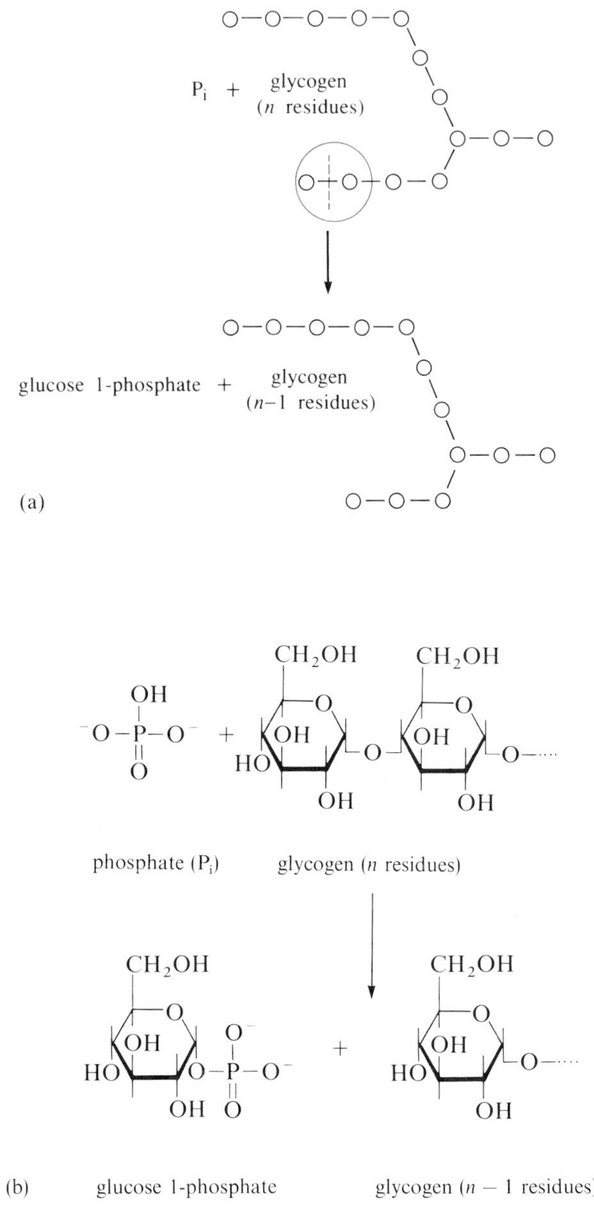

Figure 5.15 Release of glucose residues from glycogen by reaction with phosphate (P$_i$) catalysed by glycogen phosphorylase. For each cycle of the reaction one glucose residue is released as glucose 1-phosphate: (a) simplified scheme, showing 'release' of the glucose residue to the left of the dashed blue line; (b) the chemistry of the reaction occurring within the blue circle in (a). In addition to glycogen phosphorylase, different enzymes are needed to break down the branch points in the glycogen molecule.

In skeletal muscle, phosphorylase exists in two *interconvertible* forms known as (glycogen) phosphorylase *a* and phosphorylase *b*. Under most conditions inside the cell, phosphorylase *b* is inactive, whereas phosphorylase *a* is active. Thus the cell's capacity for releasing glucose from glycogen will vary according to the relative amounts of phosphorylase present in the *a* and *b* forms.

In structural terms the difference between *a* and *b* rests on the presence or absence of a *phosphate group*. To be more precise: phosphorylase is composed of two identical subunits each of M_r 97 000; in the inactive *b* form each subunit has a serine residue at position 14, in the active *a* form this residue (in each subunit) carries a phosphate group. The interconversion of *a* and *b* forms is itself catalysed by two different enzymes: *b* to *a* by phosphorylase kinase, and *a* to *b* by protein phosphatase 1 as shown below. Notice that here (and elsewhere) the active form of the enzyme is shown in blue and the inactive (or less active) form in black.

$$
\text{phosphorylase } b \;+\; \text{ATP} \xrightarrow{\;\;\text{phosphorylase kinase}\;\;} \text{phosphorylase } a \;+\; \text{ADP} + \text{H}^+
$$

phosphorylase *b*   +   ATP   ········▶   phosphorylase *a*   +   ADP + H$^+$

|
Ser—OH   Ser—O—PO$_3^{2-}$

(inactive)   (active)

phosphorylase *a*   +   H$_2$O   ────protein phosphatase 1────▶   phosphorylase *b*   +   P$_i$

Ser—O—PO$_3^{2-}$ Ser—OH

(active)   (inactive)

Thus the relative proportions of phosphorylase *a* and *b* depend on the relative activity of the two enzymes involved in the interconversion of *a* and *b* (phosphorylase kinase and protein phosphatase 1). As the catalytic activity of these two enzymes is *also* subject to regulation and control, the story gets more and more complicated! As alluded to above, level upon level of control exists—one system controlling another (Section 5.7). This, however, goes way beyond our current concern with the role of reversible covalent modification of phosphorylase.

In short, the activity of phosphorylase is regulated by either adding a phosphate to the Ser-14 in each subunit of the enzyme (giving *active* phosphorylase *a*) or removing this phosphate (giving *inactive* phosphorylase *b*). Hence the regulation and control of glycogen phosphorylase involves a *reversible covalent modification* of the enzyme protein.

5.5.2 Reversible covalent modification involving a nucleoside monophosphate group

As with (glycogen) phosphorylase, there are a number of enzymes in which reversible covalent modification involves a phosphate group. However other types of chemical group can be involved. One such example, where the group attached and detached is a *nucleoside monophosphate*, involves what is by any standards a most remarkable enzyme, *glutamine synthetase* from *E. coli*.

Glutamine synthetase is a large enzyme comprising *12* identical subunits each of M_r 50 000. As mentioned in Section 5.2, it catalyses the conversion of the amino acid glutamate (i.e. glutamic acid) to another amino acid, glutamine:

$$
\text{glutamate} + \text{NH}_4^+ + \text{ATP} \xrightarrow{\;\;\text{glutamine synthetase}\;\;} \text{glutamine} + \text{ADP} + \text{P}_i + \text{H}^+
$$

As the enzyme catalysing the formation of glutamine, glutamine synthetase has a key role in the 'flow of nitrogen' in a number of metabolic pathways. This derives from the fact that the amide group of glutamine ($-CONH_2$) is a source of nitrogen in the synthesis of a wide variety of substances such as histidine, carbamoyl phosphate, glucosamine 6-phosphate, tryptophan, CTP and AMP.

◇ Given this key role, what might you anticipate about the reaction catalysed by glutamine synthetase?

⬧ It is a *control point* for a number of pathways involving nitrogen metabolism, and hence glutamine synthetase is a key *regulatory enzyme*.

You have already seen in Section 5.2 that glutamine synthetase is inhibited by one of these end-products, glucosamine 6-phosphate. In fact glutamine synthetase is inhibited by *each* of the six end-products named above (including glucosamine 6-phosphate), as well as by two more amino acids, glycine and alanine! When all eight substances are at high levels, the catalytic activity of glutamine synthetase is almost completely inhibited—a highly complex example of *allosteric regulation*.

◇ What type of (allosteric) feedback inhibition does this represent?

⬧ Multiproduct feedback inhibition (Figure 5.6).

This multiproduct inhibition does not itself involve any covalent modification of the enzyme, *but* glutamine synthetase is *also* subject to reversible covalent inhibition.

Like (glycogen) phosphorylase, glutamine synthetase can exist in two forms, but here the distinguishing feature is not a phosphate group, but the presence or absence of a *covalently bound* adenosine monophosphate (AMP) group. (Do not confuse this with the *non*-covalent binding of AMP involved in the multiproduct feedback inhibition of the enzyme.) In the adenylated form of the enzyme (i.e. the form with the AMP covalently bound) the AMP is bound to a particular tyrosine side chain in each of the 12 subunits. This adenylated form of the enzyme is *more sensitive to multiproduct feedback inhibition* than is the non-adenylated form. Thus this reversible covalent modification, by adenylating or de-adenylating the enzyme, can either enhance or reduce its sensitivity to the other form of inhibition (allosteric) to which it is susceptible; one type of control is thus superimposed directly on the other (Figure 5.16).

As Figure 5.16 indicates, the addition or removal of the covalently bound AMP group is catalysed by an enzyme; here, unlike phosphorylase (Section 5.5.1), a *single* enzyme, adenylyl transferase, governs whether the covalently bound group (AMP) is added or removed. But, as you might by now expect, the direction in which adenylyl transferase works (adding *or* subtracting AMP) is itself subject to regulation and control! We shall not go further into the story. Suffice it to say that, as with our abridged treatment of phosphorylase, there are further levels of control built on top of each other.

Figure 5.16 Reversible covalent modification of glutamine synthetase. Adenylyl transferase catalyses the addition and removal of a covalently bound AMP to glutamine synthetase. The adenylated enzyme is more sensitive to multiproduct feedback inhibition than the de-adenylated enzyme.

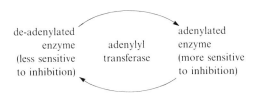

de-adenylated enzyme (less sensitive to inhibition)   adenylyl transferase   adenylated enzyme (more sensitive to inhibition)

5.6 COMPARTMENTATION

So far in our discussion of regulation and control of metabolism we have treated the cell as if it were a single entity, a single 'pot', in which a large number of different reactions occur simultaneously and in which regulation and control is focused on key individual enzymes. But, though this *may* be a reasonable view of a prokaryote cell, for eukaryote cells it is an incomplete picture. For, as you are well aware from Chapter 2, a eukaryote cell comprises a number of types of membrane-bound organelles (e.g. mitochondria, lysosomes, chloroplasts, etc.) and to this extent the cell comprises a set of separate *compartments*. The mechanisms of regulation and control that we have discussed so far, such as allosteric inhibition and activation and reversible covalent modification, can, in principle, apply equally to 'single compartment' prokaryote cells and to individual compartments within compartmentalized eukaryote ones; and indeed these mechanisms are known to occur in virtually all cells. But the **compartmentation** of eukaryote cells opens up further opportunities for regulation and control of metabolism.

Above all else, a set of different compartments (organelles) allows the cell to keep certain reactions or entire metabolic pathways separate from each other. This can help alleviate the 'confusion' that might otherwise arise where two enzymes/pathways compete for the same substrate (Chapter 4, Section 4.1). Physical separation of pathways can simplify the problem of regulation and control; subdivide a large and complex system into a set of smaller, and thereby less complex, systems and these can be regulated and controlled, largely independently.

A eukaryote cell seems to exploit this sort of system. Some metabolic pathways take place entirely in the cytosol—glycolysis and fatty acid synthesis, for example—while others occur exclusively in the mitochondrion—the TCA cycle and oxidative phosphorylation, for instance. As you will see shortly when you consider some of these pathways in Chapter 6, they do indeed have their own independent systems of regulation and control.

Yet for the cell overall, *total* independence of compartments can make no sense. Different cellular compartments must communicate with each other, at the very least. For example, for energy production the rate of glycolysis in the cytosol must be coordinated with the rate of operation of the TCA cycle in the mitochondrion. Sometimes an individual metabolic pathway does not even occur entirely within one distinct compartment: both the urea cycle (whereby excess nitrogen from the breakdown of amino acids is converted into urea) and gluconeogenesis (by which glucose is synthesized from pyruvate) are pathways where some reactions occur in the cytosol and some others in the mitochondrion. (Actually one step in gluconeogenesis occurs inside the lumen of the endoplasmic reticulum (Chapter 2, Section 2.4), so in fact three compartments—cytosol, mitochondrion and lumen of the endoplasmic reticulum—are involved in this single metabolic pathway.) Thus some intermediates in these pathways must be able to cross specific compartment boundaries, i.e. that between mitochondrion and cytosol, or cytosol and lumen of the endoplasmic reticulum.

If you think back to Chapter 4 you will appreciate another situation where the boundary between mitochondrion and cytosol must be capable of being traversed. In Section 4.1 we considered the apparent problem for a cell when the same substance was capable of being involved in more than one enzyme-catalysed reaction. The example we chose was *pyruvate*, which has four alternative fates in a eukaryote cell (Figure 5.17). One strategy by which the cell appears to avoid 'confusion' is to 'compartmentalize the problem'—execute two of the fates in the cytosol and two in the mitochondrion.

But in avoiding one problem this opens up another. For example, pyruvate *production*, via glycolysis, occurs in the cytosol whereas its *utilization* in forming acetyl CoA occurs in the mitochondrion (Figure 5.17). Acetyl CoA is in a sense the starting substance for a large set of reactions located in the mitochondrion, the TCA cycle; it is the means by which the pyruvate of glycolysis is fed into the cycle. So for pyruvate, the end-product of one degradative pathway (glycolysis), to be further degraded (via acetyl CoA and the TCA cycle), it must first cross from the cytosol to the inside of the mitochondrion.

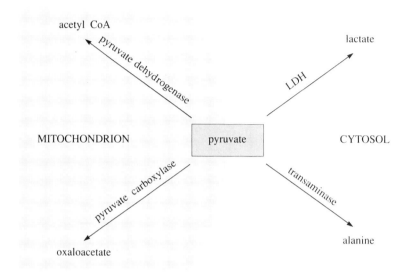

Figure 5.17 Four alternative fates for pyruvate and where they take place.

We have thus seen that the price to be paid for the benefits of compartmentation (subdividing a complex system) is that mechanisms must exist that allow *selective* breaching of the compartment barriers.

◇ Why must compartment barriers be selective?

If the barriers were not selective and anything could get through, they would not be barriers! Without selective barriers 'compartmentation' might still have a physical presence (i.e. organelles could exist) but could well be meaningless in any functional sense.

Physically the barriers between compartments are *membranes* (Chapter 2) and it is to the detailed structure of membranes that these intercompartment barriers owe their selectivity. The structural basis for this selective permeability of membranes forms much of the subject of Chapter 7, so we need not dwell on it here. But for our present purposes, it is important to note that this selectivity provides the eukaryote cell with one more focus of regulation and control—one more type of control point.

In essence, by controlling the flow of specific substances across the membranes that separate compartments, the cell can control the rate of some of the metabolic pathways in which those substances participate. For example, suppose that a substance produced in the mitochondrion is a vital component of a particular pathway that is in the cytosol. By controlling the rate of exit of the substance from the mitochondrion, the cell can control the rate of operation of the pathway in the cytosol.

By analogy with human affairs, it is like controlling and regulating the import (or export) of goods between countries: decide on the types of goods involved and control their import by surveillance of the borders (customs control = selective membranes). The level of allowed import can be varied as desired (regulated) and this will have consequences for particular production processes (metabolic pathways) within the importing country. So cut UK import levels of wood pulp from Finland and you reduce the rate of production of paper for Open University courses, the university can take fewer students each year and this slows down the rate of graduation in the UK ('UK graduate production from wood pulp' = metabolic pathway in one compartment)! The critical control point is at the borders; for the cell, the intercompartment membranes. And, as Chapter 7 will show, a similar argument holds for the cell's interaction with the outside world, its surrounding environment: here the *cell membrane* (Chapter 2, Section 2.2) is the key boundary where selective permeability operates.

5.7 LEVELS OF REGULATION AND CONTROL

One fact that has doubtless emerged by now is that frequently the same metabolic pathway is subject to more than one control. Where this is the case, these different controls may be all by the same type of mechanism (e.g. all by allosteric control) or by different types (e.g. even a single regulatory enzyme, such as glutamine synthetase, can be subjected to two different types of control). We have also seen how the same pathway, or enzyme within it, can be subjected to several levels of regulation and control—the controls themselves are controlled (e.g. phosphorylase and glutamine synthetase, Section 5.5).

Such level upon level of control of particular key enzymes has been likened to a series of waterfalls, a cascade, and indeed sometimes goes under the label of 'reaction cascade'; not unreasonable imagery, as Figure 5.18 shows.

◇ The cascade in Figure 5.18 is a purely hypothetical one. But if you used it to represent glycogen phosphorylase placed at the lowest level in the cascade, what would enzyme 3 (blue), enzyme 3 (black) and enzyme 2 (blue) represent?

◆ Remember our convention: blue forms are active and black inactive. Therefore, enzyme 3 (black) would represent glycogen phosphorylase *b*, enzyme 3 (blue) glycogen phosphorylase *a* and, as enzyme 2 (blue) activates/converts *b* to *a*, it would represent phosphorylase kinase (Section 5.5.1).

Such systems are not purely a 'belt and braces' approach to regulation and control. Having several different levels of control increases the number of potential *control points*. This in turn means that more influences (e.g. activators, inhibitors) can be brought to bear on the system, thus increasing the subtlety with which regulation and control of the various metabolic pathways can be achieved. And, most importantly, it allows *interconnections* to be built up between different levels of control and between systems controlling different pathways. This importance of interconnections applies to all levels of control, at the level of individual enzymes, whole pathways, several pathways in individual organelles and, ultimately, to the cell as a whole.

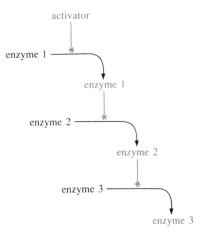

Figure 5.18 A hypothetical control system involving a 'cascade'. Active forms of enzymes are shown in blue, inactive in black. We use blue lines and asterisks to denote the process of activation (Section 5.2); black arrows again indicate reactions, but are curved here to enhance the 'aesthetic' sense of water flowing. Active and inactive forms need not necessarily all be enzymes (as here); they could, for example, be small molecules where one substance (black) is converted into another which is an activator (blue) of the reaction at the next level down the cascade. (The terms *activator* and *activation* are here used in their *general* sense not implying that the mechanism of activation is necessarily by *allosteric* activation, though in many cases it may be.)

Interconnections permit integration; without integration, though individual systems (e.g. metabolic pathways, whole organelles) might be well controlled, the cell as a whole might not be. It is a little like human governments. While it makes good sense to divide responsibilities both geographically and by subject (i.e. local government and ministries), the various divisions must be integrated and this they can only be if they are interconnected, at the level of communicating, at the very least.

But not only must a cell integrate its various internal systems of regulation and control, this must be in a way that makes them responsive to the surrounding environment (Section 5.1), an environment that in multicellular organisms will contain other cells. Though each cell is to some extent an individual, in multicellular organisms there must be levels of regulation and control that transcend the cellular level, at the supracellular level, that allow integration of the activities of millions of similar cells in tissues, the integration of different tissues, of organs and the organism as a whole. The study of such levels of regulation and control is the province, not of the cell biologist, but of the physiologist and the developmental biologist. However, when you consider these subjects you will find some of the same motifs, the same basic concepts, as for regulation and control of intermediary metabolism: the importance of a correspondence between structure and function; feedback control; and, different levels of regulation and control.

Summary of Sections 5.5 to 5.7

1 The catalytic activity of certain regulatory enzymes can be controlled by reversible covalent modification.

2 (Glycogen) phosphorylase, the enzyme catalysing the 'release' of glucose residues from glycogen, can exist in two forms, *a* and *b*. Under most conditions inside skeletal muscle, phosphorylase *a* is active, while phosphorylase *b* is inactive.

3 Phosphorylase *a* bears a covalently attached phosphate group on each of the enzyme subunits; phosphorylase *b* does not. The two forms are interconvertible. The interconversion is catalysed by two other enzymes: *b* to *a* by phosphorylase kinase, and *a* to *b* by protein phosphatase 1. Phosphorylase kinase and protein phosphatase 1 are themselves subject to control.

4 Glutamine synthetase is a highly complex regulatory enzyme that is subject to both allosteric inhibition and reversible covalent modification. In reversible covalent modification, the chemical group covalently attached (and detached) is AMP; i.e. the enzyme can be in an adenylated or a non-adenylated form. The adenylated enzyme is more sensitive to allosteric inhibition than is its non-adenylated counterpart.

5 The covalent attachment and detachment of the AMP is catalysed by another enzyme, adenylyl transferase, which itself is subject to control.

6 In eukaryotes the division of the cell into different compartments affords yet another means of regulation and control. Different metabolic pathways, or parts of pathways, can be in different compartments; by controlling the flow of substances between compartments the cell may regulate the rate of output of metabolic pathways.

7 Metabolic pathways can be subject to a number of different levels of control. The controls operating on one enzyme, for example, can themselves be subject to controls; in turn, these controls can themselves be controlled, and so on. The net result is a cascade of controls. This helps a cell bring several different controls to bear on a single pathway and also increases the opportunity to integrate control systems operating on different pathways.

8 In a multicellular organism, other cells are part of an individual cell's environment. Thus in controlling and regulating its metabolism, each cell must be able to coordinate with other cells, to some extent at least. The study of these controls, operating at the supracellular level of tissues, organs and the organism as a whole, fall within the provinces of physiology and developmental biology, not that of cell biology. Nevertheless, some of the general principles underlying regulation and control that operate within a cell, such as feedback inhibition, also apply at these supracellular levels too.

Question 8 (*Objectives 5.1, 5.8 and 5.10*) Which of the following statements are true, and which are false?

(a) Reversible covalent modification requires two different enzymes; one to catalyse the addition of the modifying group, the other to catalyse its removal.

(b) An enzyme can be subject to either reversible covalent modification or allosteric control but not both.

(c) Enzymes catalysing the reversible covalent modification of other enzymes can themselves be subject to regulation and control.

Question 9 (*Objectives 5.1 and 5.9*) Which of the following statements are true and which are false?

(a) A particular metabolic pathway need not necessarily be solely within one compartment.

(b) If a metabolic pathway is split between two compartments, then regulation and control of the pathway can only occur at the compartment boundary.

(c) A metabolic pathway contained entirely within one compartment is only affected by regulation and control operating within that compartment.

SUMMARY OF CHAPTER 5

1 Cells are capable of regulating their metabolism to adjust to changes in their environment.

2 Regulation of metabolism overall depends on regulation and control of individual metabolic pathways.

3 Enzymes are the key targets for control in metabolic pathways.

4 The rate of an enzyme-catalysed reaction can be altered either by altering the amount of enzyme or by altering the catalytic activity of a fixed amount of enzyme.

5 Mechanisms that regulate and control the catalytic activity of a fixed amount of enzyme work much more quickly than those that alter the amount of enzyme.

6 In pathways subject to feedback inhibition the first step of the pathway is generally the control point and in branched pathways there are additional control points after the branch-point. The controls operate on the enzymes catalysing the reactions at these points; so-called regulatory enzymes.

7 Regulatory enzymes sensitive to allosteric inhibition and allosteric activation are known as allosteric enzymes.

8 Allosteric control involves an effector (allosteric activator or inhibitor) that binds to a separate site from the enzyme active site, the allosteric site.

9 Some enzymes, such as ATCase, can be sensitive to both allosteric activation and inhibition.

10   Allosteric enzymes also exhibit cooperation between their multiple active sites.

11   Both cooperativity and allosteric control depend on long-range interactions between sites, which, if all allosteric enzymes are like ATCase, are mediated by large-scale changes in the tertiary and/or quaternary structure of the enzyme.

12   Reversible covalent modification, which involves addition and removal of a chemical group from the enzyme protein, is another different mechanism of regulating and controlling the activity of an enzyme.

13   Compartmentation, whereby different metabolic pathways or parts of pathways are split between different eukaryote cell compartments, affords another means of regulating and controlling pathways. Notably the rate of flow of needed substrates (intermediates) across organelle membranes may be subject to regulation and control.

14   Regulation and control exists at several levels inside the cell and one level of control can control other levels of control. Integration of the different systems and the controls operating upon them is very important to the cell as a whole.

15   Other levels of control exist at the supracellular level—between groups of cells in tissues, tissues in organs and so on.

OBJECTIVES FOR CHAPTER 5

Now that you have completed this chapter you should be able to:

5.1 Define and use, or recognize definitions and applications of, each of the terms printed in **bold** in the text.

5.2 Recognize where a 'designer' approach is being used and what are the strengths and weaknesses of such an approach. (*Question 1*)

5.3 Distinguish between the two main ways of regulating and controlling metabolic pathways: altering the amount of enzyme present and altering the catalytic activity of a fixed amount of enzyme. (*Questions 2 and 3*)

5.4 Predict the control points for feedback inhibition or enzyme activation when shown linear or branched metabolic pathways. (*Question 4*)

5.5 Given metabolic pathways and their control points for feedback inhibition or enzyme activation, predict levels of various metabolites under a variety of given circumstances. (*Question 5*)

5.6 Distinguish the main features of allosteric control. (*Questions 6 and 7*)

5.7 Given data about real or hypothetical allosteric enzymes, distinguish between valid and invalid conclusions. (*Question 7*)

5.8 Distinguish between allosteric control and control by reversible covalent modification. (*Question 8*)

5.9 Identify the main features, consequences and advantages to a cell of compartmentation. (*Question 9*)

5.10 Define or describe or distinguish between types of regulation and control that operate at different levels. (*Question 8*)

ENERGY FROM GLUCOSE ◆ CHAPTER 6 ◆

6.1 INTRODUCTION

Even the simplest of organisms, such as, say, the single-celled bacterium *E. coli*, carries out many hundreds of different chemical reactions. Faced with such a display of chemical virtuosity, it is easy to be overwhelmed by the complexity of intermediary metabolism (Chapter 4, Section 4.1). Fortunately, some patterns can be seen within this complexity.

Firstly, despite the number of different reactions, all of them must obey the standard laws of organic chemistry; knowing these we can classify each of the reactions as belonging to one or other of a relatively few different types or classes of reaction.

Secondly, rather than viewing the reactions of intermediary metabolism as a complex hotch-potch, we can see how they in fact form a number of distinct sequences of reactions, a number of *metabolic pathways* (Chapter 4, Section 4.1) in which the product of one reaction is the reactant (substrate) of the next. Each metabolic pathway can thus be seen to be the way in which the cell can precisely manoeuvre the chemical bonds of a substance, step-by-step—via a sequence of intermediate substances—to produce, ultimately, the end-product of the pathway.

Thirdly, by considering metabolic pathways rather than isolated individual reactions, we can see that these pathways can themselves be classified into types. The simplest division distinguishes between *catabolic* and *anabolic* pathways. Catabolic pathways are degradative, breaking down complex molecules into simpler ones. Anabolic pathways are synthetic, treating simple molecules as precursors from which more complex molecules can be produced. The cell maintains a delicate balance between the two types of pathway, also in keeping with its interaction with its surrounding environment (e.g. availability of nutrients).

Finally, each reaction is enzyme-catalysed, this enabling the cell to carry out intermediary metabolism despite the mild conditions within the rapid time-frame in which cells operate, *and* avoiding chaos! It is their specificity, catalytic power and sensitivity to regulation and control that make enzymes the key to intermediary metabolism. And, as we have argued at some length (Chapters 4 and 5), by understanding these features of enzymes, we can understand how intermediary metabolism in general operates. Having invested some considerable time and space in elaborating this claim, it is now time to put it to some sort of test—to look at some actual metabolic pathways and show how an analysis of the enzymes within them helps us understand the ways in which these pathways work. It is this that constitutes one of the two main aims of this chapter.

The pathways we shall look at are catabolic ones, pathways that provide the cell with building blocks and, the main focus of our treatment, *energy* in the form of ATP. The building blocks and energy are both needed for synthetic, anabolic, pathways. Energy is also needed for a number of other cellular activities—for example, pumping of ions across membranes, muscular contraction, and, more exotic perhaps, bioluminescence (as in fireflies) or electric discharges (as in electric eels). The catabolic pathways can be seen to begin with sugars, fatty acids and amino acids which, in animals, are often derived from the digestion of polysaccharides, fats and proteins, respectively (as shown in Figure 6.1), obtained in the animal's diet. Plants and micro-organisms can synthesize organic building blocks from simple inorganic materials such as carbon dioxide (CO_2), and nitrogen (N_2) or nitrate (NO_3^-) or ammonium (NH_4^+) ions.

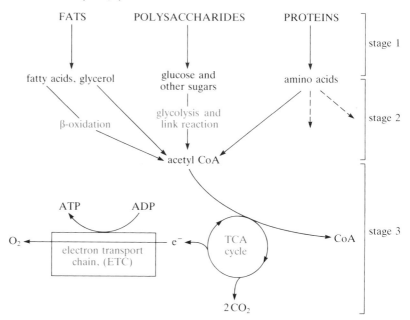

Figure 6.1 Outline of the three main stages in the extraction of energy from foodstuffs.

As Figure 6.1 shows, the first stage of energy (ATP) generation is the breaking down of large molecules in food into smaller units (sugars, fatty acids and amino acids). In the second stage, these small molecules, which we can call 'fuel' molecules, are degraded to a few simple units that play a central role in intermediary metabolism. The most significant of these is the acetyl unit of **acetyl CoA**, a very important intermediate in metabolism.

We are concerned with just stages 2 and 3. We shall trace the pathway taken from just one fuel molecule—arguably the most important—*glucose*, all the way to its final products, carbon dioxide and water. The first part of this journey is from glucose to acetyl CoA (stage 2 in Figure 6.1). From then on (stage 3 in Figure 6.1) we shall follow from acetyl CoA via the tricarboxylic acid (TCA) cycle and the electron transport chain (ETC)—the pathways common to the oxidation of all fuel molecules in aerobic cells.

Remember that our aim is not to consider all the details, but rather to see how the concepts developed in Chapters 4 and 5 concerning enzymes help an understanding of the general ways in which metabolic pathways operate, and these energy-generating pathways in particular. It is the concepts of Chapters 4 and 5, particularly those related to regulation and control, that you should look out for in our treatment of how energy is obtained from glucose. Last, but not least, the other aim of this chapter is to show how cells actually obtain energy from food in general, glucose in particular, and how this is carefully geared to the 'fuel' available and the cell's energy needs.

We start, then, with the *first* part of the route from glucose to carbon dioxide and water—that is, the conversion of glucose to pyruvate via the pathway known as **glycolysis**.

6.2 GLUCOSE TO PYRUVATE: GLYCOLYSIS

Glycolysis means literally 'dissolution of sugar'. All the reactions of glycolysis occur in the cytosol, whereas the subsequent metabolism of the end-product of glycolysis, pyruvate, occurs in the mitochondrion (Sections 6.3 to 6.5). First, we shall look at the reactions of glycolysis and the enzymes that catalyse them.

6.2.1 The reactions and enzymes of glycolysis

The reactions of glycolysis, together with the enzymes catalysing them, are summarized in Figure 6.2.

Once again, as we do not wish to dwell on the details of the reactions, it is convenient to treat the ten reactions of glycolysis as members of a smaller

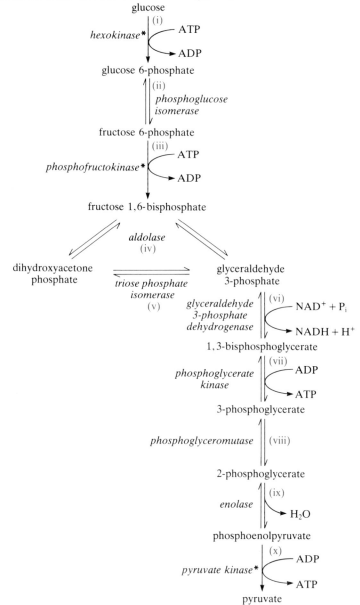

Figure 6.2 Summary of the glycolytic pathway for the conversion of glucose to pyruvate. Control points, where regulatory enzymes have their effect, are marked with a *black* asterisk.

number of types of reaction. As we mentioned in Chapter 4 (Section 4.2), enzymes are named after the reactions that they catalyse. So we can exploit this here and, in a sense, reverse the process to gain some clues as to the types of reactions occurring in glycolysis by examining the names of some of the enzymes shown in Figure 6.2. For example, a glycolytic enzyme may be a *phosphotransferase*, an *oxidoreductase* or an *isomerase*.

Phosphotransferases are enzymes that catalyse the transfer of phosphate groups. Often, ATP or ADP is required as phosphate (P_i) donor or acceptor, respectively. An enzyme that transfers a phosphate group from ATP to an acceptor molecule is called a **kinase**. You have already met the first kinase of glycolysis, hexokinase (Chapter 4, Section 4.3.3); the other three are phosphofructokinase, phosphoglycerate kinase and pyruvate kinase (reactions i, iii, vii and x in Figure 6.2).

Oxidoreductases are enzymes that catalyse oxidation–reduction reactions. Oxidation involves the addition of oxygen atoms or the removal of hydrogen atoms. However, the term oxidation is used in a more general sense to describe the removal of electrons, for example, the change of an Fe(II) ion to an Fe(III) ion. Reduction is the opposite of oxidation. Thus, reduction describes the removal of oxygen from a substance, or the addition of hydrogen to it or, more generally, any reaction in which an atom gains electrons. Oxidoreductases are known generally as **dehydrogenases** when hydrogen atoms are added to or removed from a coenzyme. As you can see from Figure 6.2 there is one dehydrogenase enzyme in the glycolytic pathway, glyceraldehyde 3-phosphate dehydrogenase (reaction vi). (Oxidoreductases are known as *oxidases* when molecular oxygen is directly involved; see Section 6.5.)

Isomerases are enzymes that catalyse isomerization reactions, that is, reactions involving molecular rearrangements but no net addition or removal of atoms. The three isomerases involved in glycolysis are phosphoglucose isomerase, triose phosphate isomerase and phosphoglyceromutase (catalysing reactions ii, v and viii respectively in Figure 6.2).

As glycolysis is a metabolic pathway, the enzymes act in sequence, with the product of the first enzyme-catalysed reaction becoming the substrate of the next, and so on. Glycolysis takes place in the cytosol (Chapter 2, Section 2.1). The enzymes involved are not grouped or arranged within a membrane-bound organelle, as is the case for some enzyme systems such as those responsible for the TCA cycle and photosynthesis (which are located in the mitochondria and chloroplasts respectively). It should be noted that most of the reactions of glycolysis are freely *reversible*; they also operate in the reverse direction from glycolysis in the pathway by which glucose is synthesized from pyruvate. However there are three reactions occurring during the breakdown of glucose to pyruvate that are regarded as essentially *irreversible*, namely those catalysed by hexokinase, phosphofructokinase and pyruvate kinase and these are important in the control of glycolysis (see Section 6.2.5). (When pyruvate is converted to glucose, in gluconeogenesis, these essentially irreversible steps of glycolysis are bypassed by the action of other enzymes.)

6.2.2 The role of the phosphate group in glycolysis

You will have noticed from Figure 6.2 that all of the intermediates in the glycolytic pathway contain phosphate groups; only the initial reactant, glucose, and the final product, pyruvate, are not phosphorylated compounds (phosphates). The phosphate group has a special significance because it is the group required for generating ATP from ADP. However another important property of phosphorylated compounds is that they cannot penetrate through cell membranes. Thus phosphates are in effect locked inside the cell and cannot diffuse out. If such phosphorylated intermediates could escape from the cell, the rate of glycolysis might become very small. Moreover, ATP and ADP, being phosphates, are also unable to pass through the cell membrane. In general, all phosphorylated compounds have a high density of (negative) electrical charge on their phosphate groups. The lipid-rich cell membrane resists passage of such highly charged solute molecules. It would seem, then, that the selection of phosphate compounds as metabolic intermediates during the course of evolution could have been related to both of their key properties; that is, their inability to cross membranes *and* their facilitation of the interconversion of ADP and ATP.

6.2.3 The role of nicotinamide adenine dinucleotide (NAD$^+$) in glycolysis

Cells derive energy from the oxidation of fuel molecules and in aerobic organisms the *ultimate* electron acceptor is oxygen. These fuel molecules transfer electrons to special carriers, and the reduced forms of these carriers then transfer electrons to oxygen via the electron transport chain (ETC) located in the inner membrane of the mitochondria (Section 6.5). **Nicotinamide adenine dinucleotide (NAD$^+$)** is an important coenzyme that acts as a major electron acceptor, or carrier, in the oxidation of fuel molecules. On oxidation the substrate gives up two hydrogen atoms. One of these is transferred to the nicotinamide ring of the coenzyme, the acceptor, NAD$^+$ (Figure 6.3), as are *two* electrons (i.e. both the electrons from the *two* hydrogen atoms lost by the substrate). The reduced coenzyme thus produced can be represented as NADH. The other hydrogen atom from the substrate, *minus* its electron (so it is a hydrogen ion or proton, H$^+$) is transferred direct to the solvent.

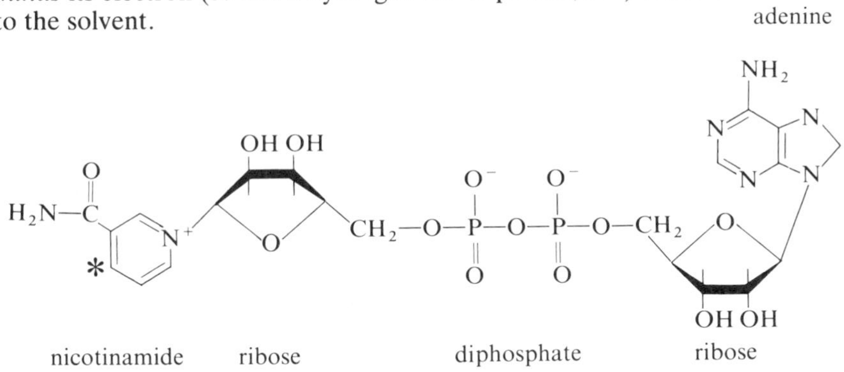

Figure 6.3 The structure of NAD$^+$, nicotinamide adenine dinucleotide. When NAD$^+$ is reduced to NADH, a hydrogen atom becomes attached at C-4 (marked *) of the nicotinamide ring. The other hydrogen atom donates its *electron* to the ring, allowing the N$^+$ to become N.

As a coenzyme, NAD$^+$ is the electron acceptor in many reactions of the type:

$$XH_2 + NAD^+ \rightleftharpoons X + NADH + H^+ \qquad (6.1)$$

NAD$^+$ occurs in all cells and its role is analogous to that of ATP, ATP being a phosphate carrier and NAD$^+$ an electron carrier. Note that as one of its building blocks NAD$^+$ contains the substance nicotinamide (see Figure 6.3). Nicotinamide is the amide of nicotinic acid, a vitamin of the B group, which must be present in the diet of vertebrates. Its absence from the human diet results in the deficiency disease pellagra (Chapter 4, Section 4.5.1).

◇ Can you see a possible reason for disease resulting from a lack of nicotinic acid in the diet?

◈ If this vitamin is not supplied then the cells are unable to form the complete NAD^+ molecule. As a result, in pellagra, there is a defect in certain enzymic reactions involving electron transfers to NAD^+ and from NADH.

In glycolysis, NAD^+ is the electron acceptor in the oxidation of glyceraldehyde 3-phosphate to 1,3-bisphosphoglycerate (reaction vi, Figure 6.2) and NADH is formed. So NAD^+ must be regenerated for glycolysis to proceed. In aerobic organisms this is achieved later, when the NADH transfers its electrons to oxygen via the electron transport chain (Section 6.5) which thereby regenerates NAD^+. Under anaerobic conditions, when the ETC is inoperative, NAD^+ is generated by the reduction of pyruvate to lactate. In some micro-organisms, NAD^+ is regenerated by the synthesis of lactate or ethanol from pyruvate, processes known as **fermentations**. These processes are described in more detail in Section 6.6.

6.2.4 Energy made available by the glycolytic pathway

The details of the reactions involved in the glycolytic pathway are shown in Figure 6.4. *In terms of carbon intermediates*, the net result of glycolysis is the conversion of glucose to pyruvate, but there are two other important conversions, that are linked to the catabolism of glucose to pyruvate.

◇ What are they (look at Figure 6.4)?

◈ The conversion of ADP to ATP and the conversion of NAD^+ to NADH.

The *net* reaction in the transformation of glucose into pyruvate is thus

$$\text{glucose} + 2P_i + 2ADP + 2NAD^+ \longrightarrow$$
$$2 \text{ pyruvate} + 2ATP + 2NADH + 2H^+ + 2H_2O \qquad (6.2)$$

To calculate the total number of ATP molecules that result from glycolysis, we need to add up the results of all the kinase-catalysed steps in the pathway, as shown in Table 6.1. The reason why the tally of NADH molecules resulting from the glycolytic pathway is also included in Table 6.1 will become evident later. In this and subsequent tables in this chapter, the reduced coenzymes (e.g. NADH) formed are shown in *blue*.

Table 6.1 ATP and NADH yields from glycolysis

Reaction	ATP yield per glucose
glucose → glucose 6-phosphate	−1
fructose 6-phosphate → fructose 1,6-bisphosphate	−1
2 (1,3-bisphosphoglycerate → 3-phosphoglycerate)	+2
2 (phosphoenolpyruvate → pyruvate)	+2
2 NADH formed by oxidation of 2 molecules of glyceraldehyde 3-phosphate	
net yield of ATP	+2

Note: Because *two* 'interconvertible' 3-carbon intermediates result from the aldolase-catalysed reaction (reaction iv in Figures 6.2 and 6.4), each reaction after this point occurs *twice* for every glucose molecule degraded. Thus the net yield of ATP from glycolysis is two molecules of ATP (i.e. $+4-2$) for each glucose molecule catabolized.

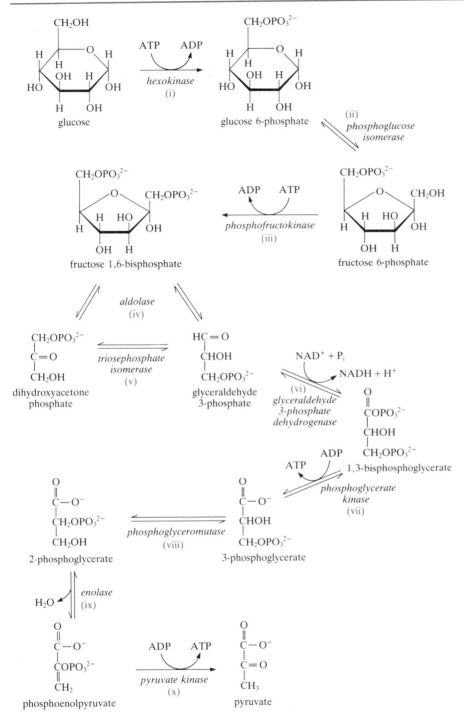

Figure 6.4 Details of the glycolytic pathway, showing the kinds of chemical transformation occurring at each step.

It may seem anomalous that during glycolysis two molecules of ATP should be cleaved to form ADP (reactions i and iii in Figures 6.2 and 6.4) for each molecule of glucose catabolized to pyruvate, when the importance of this pathway in *generating* ATP has been stressed. However, the input of two ATP molecules is simply a means of priming the system and four ATPs are formed later in the pathway, giving a *net* yield of two ATPs. Also, you may think that the glycolytic sequence is unnecessarily complicated when considering the end result. However, there are many reasons for concluding that it represents the simplest possible way in which the glucose molecule can be degraded at pH 7.0 in dilute solution, at normal atmospheric pressure and temperatures compatible with life, so that some of its energy content can be conserved in the form of ATP. The glycolytic pathway inherently contains, presumably through the 'trial and error' of evolution, highly developed chemical and engineering 'wisdom', which biochemists have really just begun to understand and appreciate.

Glycolysis involves two examples of *substrate level phosphorylation*, as distinct from *oxidative phosphorylation*. **Substrate level phosphorylation** describes the direct synthesis of ATP (or analogous compounds such as GTP; see Section 6.4.1) by transfer of a phosphate group (P_i) to ADP, *not* involving the electron transport chain (ETC). **Oxidative phosphorylation** is the formation of ATP coupled to the oxidation of NADH and reduced flavin adenine dinucleotide ($FADH_2$) by molecular oxygen via the electron transport chain (see Section 6.5).

◇ Which reactions of glycolysis involve substrate level phosphorylation?

◈ The conversion of 1,3-bisphosphoglycerate to 3-phosphoglycerate (reaction vii, Figure 6.4) and the conversion of phosphoenolpyruvate to pyruvate (reaction x, Figure 6.4).

6.2.5 Regulation and control of glycolysis

One role of glycolysis is the degradation of glucose to provide energy in the form of ATP. In aerobic cells, yet more ATP is synthesized when the end-product of glycolysis, pyruvate, is broken down further via the link reaction, TCA cycle and ETC, of which more later (Sections 6.3 to 6.5). Thus glycolysis is itself a producer of ATP *and* the starting point for other major ATP-producing pathways. As such a key element in energy production, it is not surprising that the rate of glycolysis is responsive to cellular energy needs—that the rate of glycolysis is geared to the general *energy level* in the cell.

One reflection of the energy level in the cell is the relative amount of ATP in the cell. When ATP is used up in energy-consuming processes its level naturally falls and that of ADP, the usual product of ATP utilization, rises. This decline in the cell's energy level is thus reflected in a fall in the ratio of ATP/ADP.

◇ Can you name some cellular energy-consuming processes?

◈ Numerous syntheses of chemicals and larger cellular components, performance of mechanical work by muscles, transport of ions and other compounds against concentration gradients (of which more in Chapter 7) or somewhat more exotic examples such as generation of bioluminescence (e.g. in fireflies and glow-worms) are among the many energy-consuming reactions that you may have considered.

However, a number of reactions are energy-yielding and generate ATP, helping raise the ATP/ADP ratio in the cell.

◇ Can you recall the energy-yielding (ATP-generating) reactions of glycolysis?

◈ Those catalysed by phosphoglycerate kinase and by pyruvate kinase (reactions vii and x, respectively, in Figures 6.2 and 6.4).

So ATP is recycled; the energy-consuming reactions *de*phosphorylate ATP to ADP, while those producing ATP do so by phosphorylating ADP. Some energy-consuming reactions convert ATP to AMP (adenosine *mono*phosphate) rather than to ADP. As AMP, ADP and ATP are interconvertible, ATP can also be recycled from AMP. So, in essence, the ratio of ATP/AMP reflects the general energy level in the cell, as does that of ATP/ADP.

◇ Do relatively low ATP/ADP and ATP/AMP ratios reflect a relatively high energy level in the cell or a relatively low one?

◆ A relatively low energy level.

By monitoring either the ratio of ATP/ADP or that of ATP/AMP the cell can, as it were, measure its energy level; further, by having control systems that respond to these ratios, the cell can act appropriately so as to maintain overall a relatively steady energy level (i.e. within a narrow range).

◇ Wearing your 'designer' hat (Chapter 5, Section 5.1), can you suggest what an appropriate response to a relatively low ATP/ADP or ATP/AMP ratio would be? And what would be appropriate for a relatively high ratio?

◆ A relatively low ATP/ADP or ATP/AMP ratio indicates that the energy level in the cell is relatively low; an appropriate response would be to step-up the rate of energy (i.e. ATP) production, by increasing the rate of operation of energy-yielding pathways. A relatively high ratio indicates that the cell is relatively energy-'rich'; it would therefore seem appropriate to decrease the rate of operation of energy-yielding pathways.

The regulation of glycolysis, gearing it to cellular energy level, essentially works in this way—a low cellular energy level (low ATP/ADP and ATP/AMP ratios) leading, directly or indirectly, to an increased rate of glycolysis, hence ATP synthesis and a restoration of the ATP/ADP and ATP/AMP ratios to within the maintenance range. Conversely, high ATP/ADP and ATP/AMP ratios tend to lead to a decrease in the rate of glycolysis. Subsequent usage of ATP, in energy-requiring processes, leading to a lowered ATP/ADP or ATP/AMP ratio, would once again stimulate glycolysis and effect increased ATP synthesis.

Though the precise details of how glycolysis is regulated in response to the ATP/ADP or ATP/AMP ratios are complex and vary somewhat among different organisms (notably between bacteria and higher organisms), many aspects are fairly universal. We shall concentrate on just some of the main aspects, chiefly as they operate in mammals. In particular, there are three *control points* of glycolysis and it is these that we shall examine.

◇ What would you expect those control points to be?

◆ From Chapter 5 (Sections 5.1 and 5.2) you might anticipate that they are enzymes.

The three enzyme-catalysed reactions involved in the regulation of glycolysis in response to the cell's energy level, that is the control points, those catalysed by the *regulatory enzymes*, are hexokinase, phosphofructokinase and pyruvate kinase (reactions i, iii and x, respectively, in Figures 6.2 and 6.4). As will become evident, though these enzymes operate in the order indicated (i, iii and x), it is logical to consider phosphofructokinase first.

Phosphofructokinase (PFK, for short) is the most important of the three regulatory enzymes. It catalyses the phosphorylation of fructose 6-phosphate to fructose 1,6-bisphosphate (reaction iii in Figures 6.2 and 6.4):

$$\text{fructose 6-phosphate} + \text{ATP} \xrightleftharpoons{\text{PFK}} \text{fructose 1,6-bisphosphate} + \text{ADP} \qquad (6.3)$$

Curiously, at high levels, ATP, which is a substrate of the enzyme, is also an *inhibitor* of the enzyme. So when the ATP level in the cell is high (i.e. ATP/ADP and ATP/AMP ratios are high) PFK is inhibited. Inhibition occurs by ATP binding to a site distinct from the active site (where it binds as a substrate).

◇ What form of inhibition does this seem to involve?

◈ Allosteric inhibition (Chapter 5, Section 5.4); the site where ATP binds as an inhibitor is the allosteric site.

So, when ATP is plentiful in the cell (high ATP/ADP and ATP/AMP ratios), PFK is inhibited, glycolysis is thus slowed and further ATP synthesis via glycolysis is therefore diminished.

However the inhibitory action of ATP on PFK is reversed by AMP, an *activator* of the enzyme (Chapter 5, Sections 5.2 and 5.4); ATP and AMP are antagonistic. So when the ATP/AMP ratio is lowered (indicating a lowering in cellular energy level), the activity of PFK increases, glycolysis steps up and ATP synthesis is increased. PFK thus effectively 'monitors' the energy level of the cell and responds to this accordingly, the two things being achieved by its, opposing, sensitivity to both ATP and AMP—that is, it responds overall to the ATP/AMP ratio.

The regulation of PFK also illustrates one other aspect of glycolysis and its control. As well as being important in provision of ATP, glycolysis helps provide carbon compounds that act as precursors for synthesis in other pathways. So we might anticipate that glycolysis is also regulated by signals that indicate whether such precursors are plentiful or not and that such regulation helps adjust the state of supply of those building blocks. Indeed PFK is inhibited by *citrate*, an intermediate of the TCA cycle (Section 6.4). This citrate inhibition of PFK enhances that by ATP. Note that the end-product of glycolysis, pyruvate, is fed into the TCA cycle (via acetyl CoA; Figure 6.1, and Sections 6.3 and 6.4 later). Hence a high level of citrate indicates that the TCA cycle, an important source of carbon-compound building blocks, is active and hence the cell need not degrade yet more glucose, via glycolysis, for their synthesis.

Thus overall PFK is most active when the cell is in need of both energy and building blocks, as signalled by a low ATP/AMP ratio and a low level of citrate. The enzyme is moderately active when either further building blocks or energy are needed. When both are abundant, the activity of PFK is almost switched off.

Hexokinase, the first enzyme of glycolysis (reaction i in Figures 6.2 and 6.4), is also a regulatory enzyme. It catalyses the phosphorylation of glucose by ATP, and its activity is inhibited by glucose 6-phosphate.

◇ What other relationship does glucose 6-phosphate have to hexokinase?

◈ It is a product of the reaction catalysed by hexokinase:

$$\text{glucose} + \text{ATP} \xrightarrow{\text{hexokinase}} \text{glucose 6-phosphate} + \text{ADP} \qquad (6.4)$$

Thus high levels of the product of the reaction, glucose 6-phosphate, inhibit the enzyme. Elevated levels of glucose 6-phosphate are likely to occur as a consequence of the inhibition of PFK.

◇ Under what circumstance will that happen, at least as concerns energy metabolism?

◈ When the cellular ATP/AMP ratio is high; i.e. ATP is abundant.

◇ Why should inhibition of PFK lead to high levels of glucose 6-phosphate?

◈ Inhibition of PFK leads to a build-up of its substrate, fructose 6-phosphate (see Equation 6.3, and reaction iii in Figures 6.2 and 6.4). But fructose 6-phosphate is the *product* of the preceding reaction in the pathway (reaction ii in Figures 6.2 and 6.4), that catalysed by phosphoglucose isomerase:

$$\text{glucose 6-phosphate} \underset{\text{isomerase}}{\overset{\text{phosphoglucose}}{\rightleftharpoons}} \text{fructose 6-phosphate} \qquad (6.5)$$

A build-up of fructose 6-phosphate causes a rise in the level of the substrate of that reaction, *glucose 6-phosphate* (see Equation 6.5), as the equilibrium of this readily reversible reaction shifts towards the left.

This rise in the level of glucose 6-phosphate, in turn, causes inhibition of hexokinase. Thus hexokinase responds to the cellular energy level (as indicated by the ATP/AMP ratio) indirectly, the direct response being an effect on the activity of PFK.

The last enzyme of the glycolytic pathway, *pyruvate kinase* (reaction x in Figure 6.2 and 6.4), is also a regulatory one. It catalyses the conversion of phosphoenolpyruvate to pyruvate, a step that, importantly, is also an ATP-yielding one:

$$\text{phosphoenolpyruvate} + \text{ADP} \overset{\overset{\text{pyruvate}}{\text{kinase}}}{\longrightarrow} \text{pyruvate} + \text{ATP} \qquad (6.6)$$

The regulation of this enzyme is complex, there being (in mammals) three isoenzymes with differing means of control (Chapter 4, Section 4.8 and Chapter 5, Section 5.3.1). To keep matters relatively simple, we shall just consider one of these isoenzymes, PK1, that found mainly in the liver.

PK1 is allosterically inhibited by ATP.

◇ How does this tie in with the 'logic' of regulating energy metabolism?

◈ When the cellular energy level is high (high ATP/ADP and ATP/AMP ratios), the high level of ATP inhibits PK1, slows glycolysis and hence further ATP synthesis.

On the other hand, fructose 1,6-bisphosphate, the product of an earlier step in glycolysis (that catalysed by PFK; Equation 6.3 and reaction iii in Figures 6.2 and 6.4), *activates* PK1. This thereby ensures that, if glycolysis is proceeding very actively, PK1 is activated, enabling it to cope with the high level of intermediates from the earlier steps.

PK1 is also allosterically inhibited by alanine, a building block for protein synthesis derived from pyruvate.

◇ What is the 'logic' of this inhibition?

◈ As with the inhibition of PFK by citrate, the inhibition of PK1 by alanine serves to slow down glycolysis when building blocks are plentiful.

Thus this isoenzyme of pyruvate kinase (PK1), like PFK, is under dual control of both energy level and levels of building blocks. Not only that, but PK1 is also subject to reversible covalent modification by phosphorylation (Chapter 5, Section 5.5.1), this in turn being under hormonal control and linked thereby to the level of glucose in the blood. And PK1 is only one of three pyruvate kinase isoenzymes!

Given that we have discussed only some of the major aspects of regulation and control of glycolysis, you can appreciate how complex the variety of control systems affecting that pathway are. And we probably do not even know all the subtleties involved. This is borne out by studies on 'reconstructing' glycolysis outside the living cell. Each of the enzymes involved in glycolysis has been crystallized, and by using these highly purified enzymes it has been possible to reproduce parts of the glycolytic sequence *in vitro*. Such reconstructive approaches have enabled the mechanism and dynamics of glycolysis to be examined in detail. More recently, metabolic pathways have also been 'reconstructed' by using computer modelling, and the results suggest that the control of glycolysis is very complex; much more complex than has been our description.

However, overall, several important general points should be evident. Firstly, some degree of control of the rate of glycolysis will be afforded by the amount of glucose available to the pathway—the rate of the first, hexokinase-catalysed, reaction will depend on the amount of glucose present, as will subsequent steps depend on the concentrations of their substrates. This 'passive' control, merely acting out the so-called *law of mass action* (that the rate of a reaction depends on the concentration of reactants), is however overlayed by more active, specific and subtle controls dependent on such mechanisms as allosteric activation and inhibition and reversible covalent modification. As you have seen, these make the rate of glycolysis sensitive to the cellular energy level (as reflected in ATP/ADP and ATP/AMP ratios) and the cellular levels of important precursors. Furthermore, these active controls also serve to integrate the rate of glycolysis with that of other pathways important in production or utilization of energy and of building blocks (e.g. citrate produced in the TCA cycle affects the rate of glycolysis via its effect on PFK). As glycolysis is such a central pathway in the provision of both energy and precursors, this integration is obviously advantageous. As the example of the hormonal control of pyruvate kinase (making it responsive to blood sugar level) shows, this integration of glycolysis with other systems can extend beyond the confines of the individual cell, demonstrating yet again the vital role of energy metabolism and its control—and the central place of glycolysis in this—to the organism as a whole.

Summary of Section 6.2

1 In aerobically metabolizing cells, glucose is completely oxidized to carbon dioxide and water. This occurs via glycolysis, the link reaction, the tricarboxylic acid (TCA) cycle and the electron transport chain (ETC).

2 In glycolysis, which can also occur under anaerobic conditions, glucose is degraded to pyruvate. Each molecule of glucose yields two molecules of pyruvate, and there is also a net gain of energy in the form of two molecules of ATP, produced via substrate level phosphorylation.

3 As well as providing energy (as ATP), glycolysis also helps provide precursors for biosynthesis. Glycolysis is sensitive to these two 'outputs'—energy and precursors—and is regulated by the cellular level of each of them.

4 In mammals, phosphofructokinase (PFK) is the central regulatory enzyme of glycolysis: it is sensitive to the cellular energy level in the form of the ratio of ATP to AMP, and to the level of precursors, as indicated by the level of citrate (a TCA cycle intermediate). This also serves to link regulation of glycolysis to the TCA cycle.

5 Hexokinase and pyruvate kinase are also regulatory enzymes of glycolysis. The former responds to the cellular energy level indirectly, in being inhibited as a consequence of the inhibition of PFK. The regulation of the latter, pyruvate kinase, is complex. Pyruvate kinase comprises three isoenzymes one of which, PK1, is subject to both allosteric regulation, by energy level (ATP), level of inputs to glycolysis (indicated by fructose 1,6-bisphosphate) and level of precursor (alanine); and reversible covalent modification, dependent on blood glucose level.

6 Another product of glycolysis is NADH, generated from NAD^+ in the one oxidation reaction of the pathway, that catalysed by glyceraldehyde 3-phosphate dehydrogenase.

Question 1 (*Objectives 6.1, 6.2 and 6.3*) Which glycolytic enzyme is an oxidoreductase? Name the coenzyme that is reduced in the reaction catalysed by this enzyme.

Question 2 (*Objective 6.1, 6.2 and 6.10*) There is a net yield of two ATPs from the catabolism of glucose to pyruvate, whereas there are *two* reactions in the glycolytic pathway *each* of which yields two ATPs. Account for this apparent anomaly.

Question 3 (*Objectives 6.1, 6.2 and 6.5*) Which three glycolytic enzymes are subject to regulation? The rate of glycolysis can be adjusted in accordance with the cell's needs for energy and precursors for biosynthesis. Explain, with reference to the three regulatory enzymes, how this is achieved.

6.3 THE LINK REACTION

As we have mentioned, in aerobic cells, glycolysis is the first pathway in the overall oxidation of glucose to carbon dioxide and water. The next step, *linking* glycolysis to the TCA cycle, is the conversion of pyruvate to acetyl CoA. This, the so-called, **link reaction** takes place in the mitochondrial matrix (see Chapter 2, Section 2.8) and may be represented as:

$$\text{pyruvate} + \text{CoA} + NAD^+ \longrightarrow \text{acetyl CoA} + CO_2 + NADH + H^+ \tag{6.7}$$

It is by this essentially irreversible reaction that the product of glycolysis, pyruvate, is converted into the starting substance of the TCA cycle, acetyl CoA. The reaction is catalysed by pyruvate dehydrogenase. Pyruvate dehydrogenase (PDH) is not a single enzyme but is an organized assembly (a complex) of *three* different enzymes, and as such is an example of a **multi-enzyme complex** (see also Chapter 3, Section 3.4.4). The complex contains multiple copies of the three separate enzymes, each with its own specific binding sites for substrates and coenzymes. Enzyme E_a of the complex binds pyruvate and thiamine pyrophosphate (TPP), enzyme E_b binds lipoic acid and coenzyme A, and enzyme E_c binds FAD and NAD^+. The overall scheme of the PDH-catalysed reaction is shown in Figure 6.5.

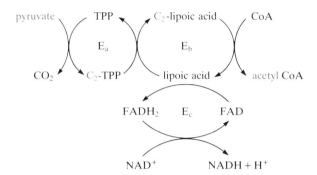

Figure 6.5 Summary of the sequence of reactions catalysed by the pyruvate dehydrogenase complex. 'C_2' denotes the 2-carbon fragment that is ultimately transferred to coenzyme A (CoA) to form acetyl CoA.

6.3.1 Role of the coenzymes TPP, lipoic acid and FAD in the link reaction

During the conversion of pyruvate to acetyl CoA there is an *oxidation* and the concomitant *removal of carbon dioxide*. This *oxidative decarboxylation* of pyruvate is catalysed by enzyme E_a of the PDH complex, when its TPP (thiamine pyrophosphate) coenzyme becomes covalently linked to pyruvate. A $C-C$ bond of pyruvate breaks, releasing carbon dioxide and leaving a 2-carbon fragment attached to TPP (see Figure 6.5). The 2-carbon fragment is removed from TPP on E_a (which then returns to its original form ready for further reaction) and moved to the coenzyme lipoic acid on E_b. From here the 2-carbon fragment passes to coenzyme A (CoA), forming acetyl CoA, which is released into the medium of the mitochondrial matrix as the major product of the PDH-catalysed reaction. The remainder of the overall reaction is concerned solely with coenzyme regeneration, that is, the oxidation of $FADH_2$ back to FAD (catalysed by enzyme E_c; Figure 6.5).

6.3.2 The multi-enzyme nature of pyruvate dehydrogenase

The PDH complex is a multi-enzyme complex that catalyses sequential steps in what may be regarded as a very short metabolic pathway. Although the individual components, E_a, E_b and E_c, do remain catalytically active when separated out from the complex, they contribute far more efficiently to the overall reaction by being linked together. The structural integration of three kinds of enzymes makes possible the coordinated catalysis of a complex reaction. This arrangement allows the product of one enzyme-catalysed reaction to be passed immediately to the next enzyme, without moving off into the medium of the mitochondrial matrix. In many respects, therefore, the PDH complex resembles a membrane-bound metabolic pathway like the mitochondrial electron transport chain (ETC) (see Section 6.5), rather than one catalysed by a sequence of soluble enzymes (e.g. glycolysis).

6.3.3 Regulation of the link reaction

The rate of production of acetyl CoA from pyruvate in the link reaction is subject to several controls, of which we shall consider just some. For example, two of the products of the reaction, acetyl CoA and NADH, both inhibit pyruvate dehydrogenase and this inhibition is overcome by CoA and

NAD^+, respectively. The enzyme is also sensitive to the energy level in the cell, since it is inhibited by GTP (which reflects the level of ATP) and activated by AMP. In addition, it is regulated by reversible covalent modification involving a phosphate group (Chapter 5, Section 5.5.1). Phosphorylation (to the inactive form of the enzyme complex) is enhanced by a high energy level (as ATP/ADP) as well as high ratios of acetyl CoA/CoA and $NADH/NAD^+$.

◇ What is the 'logic' of these controls on the degree of phosphorylation of pyruvate dehydrogenase?

◆ High ratios of ATP/ADP, acetyl CoA/CoA and $NADH/NAD^+$ are all indicative of a high cellular energy level, active glycolysis and link reaction. As these stimulate phosphorylation of pyruvate dehydrogenase, and the phosphorylated enzyme is *inactive*, this will lead to reduced activity of the link reaction, leading, in turn, to a reduction in further metabolism of pyruvate and hence slowing down of ATP production.

On the other hand, pyruvate inhibits phosphorylation of pyruvate dehydrogenase.

Thus there are several key metabolites whose regulatory effects on the rate of the link reaction ensure that acetyl CoA, the common intermediate in the catabolism of many fuel molecules, is maintained at a level consistent with the metabolic needs of the cell.

6.4 THE TRICARBOXYLIC ACID (TCA) CYCLE

The discovery of the **tricarboxylic acid (TCA) cycle** has an interesting history. In the mid-1930s Albert Szent-Györgyi found that certain organic acids occurring in cells—succinate, fumarate and malate—each of which contains four carbon atoms and two carboxyl (−COOH) groups, are vigorously oxidized by suspensions of minced skeletal muscle. He also discovered that these substances greatly stimulate the oxidation of glucose by muscle and for each molecule of organic acid added to the muscle, many molecules of glucose are oxidized. A few years later Hans Krebs surveyed a variety of other naturally occurring dicarboxylic and tricarboxylic organic acids to see whether they also showed this 'catalytic' effect on the oxidation of glucose. He made the important observation that the 6-carbon tricarboxylic acids, citrate and isocitrate, and a 5-carbon dicarboxylic acid, α-oxoglutarate, were capable of stimulating the oxidation of glucose in muscle suspensions. No other related organic acids produced this effect.

Krebs concluded that those organic acids that did show the effects must be interrelated and participate together in a metabolic sequence. He proved that each enzymic step takes place in muscle at a rate consistent with the rate of respiration. He also showed that these reactions function as a *cycle*, and thus differ from those of glycolysis which take place as a *linear* sequence. A key observation supporting the existence of a cyclic series of reactions was that oxaloacetate, the *last* intermediate in the sequence, reacted enzymically with acetyl CoA to make citrate, the *first* substance, in a step that joined the 'head' and 'tail' of the sequence:

$$\text{oxaloacetate} + \text{acetyl CoA} \longrightarrow \text{citrate} + \text{CoA} \qquad (6.8)$$

The citrate thus formed was oxidized again to oxaloacetate, which could react with another molecule of acetyl CoA to start another cycle. In each cycle two molecules of carbon dioxide appeared for the input of one molecule of acetyl

CoA (produced from pyruvate via the link reaction). The catalytic effect of the various 6-carbon, 5-carbon and 4-carbon acids on oxidation of fuel molecules could then readily be explained by assuming that they are all intermediates in the cycle. This cycle, called the tricarboxylic acid (TCA) cycle, the citric acid cycle, or simply the Krebs cycle, has been shown to occur in almost all aerobic heterotrophic and autotrophic cells, and to be the major route of oxidative metabolism of all foodstuffs. That the cycle actually operates in intact cells has been verified by means of experiments using radioactively labelled intermediates.

Figure 6.6 gives details of the TCA cycle. The cycle can be summarized as follows.

A 4-carbon compound (oxaloacetate) condenses with a 2-carbon acetyl unit (as acetyl CoA) to produce a 6-carbon tricarboxylic acid (citrate). This is reaction i in Figure 6.6. Citrate is isomerized to isocitrate (reaction ii), which is then decarboxylated (reaction iii). The resulting 5-carbon compound (α-oxoglutarate) is decarboxylated to yield the 4-carbon compound succinate (reactions iv and v). Oxaloacetate is then regenerated from succinate (reactions vi, vii and viii).

Figure 6.6 Details of the tricarboxylic acid (TCA) cycle. The light blue boxes are used to show the fate of the acetyl group as it passes once round the cycle. Notice that neither of the two molecules of carbon dioxide produced originate from this group during its first turn of the cycle. The source of each of the carbon dioxide molecules is shown by the grey boxes.

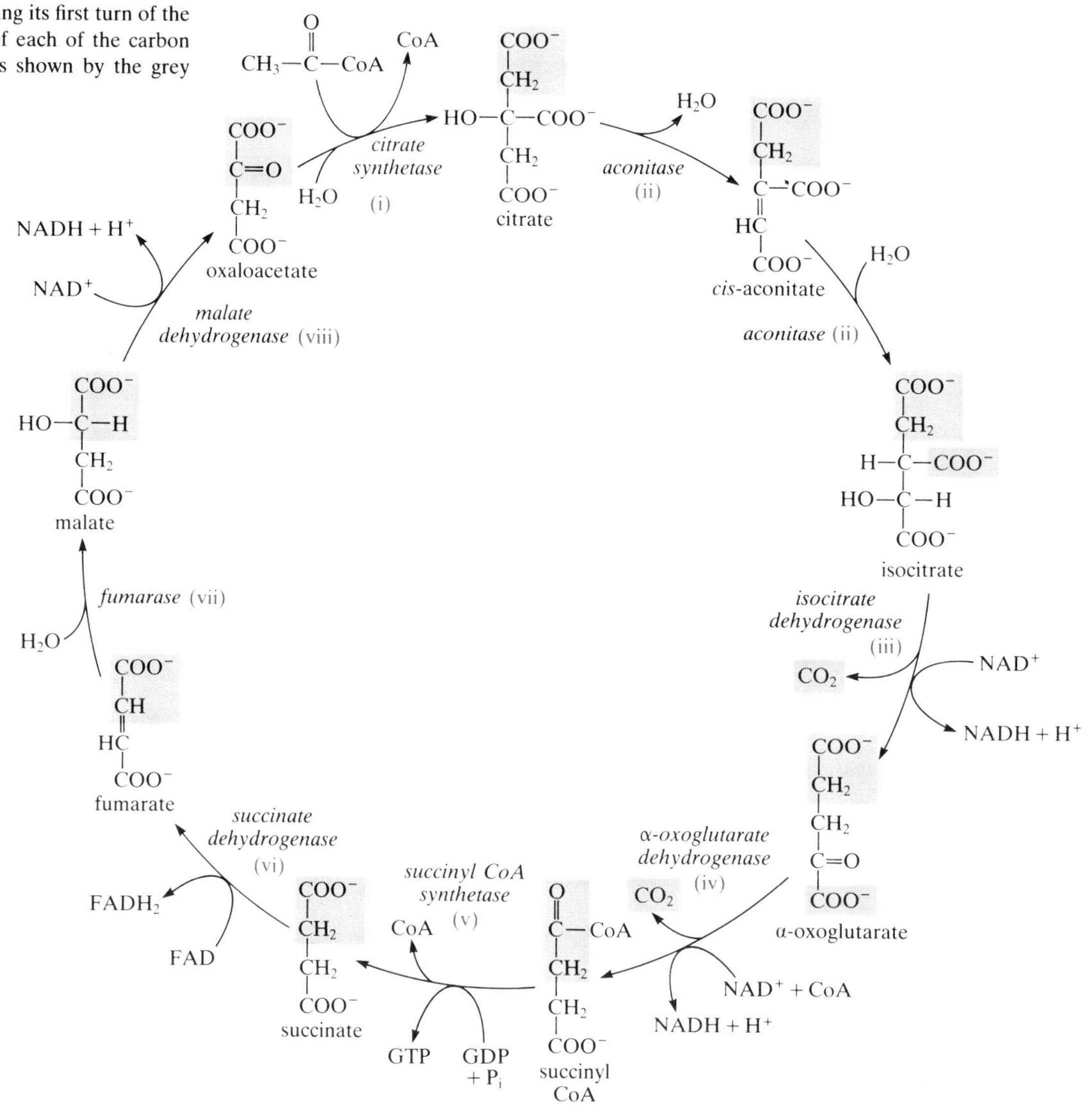

Thus, overall, two carbon atoms enter the cycle as an acetyl unit (acetyl CoA) and two carbon atoms leave the cycle in the form of two molecules of carbon dioxide. In addition three pairs of hydrogen atoms reduce three NAD^+ molecules to form three NADHs (plus three H^+ ions, of course), and one pair of hydrogen atoms is transferred to a flavin adenine dinucleotide (FAD) molecule. As you will see later, these electron carriers yield a total of eleven molecules of ATP when they are oxidized subsequently by oxygen via the electron transport chain (ETC). A further 'high-energy' phosphate bond is formed in each round of the TCA cycle itself, in the form of GTP (reaction v). These processes are summarized in Figure 6.7.

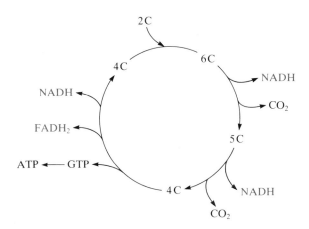

Figure 6.7 A summary diagram of the TCA cycle showing the outputs (reduced coenzymes, GTP and carbon dioxide) in relation to the sequence of reactions of the cycle.

As described in Chapter 5, Section 5.6, the metabolic patterns of eukaryotic cells are markedly affected by *compartmentation* of those cells. Glycolysis takes place in the cytosol whereas the link reaction, TCA cycle and oxidative phosphorylation occur in the mitochondria. This compartmentation of the major pathways of metabolism is summarized in Figure 6.12 (see later). Because mitochondria are the sites of the enzyme systems that participate in the oxidation of foodstuffs by molecular oxygen and the recovery of the energy of oxidation as ATP, they are often called the 'powerhouses' of eukaryotic cells. In prokaryotic cells, which do not have mitochondria or internal membranes (Chapter 2, Section 2.13), the enzymes that catalyse the oxidation of foodstuffs are located in the cell membrane.

6.4.1 Reactions of the TCA cycle and the enzymes involved

In the first step of the TCA cycle, the acetyl group from acetyl CoA becomes linked to oxaloacetate (Equation 6.8) in a reaction catalysed by citrate synthetase. In this reaction free CoA is regenerated and the 6-carbon tricarboxylic acid citrate is formed. In the next reaction (reaction ii, Figure 6.6), citrate is isomerized into isocitrate by the interchange of a hydrogen atom and an hydroxyl ($-OH$) group. As shown in Figure 6.6, this reaction occurs in two steps and is catalysed by the enzyme aconitase.

The next reaction is the first of the four dehydrogenation reactions in the TCA cycle. This is the oxidative decarboxylation (Section 6.3.1) of isocitrate to α-oxoglutarate (reaction iii, Figure 6.6), catalysed by isocitrate dehydrogenase:

$$\text{isocitrate} + NAD^+ \xrightleftharpoons[\text{dehydrogenase}]{\text{isocitrate}} \alpha\text{-oxoglutarate} + CO_2 + NADH + H^+ \qquad (6.9)$$

The molecule of carbon dioxide generated in this reaction is the first of the two that are produced in each turn of the TCA cycle. There are in fact two isoenzymes (Chapter 4, Section 4.8) of isocitrate dehydrogenase; one specific for NAD^+, the other for the related coenzyme, nicotinamide adenine dinucleotide phosphate ($NADP^+$). The NAD^+-specific enzyme is confined to the mitochondria and is the important one for the TCA cycle (Equation 6.9). However, the $NADP^+$-specific enzyme is present both in mitochondria *and* in the cytosol and has a different metabolic role.

There then follows the second oxidative decarboxylation reaction of the cycle—the formation of succinyl CoA from α-oxoglutarate (reaction iv, Figure 6.6).

$$\text{α-oxoglutarate} + NAD^+ + CoA \xrightleftharpoons[\text{dehydrogenase}]{\text{α-oxoglutarate}} \text{succinyl CoA} + CO_2 + NADH + H^+ \quad (6.10)$$

This reaction is catalysed by the α-oxoglutarate dehydrogenase complex, a multi-enzyme complex of three different enzymes. This is a very similar reaction to the conversion of pyruvate into acetyl CoA (Section 6.3). It involves the same coenzymes: NAD^+, CoA, TPP, lipoic acid and FAD, and the same sequence of steps.

◇ From Figure 6.6, compare the nature of the groups attached to CoA in the PDH reaction and in the α-oxoglutarate dehydrogenase reaction by writing out in full the RCO− parts of the two RCO−CoA formulae.

▸ In the PDH reaction the product is acetyl CoA, $CH_3-CO-CoA$; in reaction iv of the TCA cycle the product is succinyl CoA, $^-OOC-CH_2-CH_2-CO-CoA$.

The subsequent cleavage of succinyl CoA to form succinate and CoA is coupled to the formation of a molecule of guanosine triphosphate (GTP) from guanosine diphosphate (GDP) and P_i (reaction v, Figure 6.6).

$$\text{succinyl CoA} + P_i + GDP \xrightleftharpoons[\text{synthetase}]{\text{succinyl CoA}} \text{succinate} + GTP + CoA \quad (6.11)$$

This reaction is catalysed by succinyl CoA synthetase. The terminal phosphate group in GTP can be subsequently transferred to ADP to form ATP. This reaction is catalysed by nucleoside diphosphokinase:

$$GTP + ADP \xrightleftharpoons[\text{diphosphokinase}]{\text{nucleoside}} GDP + ATP \quad (6.12)$$

The production of GTP in Equation 6.11 is the one example of substrate level phosphorylation in the TCA cycle. You have already met substrate level phosphorylation in two reactions of glycolysis—the conversion of 1,3-bisphosphoglycerate to 3-phosphoglycerate and the conversion of phosphoenolpyruvate to pyruvate (Section 6.2.4). As you will recall, the contrasting process is oxidative phosphorylation—ATP synthesis coupled to the oxidation of NADH and $FADH_2$ by oxygen.

The last three reactions of the TCA cycle are all transformations involving 4-carbon intermediates (reactions vi, vii and viii, Figure 6.6):

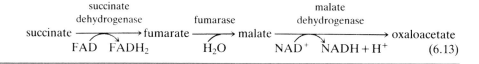

$$(6.13)$$

Succinate is oxidized to fumarate by succinate dehydrogenase. Notice that FAD is the hydrogen acceptor in this reaction, not NAD^+, which is used in the other three oxidation reactions in the cycle. Succinate dehydrogenase differs from the other TCA cycle enzymes in being an integral part of the inner mitochondrial membrane. (The others are all soluble enzymes within the mitochondrial matrix.) In fact, this enzyme is directly linked to the electron transport chain (Section 6.5). The $FADH_2$ produced by the oxidation of succinate remains bound to the enzyme (in contrast with NADH, which is freely soluble), and two electrons from $FADH_2$ are thereby transferred *directly* to the electron transport chain. (The ultimate acceptor of electrons from both NADH and $FADH_2$ is molecular oxygen, as will be discussed in Section 6.5.) The penultimate step in the TCA cycle is addition of $-H$ and $-OH$, i.e. water, to fumarate to form malate. This reaction is catalysed by fumarase. The final reaction is the oxidation of malate to oxaloacetate—the starting point of the cycle. This reaction is catalysed by malate dehydrogenase and NAD^+ is the hydrogen acceptor (Equation 6.13).

The overall reaction of the TCA cycle, *plus* the link reaction, may be summed up as shown in Figure 6.8.

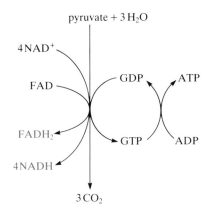

Figure 6.8 The overall result of the link reaction plus the TCA cycle.

◇ What is the total yield of reduced coenzymes per acetyl group passing round the TCA cycle?

◆ Three molecules of NADH and one molecule of $FADH_2$. (Remember that a *further* molecule of NADH comes from the link reaction.)

6.4.2 Energy made available via the TCA cycle

The overall reaction of the TCA cycle (without the link reaction) is

$$\text{acetyl CoA} + 3NAD^+ + FAD + P_i + GDP + 2H_2O \longrightarrow$$
$$2CO_2 + 3(NADH + H^+) + FADH_2 + GTP + CoA \quad (6.14)$$

It may be summarized in words as follows:

1 Two carbon atoms enter the cycle when an acetyl unit (from acetyl CoA) combines with oxaloacetate. Two carbon atoms are lost from the cycle, as carbon dioxide, via two decarboxylations.

◇ What are the names of the enzymes that catalyse these decarboxylation reactions? (Refer to Figure 6.6.)

◆ Isocitrate dehydrogenase (reaction iii, Figure 6.6) and α-oxoglutarate dehydrogenase (reaction iv, Figure 6.6).

2 One molecule of ATP is synthesized by a substrate level phosphorylation (via the formation of GTP) from succinyl CoA (reaction v, Figure 6.6).

3 Four pairs of hydrogen atoms are removed from the cycle in four oxidation–reduction reactions.

◇ What are the coenzymes involved in these oxidation–reduction reactions? (Refer to Figure 6.6.)

◈ One NAD^+ molecule is reduced (to NADH) in each of the oxidative decarboxylations, of isocitrate and α-oxoglutarate (reactions iii and iv, Figure 6.6), one FAD molecule is reduced (to $FADH_2$) in the oxidation of succinate (reaction vi, Figure 6.6), and one NAD^+ molecule is reduced in the oxidation of malate (reaction viii, Figure 6.6).

4 Two water molecules are used: one in the synthesis of citrate, and the other in the formation of malate from fumarate (reactions i and vii, respectively, in Figure 6.6).

The yields of ATP, NADH and $FADH_2$ from the passage of one molecule of glucose through the glycolytic pathway, the link reaction, and round the TCA cycle are shown in Table 6.2. As in Table 6.1, yields of reduced coenzymes (here NADH and $FADH_2$) are shown in blue.

Table 6.2 ATP, NADH and $FADH_2$ yields from glycolysis, the link reaction and the TCA cycle

Reaction	ATP yield per glucose
Glycolysis (in cytosol)	
glucose → glucose 6-phosphate	−1
fructose 6-phosphate → fructose 1,6-bisphosphate	−1
2(1,3-bisphosphoglycerate → 3-phosphoglycerate)	+2
2(phosphoenolpyruvate → pyruvate)	+2
2NADH formed by oxidation of 2 molecules of glyceraldehyde 3-phosphate	
Link reaction (in mitochondria)	
2NADH formed by conversion of pyruvate to acetyl CoA	
TCA cycle (in mitochondria)	
2 molecules of GTP formed from 2 molecules of succinyl CoA (convertible to ATP: Equation 6.12)	+2
6NADH formed in the oxidation of 2 molecules each of isocitrate, α-oxoglutarate and malate	
2FADH$_2$ formed in the oxidation of 2 molecules of succinate	
net yield of ATP	+4

Recall that each molecule of glucose yields two molecules of pyruvate (Section 6.2.4), and hence two of acetyl CoA.

6.4.3 The TCA cycle as a source of biosynthetic intermediates

We have seen how the TCA cycle enables the *energy* from the oxidation of acetyl units (from acetyl CoA) to be stored in the form of the reduced coenzymes NADH and $FADH_2$ (prior to their oxidation via the ETC to generate ATP). However, another important role of the TCA cycle is to provide *intermediates* for biosynthesis, as shown in Figure 6.9. For example,

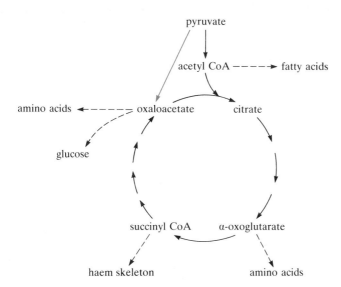

Figure 6.9 Biosynthesis via the TCA cycle. Intermediates drawn off for use in biosynthesis (shown by dashed arrows) can be replenished by converting pyruvate to oxaloacetate (blue arrow).

many amino acids are derived from α-oxoglutarate and oxaloacetate. But the intermediates of the TCA cycle are not generated (or consumed) by the reactions of the cycle itself. As Krebs found, they act in a catalytic manner. So *if* any are drawn off for biosynthesis they must be replenished.

Consider, for example, the consequences of the conversion of oxaloacetate into amino acids for protein synthesis. The TCA cycle would soon cease to operate unless oxaloacetate (or any other TCA cycle intermediate, which can regenerate oxaloacetate) is formed *de novo*, because acetyl CoA cannot enter the cycle unless it combines with oxaloacetate. But how is oxaloacetate replenished? Mammals are unable to convert acetyl CoA into oxaloacetate, or into any other TCA cycle intermediate. Instead, oxaloacetate is formed by the carboxylation of the end-product of glycolysis, pyruvate. This reaction is catalysed by pyruvate carboxylase:

$$\text{pyruvate} + CO_2 + ATP + H_2O \xrightarrow[\text{carboxylase}]{\text{pyruvate}} \text{oxaloacetate} + ADP + P_i + 2H^+ \qquad (6.15)$$

In this way, TCA cycle intermediates are kept 'topped up' and not depleted by biosynthesis.

6.4.4 Control of the TCA cycle

The rate of the TCA cycle is precisely controlled in accordance with the cell's needs for ATP and is regulated at three points. The three enzymes involved in the regulation of the TCA cycle are citrate synthetase, isocitrate dehydrogenase and α-oxoglutarate dehydrogenase (reactions i, iii and iv in Figure 6.6).

1 *Citrate synthetase* is allosterically inhibited by ATP. Thus, as the level of ATP in the cell increases, this enzyme becomes less active and so less citrate is formed.

2 *Isocitrate dehydrogenase* is regulated allosterically: ADP is an activator, while NADH and ATP are inhibitors (in some tissues, just ATP is inhibitory).

◇ How do these properties of isocitrate dehydrogenase fit in with the cell's regulation of its energy level?

◈ A high concentration of NADH indicates that there is active metabolism of glucose, as several steps require the participation of NAD$^+$ which is reduced to NADH in the process. If glucose metabolism is active, then presumably there is a high cellular level of ATP. Inhibition of the enzyme by NADH will slow down the TCA cycle and thus further synthesis of ATP. A high level of ATP is a more *direct* signal that the energy level in the cell is high, and ATP being an inhibitor of the enzyme means that this signal is responded to appropriately.

On the other hand, a high level of ADP indicates a low energy level (low ATP/ADP ratio) in the cell. ADP being an activator of the enzyme tends to increase TCA activity, and hence ATP synthesis and the ATP/ADP ratio.

3 Certain features of the control of the third enzyme, *α-oxoglutarate dehydrogenase*, are like those of the PDH complex (Section 6.3.3). This enzyme is inhibited by succinyl CoA and NADH, the products of the reaction it catalyses. A high cellular energy level also inhibits it.

Remember also that the rate of formation of acetyl CoA in the link reaction, catalysed by pyruvate dehydrogenase, is also subject to similar controls (Section 6.3.3). Thus, the rate of production of the 2-carbon 'substrate' of the TCA cycle—*and* the rate of operation the cycle itself—are reduced when the ATP/ADP and NADH/NAD$^+$ ratios are high. The regulatory enzymes of the TCA cycle are indicated in Figure 6.10.

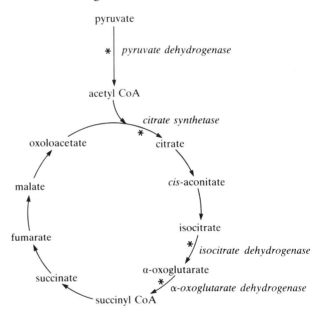

Figure 6.10 Control of the TCA cycle and the link reaction. Regulatory enzymes are indicated by *black* asterisks. (See text for details.)

Summary of Sections 6.3 and 6.4

1 The link reaction is the conversion of pyruvate to acetyl CoA—so-called because it links glycolysis with the TCA cycle.

2 The link reaction is catalysed by a multi-enzyme complex—comprising three enzymes acting in sequence—called the pyruvate dehydrogenase (PDH) complex.

3 In this reaction, in addition to acetyl CoA, NADH and carbon dioxide are produced.

4   Pyruvate dehydrogenase is regulated by the energy level of the cell and by acetyl CoA/CoA and NADH/NAD$^+$ ratios.

5   The link reaction and the TCA cycle take place in the mitochondria, in contrast to glycolysis, which occurs in the cytosol.

6   The TCA cycle is a cyclic sequence of reactions in which the two carbon atoms of the acetyl group of acetyl CoA are converted to carbon dioxide with the concomitant reduction of the coenzymes NAD$^+$ and FAD. The energy thus 'trapped' as NADH and FADH$_2$ is subsequently used to generate ATP via the electron transport chain (oxidative phosphorylation).

7   The source of the acetyl CoA can be a variety of fuel molecules including pyruvate (via the link reaction), fatty acids (via β-oxidation) as well as certain amino acids.

8   The first step in the TCA cycle is the joining of acetyl CoA to oxaloacetate to form citrate.

9   In each turn of the cycle, citrate is converted, via a series of 6-carbon, 5-carbon and 4-carbon intermediates, back to oxaloacetate.

10   One reaction of the TCA cycle involves a substrate level phosphorylation; initially GTP is formed, and this can then be converted to ATP.

11   The TCA cycle is an important source of precursors for biosynthesis. TCA cycle intermediates drawn off for biosynthesis must be replenished; for example, by carboxylation of pyruvate to oxaloacetate.

12   The TCA cycle is controlled directly at three points: the enzymes citrate synthetase, isocitrate dehydrogenase and α-oxoglutarate dehydrogenase are all subject to regulation by ATP/ADP and/or NADH/NAD$^+$ ratios. These effectors also control the link reaction enzyme, pyruvate dehydrogenase, and thus the rate of production of acetyl CoA, the starting substance of the TCA cycle—this serving as a further, indirect, control on the cycle.

Question 4 (*Objectives 6.1 and 6.4*) Give two examples of catalysis by multi-enzyme complexes in metabolism, indicating the advantages of such systems.

Question 5 (*Objectives 6.1 and 6.2*) Give a brief description of the importance to the cell of the TCA cycle in terms of biosynthesis and energy production.

Question 6 (*Objectives 6.1 and 6.2*) The following statements relate to the TCA cycle. Are they true or false? If false, state why.

(a) Acetyl CoA and oxaloacetate react to form citrate.

(b) The TCA cycle regenerates oxaloacetate.

(c) The two carbon dioxide molecules evolved in one turn of the cycle originate directly from the two carbon atoms of the acetyl group in acetyl CoA.

Question 7 (*Objectives 6.1 and 6.3*) Which of the following coenzymes are involved in oxidation–reduction reactions?

(a) Thiamine pyrophosphate (TPP)

(b) NAD$^+$

(c) FAD

Question 8 (*Objectives 6.1 and 6.2*) Can *net* synthesis of TCA cycle intermediates be obtained by adding acetyl CoA to a cell extract that contains all the enzymes and coenzymes of the TCA cycle? Explain your answer.

6.5 THE ELECTRON TRANSPORT CHAIN (ETC)

We have seen that each turn of the TCA cycle involves four dehydrogenation steps. In three of these, NAD^+ serves as the electron acceptor and in the fourth the pair of electrons removed from succinate is accepted by the coenzyme of succinate dehydrogenase, namely **flavin adenine dinucleotide (FAD)**. This coenzyme is a derivative of the water-soluble vitamin B_2, otherwise known as riboflavin. The reduced coenzymes thus formed in the TCA cycle—three of NADH and one of $FADH_2$—then donate the electrons they are carrying to a series of electron carriers that constitute the respiratory chain or **electron transport chain (ETC)**. The ETC is the final common pathway by which all electrons derived from different fuel molecules flow to oxygen, the final oxidant—acceptor of electrons—in aerobically metabolizing cells. This stepwise transfer of electrons, to the final acceptor, oxygen, is accompanied by the 'release' of a large quantity of energy. Some of this energy can be used to generate ATP.

6.5.1 Energy made available via the electron transport chain

Oxidative phosphorylation is the process in which ATP is formed as electrons are transferred from NADH or $FADH_2$ to oxygen via a series of electron carriers. This is the major source of ATP production in aerobically metabolizing organisms. In fact, oxidative phosphorylation generates 32 of the 36 molecules of ATP that are formed when glucose is completely oxidized to carbon dioxide and water. The important features of oxidative phosphorylation are listed below.

1 Oxidation of reduced coenzymes is *coupled* to oxidative phosphorylation. The oxidation of one molecule of NADH yields *three* molecules of ATP, whereas the oxidation of one molecule of $FADH_2$ yields *two* molecules of ATP.

2 Oxidative phosphorylation is carried out by *respiratory complexes* in the inner membrane of the mitochondrion.

3 The respiratory complexes each contain a sequence of electron carriers, and so are often referred to as *electron-transfer complexes*. Electrons are transferred from NADH or $FADH_2$ to oxygen via these complexes. This

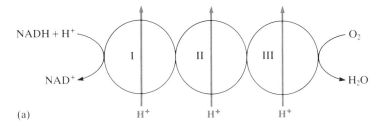

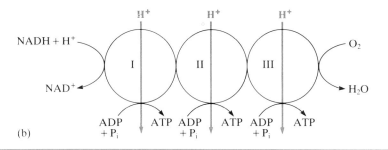

Figure 6.11 Simplified diagrams to show how electron transport (from NADH) is coupled to ATP synthesis (oxidative phosphorylation) via the generation of a proton gradient across the inner mitochondrial membrane. (a) Protons are pumped out of the mitochondiral matrix across the inner membrane by three electron-transfer complexes, here labelled I, II and III. (b) Protons flow back into the matrix, down their concentration gradient, and ATP is synthesized at each of the three complexes.

electron transfer causes protons to be pumped out of the mitochondrial matrix and a membrane potential, called the *proton-motive force*, is thereby generated. The protons are pumped out by *three* kinds of electron-transfer complexes acting in sequence (Figure 6.11a).

4 When protons flow *back* into the mitochondrial matrix, down their concentration gradient, ATP is synthesized (Figure 6.11b). Thus oxidation (electron-transfer) is coupled to ATP formation via a proton gradient across the inner mitochondrial membrane.

6.5.2 Mitochondrial structure in relation to the ETC.

As discussed in Chapter 2, (Section 2.8), mitochondria are organelles with two membrane systems, an outer membrane and an extensive, greatly folded inner membrane (see also Plate 2.5). Hence there are effectively two fluid compartments in the mitochondrion—the intermembrane space between the inner and outer membranes, and the matrix, bounded by the inner membrane. The electron-transfer complexes are an integral part of the inner mitochondrial membrane, whereas the link reaction, the TCA cycle and fatty acid oxidation all occur in the mitochondrial matrix (see Figure 6.12).

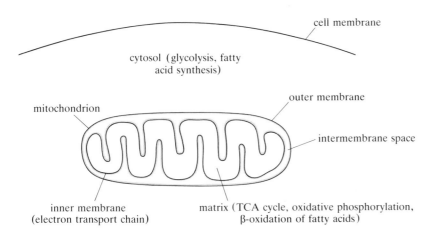

Figure 6.12 Compartmentation of some of the major metabolic pathways.

Most small molecules can pass through the outer mitochondrial membrane via specific transport systems (Chapter 7). But most substances are unable to cross the inner membrane. However, certain vital substances (ADP, for example) can cross it, carried by specific transport molecules.

6.5.3 The electron carriers of the ETC

Electrons are transferred from NADH and $FADH_2$ to oxygen via a series of **electron carriers**: flavin coenzymes, iron–sulphur proteins, quinones and cytochromes (see Figure 6.13). So, chemically, the electron carriers of the inner mitochondrial membrane are a diverse collection of molecules. However, they can all be readily reduced and re-oxidized, that is, accommodate temporarily one or two extra electrons within their molecular structure. Most of the protein components of the ETC contain iron (Fe) and undergo oxidation–reduction reactions by a valency change from Fe(II) (reduced) to Fe(III) (oxidized). Examples are the cytochromes, all of which have an iron-containing haem group, and the iron–sulphur (Fe–S) proteins.

The first reaction in the ETC is the oxidation of NADH by NADH dehydrogenase, one of the two components of the *NADH–Q reductase complex* (see Figure 6.13). NADH reduces the flavin mononucleotide (FMN) coenzyme of NADH dehydrogenase to $FMNH_2$:

$$\text{NADH} + \text{H}^+ + \text{FMN} \xrightleftharpoons[\text{dehydrogenase}]{\text{NADH}} \text{FMNH}_2 + \text{NAD}^+ \tag{6.16}$$

Two electrons are then transferred from $FMNH_2$ to an Fe–S protein within the NADH–Q reductase complex (see Figure 6.13).

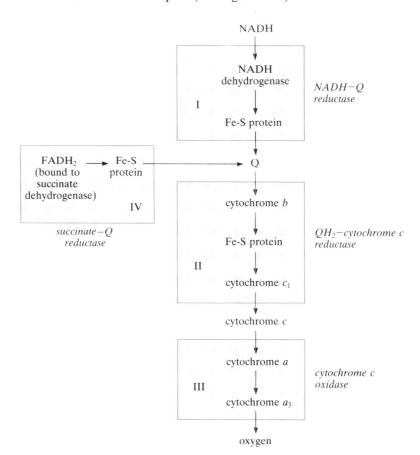

Figure 6.13 The sequence of electron carriers in the ETC. Protons are pumped by the *three* complexes between NADH and oxygen and ATP is thereby synthesized—*three* ATPs/NADH. Note that there is no proton pumping (therefore no ATP synthesis) at the succinate–Q reductase complex (IV), thus accounting for the *two* ATPs/FADH₂ observed.

It is perhaps worth emphasizing at this point that it is only the *electrons* from the hydrogen atoms involved in reducing FMN to $FMNH_2$ that are fed into the ETC. The protons also liberated in the re-oxidation back to FMN are released into the surrounding medium.

Electrons are then transferred from the Fe–S protein of NADH–Q reductase to the quinone, coenzyme Q (abbreviated as Q). The sequence of hydrogen (or electron) transfer reactions between NADH and coenzyme Q is summarized below:

$$\tag{6.17}$$

Coenzyme Q (Q) is also known as *ubiquinone* because it is ubiquitous in living organisms. It functions as a mobile carrier of electrons between the flavoproteins—NADH dehydrogenase and succinate dehydrogenase—and the cytochromes of the ETC. You will remember that $FADH_2$ is formed in the TCA cycle via the oxidation of succinate to fumarate, in the reaction catalysed by the mitochondrial membrane-bound enzyme, succinate dehydrogenase (Section 6.4.1). This enzyme is one of the two components of the *succinate–Q reductase complex*, the other being (as in NADH–Q reductase) an Fe–S protein. The electrons of $FADH_2$ in the succinate dehydrogenase component are transferred to Fe–S proteins in the complex and then to coenzyme Q for further transfer along the ETC. Thus coenzyme Q accepts electrons ultimately derived from either NADH or $FADH_2$. It is the common entry point to the ETC for both these reduced coenzymes. On accepting electrons, coenzyme Q becomes reduced. When reduced it is known as QH_2.

Apart from one Fe–S protein, the electron carriers between QH_2 and oxygen are all cytochromes. A **cytochrome** is an electron-transporting protein that contains a haem group (like haemoglobin—see Chapter 3, Figure 3.27). The haem iron in cytochromes alternates between the Fe(II) (reduced) and Fe(III) (oxidized) states during electron transport. A haem group, like an Fe–S protein, is a *single*-electron carrier, in contrast with NADH, flavins, and coenzyme Q, which are *two*-electron carriers.

◇ How many molecules of cytochrome *b* are needed to accept electrons from one molecule of QH_2?

◆ One molecule of QH_2 transfers *two* electrons to *two* molecules of cytochrome *b* (i.e. one electron per cytochrome *b*), the next member of the ETC (Figure 6.13).

There are five cytochromes between QH_2 and oxygen in the ETC:

$$QH_2 \longrightarrow \text{cyt } b \longrightarrow \text{Fe–S} \longrightarrow \text{cyt } c_1 \longrightarrow \text{cyt } c \longrightarrow \text{cyt } a \longrightarrow \text{cyt } a_3 \longrightarrow O_2 \qquad (6.18)$$

Two of these cytochromes, *b* and c_1, together with an Fe–S protein, make up the *QH_2–cytochrome c reductase complex* (see Figure 6.13). Cytochrome *c*, which acts as a mobile electron carrier on the outer surface of the inner mitochondrial membrane, transfers electrons from this to another complex, the *cytochrome c oxidase complex*, which contains the other two cytochromes, *a* and a_3 (Figure 6.13).

Thus the QH_2–cytochrome *c* reductase complex transfers electrons from QH_2 to cytochrome *c* which is thereby reduced. Reduced cytochrome *c* then transfers its electron to cytochrome *a* of the cytochrome *c* oxidase complex, and thence to cytochrome a_3. This, last, cytochrome of the ETC contains (in addition to haem iron) a copper ion, which alternates between the Cu(I) (reduced) and Cu(II) (oxidized) states as electrons are transferred from cytochrome a_3 to molecular oxygen. The reduction of a molecule of oxygen to water requires *four* electrons, whereas *single* electrons are transferred via the haem iron to the copper of cytochrome a_3. How four electrons apparently come together to reduce one molecule of oxygen (Equation 6.19) is not known.

$$O_2 + 4H^+ + 4e^- \longrightarrow 2H_2O \qquad (6.19)$$

Each electron carrier of the ETC has its own characteristic oxidation-sensitive absorption spectrum, so that absorbance can be used to distinguish between those in a reduced state and those in an oxidized state. The reduction of cytochrome *c* can be followed spectroscopically (Chapter 4, Section 4.3.1) because its absorption spectrum changes depending on whether it is in the reduced or oxidized state (see Figure 6.14). This experimental approach can be used to investigate the sequence of carriers in the ETC by using specific inhibitors that block electron transport between particular pairs of carriers.

◇ Suppose that inhibitor X blocks electron transport between cytochrome c and cytochrome a. In the presence of this inhibitor, would you expect a partially purified preparation containing cytochrome c, cytochrome b and cytochrome c_1, and cytochrome a and cytochrome a_3, to absorb more strongly at 530 nm or 550 nm when reduced coenzyme Q (QH_2) is added?

◈ Absorption will be greatest at 550 nm. In the presence of inhibitor X, electron transport will take place from Q→cytochrome b→cytochrome c_1→cytochrome c, but transfer of electrons beyond this point in the chain (to cytochrome a; Figure 6.13) cannot occur. Therefore cytochrome c remains *reduced* and its spectrum will resemble curve A in Figure 6.14, with an absorption peak at 550 nm.

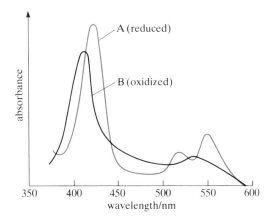

Figure 6.14 The absorption spectra of reduced (blue curve) and oxidized (black curve) cytochrome c.

A combination of investigations using fragmentation into submitochondrial particles such as inner membrane portions containing different complexes of the ETC, spectroscopy and other experimental approaches, has led to the proposed arrangement of the protein components of the ETC shown in Figure 6.15.

What has been realized only relatively recently, and is shown in Figure 6.15, is that the linear sequence is in fact physically a zigzag arrangement, with

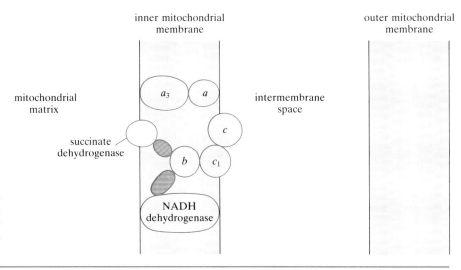

Figure 6.15 A simplified representation of the zigzag arrangement of the protein components of the mitochondrial electron transport chain (the small, mobile carrier, coenzyme Q, has been left out).

some components in contact with the mitochondrial matrix and others in contact with the intermembrane space between the inner and outer membranes. For example cytochrome c, which is a relatively loosely bound and water-soluble molecule, can be readily removed by washing the outer but not the inner surface of the inner membrane with dilute salt solution. This spatial arrangement of carriers across the inner mitochondrial membrane is of crucial importance in the linkage between oxidative phosphorylation and the passage of electrons to build up a proton gradient, as we shall now explain.

6.5.4 The chemiosmotic hypothesis

So far, we have considered the flow of electrons from the reduced coenzymes, NADH and FADH$_2$, to oxygen, whereby energy is liberated and the coenzymes are regenerated. Overall, this can be represented by:

$$NADH + H^+ + \tfrac{1}{2}O_2 \longrightarrow NAD^+ + H_2O \; (+ \text{ energy}) \qquad (6.20)$$

and

$$FADH_2 + \tfrac{1}{2}O_2 \longrightarrow FAD + H_2O \; (+ \text{ energy}) \qquad (6.21)$$

The energy liberated is used to synthesize ATP:

$$ADP + P_i \longrightarrow ATP + H_2O \qquad (6.22)$$

As discussed below, this ATP synthesis is carried out by an enzyme complex in the inner mitochondrial membrane. This enzyme complex is somewhat confusingly called *mitochondrial ATPase* because it was discovered through its catalysis of the reverse reaction, that is, ATP hydrolysis (see Chapter 2, Section 2.8).

How is the energy released by the oxidation of NADH or FADH$_2$ used to phosphorylate ADP? That is, how does oxidative phosphorylation occur?

This remained a highly controversial question for many years. Its resolution began in 1961 when Peter Mitchell put forward his **chemiosmotic hypothesis**. This proposes that electron transport and ATP synthesis are coupled by the formation of a proton (H$^+$) gradient. According to Mitchell's hypothesis, the transfer of electrons along the ETC results in the pumping of protons from the mitochondrial matrix to the outer (intermembrane space) side of the membrane. Thus the proton concentration becomes higher on the outer side of the inner membrane, and a membrane potential develops in which this, outer, side is positive (Figure 6.16). It is this *proton-motive force*, directed from the outside to the inside, that is said to drive the synthesis of ATP by the ATPase complex.

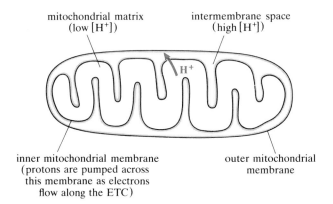

mitochondrial matrix (low [H$^+$])

intermembrane space (high [H$^+$])

H$^+$

inner mitochondrial membrane (protons are pumped across this membrane as electrons flow along the ETC)

outer mitochondrial membrane

Figure 6.16 The generation of a proton gradient (leading to a membrane potential across the inner mitochondrial membrane) as a result of electron transfer along the ETC.

Thus, according to Mitchell's hypothesis the ETC interacts with the ATP-synthesizing complex via the electrochemical proton gradient that is generated. In other words, the energy released via electron transfer along the ETC is conserved as the proton gradient which is then used to power the synthesis of ATP. This model requires the electron carriers of the ETC and the ATPase to be specifically arranged with respect to the two faces of the inner mitochondrial membrane. We know that for the ETC components this is indeed the case (Figure 6.15). In addition, the inner mitochondrial membrane cannot be freely permeable to protons, otherwise a proton gradient could not be established and maintained.

There is now a great deal of experimental evidence to support Mitchell's hypothesis that oxidation (i.e. electron transport) and phosphorylation are indeed coupled by the establishment of an electrochemical proton gradient.

Protons are pumped at *three* sites as electrons are carried along the ETC from NADH to oxygen, these sites corresponding to the three electron-transfer complexes, i.e. the NADH–Q reductase complex, the QH_2–cytochrome c reductase complex and the cytochrome c oxidase complex (see Figure 6.13). The electrochemical proton gradient generated at each site by the transfer of a pair of electrons from NADH is used to power the synthesis of one molecule of ATP, thus accounting for the three ATPs per NADH observed (Section 6.5.1).

When electrons are passed into the ETC from $FADH_2$ rather than NADH, they enter at the level of coenzyme Q (see Figure 6.13).

$\diamond$ How many molecules of ATP are formed when $FADH_2$ is the electron donor?

◆ The first site of proton pumping, and thus ATP formation, is bypassed and only *two*, rather than three, ATP molecules are formed when $FADH_2$ is the electron donor.

As already mentioned, the generation of a proton gradient by the flow of electrons through the three proton-pumping sites (i.e. electron transfer complexes) in the ETC requires that these sites be asymmetrically orientated in the membrane. These three complexes must also span the inner mitochondrial membrane so that protons can be pumped only from the matrix side to the outer (i.e. intermembrane space) side. It has been established, using a variety of experimental techniques, that all three complexes do, in fact, span the membrane and are orientated asymmetrically (see Figure 6.15). But where do the protons to be pumped come from? One might logically assume that the protons that are pumped are the ones directly involved in the oxidation–reduction reactions that occur at the three sites, e.g. when two electrons are transferred from QH_2 to the cytochrome c reductase complex, two protons must be released:

$$QH_2 \longrightarrow Q + 2e^- + 2H^+ \tag{6.23}$$

One proton per electron, or two per molecule of NADH or $FADH_2$, could be pumped by this direct mechanism at each of the proton pumping sites. However, proton concentration values *per site* of between three and four have been measured experimentally and so these results do not support this simple, direct mechanism. It appears more probable that when electrons are transferred through each of the proton pumping sites, conformational changes occur in the inner mitochondrial membrane that somehow cause protons to be transferred from the matrix to the intermembrane space side of the membrane.

Let us now consider how the electrochemical proton gradient is used to synthesize ATP. The enzymes catalysing this process appear in electron micrographs as spherical projections (knobs) on the inner surface (i.e. the matrix side) of the inner mitochondrial membrane. These knobs are called F_1 **particles** (see Plates 2.19a and 2.19b). Mitochondrial membranes that have been stripped of F_1 particles can still transfer electrons through their ETC, but they can no longer synthesize ATP. Isolated F_1 particles catalyse the hydrolysis of ATP, i.e. they act as an ATPase—the *reverse* of their physiological action—but when they are put back with the stripped membranes ATP synthesis is restored. F_1 particles constitute only part of the mitochondrial ATP-synthesizing complex. The other major unit of this enzyme complex is F_0; this component completely spans the inner mitochondrial membrane and is the channel in the inner membrane through which the protons flow *down* their electrochemical gradient (see Chapter 7, Section 7.1, on membrane channel proteins). F_0 and F_1 are joined by a 'stalk' which is the part of the complex that interacts with the antibiotic oligomycin; thus the antibiotic properties of oligomycin are due to inhibition of oxidative phosphorylation.

Having established a gradient, the flow of protons back down this gradient through the F_0 channel from the intermembrane space side to the matrix side of the inner mitochondrial membrane results in the synthesis of ATP by F_1. But how is this 'downhill' flow of protons coupled to ATP synthesis? One theory is that the proton flow may somehow induce a conformational change in the mitochondrial ATPase (F_1) which serves to activate the enzyme. As with the coupling of electron transport to proton pumping, the detailed mechanism of the energy transduction is very speculative.

6.5.5 Summary of energy yield via the electron transport chain

Our description of oxidative phosphorylation completes the survey of how glucose is completely oxidized to carbon dioxide and water. This allows us to tot up the total energy yield—in terms of ATP—of the overall process, including, most importantly, the ATPs derived from regeneration of the reduced coenzymes, NADH and $FADH_2$ (Table 6.3).

You will note from Table 6.3 that each molecule of NADH formed in glycolysis yields just *two* molecules of ATP, in contrast to the *three* molecules of ATP from the NADH of the link reaction or from that of the TCA cycle.

◇ Can you think of any possibly significant differences between these *sources* of NADH? (Consider *where* in the cell the NADH is produced.)

◆ Glycolysis occurs in the cytosol, whereas the link reaction and the TCA cycle take place in the mitochondrion.

The different sites of NADH production give rise to the *apparent* anomaly in the yields of ATP when NADH is oxidized to NAD^+ via the ETC. NADH *produced* in the mitochondrion is re-oxidized there to NAD^+, yielding three ATPs per NADH. It would seem that cytosolic NADH, from glycolysis, must be transported into the mitochondrion for oxidation, as the ETC is located there. But the mitochondrion is impermeable to NADH and NAD^+. Thus cytosolic NADH is *not* itself transported into the mitochondrion for oxidation, but, *in effect*, its two electrons are. What happens is that cytosolic NADH passes its electrons on to another cytosolic substance, dihydroxyacetone phosphate (DHAP), which you have encountered before as an intermediate in glycolysis (Figures 6.2 and 6.4). This results in oxidation of the NADH to NAD^+ and the reduction of the DHAP to glycerol 3-phosphate.

Table 6.3 Total yield of ATP from the complete oxidation of glucose

Reaction	ATP yield per glucose
Glycolysis (in cytosol)	
glucose → glucose 6-phosphate	−1
fructose 6-phosphate → fructose 1,6-bisphosphate	−1
2(1,3-bisphosphoglycerate → 3-phosphoglycerate)	+2
2(phosphoenolpyruvate → pyruvate)	+2
2NADH formed by oxidation of 2 molecules of glyceraldehyde 3-phosphate	
Link reaction (in mitochondria)	
2NADH formed by conversion of pyruvate to acetyl CoA	
TCA cycle (in mitochondria)	
2 molecules of GTP formed from 2 molecules of succinyl CoA	+2
6NADH formed in the oxidation of 2 molecules each of isocitrate, α-oxoglutarate and malate	
2FADH$_2$ formed in the oxidation of 2 molecules of succinate	
Oxidative phosphorylation (in mitochondria)	
2NADH formed in glycolysis, each yields 2ATP	+4
2NADH formed in the link reaction each yields 3ATP	+6
2FADH$_2$ formed in the TCA cycle, each yields 2ATP	+4
6NADH formed in the TCA cycle, each yields 3ATP	+18
Total yield of ATP	+36

Mitochondria *are* permeable to glycerol 3-phosphate and so this can pass across into the mitochondrion. Once inside, glycerol 3-phosphate can give up two electrons (originally obtained from NADH *in the cytosol*) to FAD, producing FADH$_2$ and regenerating DHAP. The DHAP can pass back into the cytosol where it can help 'shuttle' further electrons from more cytosolic NADH across the mitochondrial membrane. The FADH$_2$ in the mitochondrion feeds two electrons (originally from the *cytosolic* NADH) into the ETC, it being thereby re-oxidized to FAD.

◇ How many ATPs will the re-oxidation of FADH$_2$, via the ETC, yield?

✦ Two per FADH$_2$ re-oxidized.

Thus the *net* effect of this *glycerol phosphate shuttle* is the transfer of two electrons from cytosolic NADH into the ETC yielding *two* ATPs—as against direct oxidation of NADH via the ETC (worth *three* ATPs), a 'cost' to the cell of one ATP per cytosolic NADH oxidized. The glycerol phosphate shuttle operates in eukaryotes, being important, for example, in active tissues such as the flight muscles of insects. The shuttle is, however, absent in prokaryotes.

◇ Why is it not present in prokaryotes?

✦ Prokaryote cells have no mitochondria. The TCA cycle and ETC, where present, occur in the cytosol and on the cell membrane. Thus cytosolic NADH can be oxidized directly via the ETC, yielding three ATPs per molecule. The oxidation of glucose to carbon dioxide and water in prokaryotes thus yields 38 molecules of ATP per molecule of glucose.

In mammalian tissues, particularly liver and heart, there is another shuttle operating that carries no 'cost' in terms of ATP and so cytosolic NADH can yield three ATPs per molecule, just like NADH *produced* in the mitochondrion via the link reaction or TCA cycle. This *malate–aspartate shuttle* is somewhat more complex than the glycerol phosphate one, but, *in essence*, it shuttles the electrons of cytosolic NADH to malate, and back again within the mitochondrion. Put simply, in the cytosol, aspartate is converted to malate, this involving the oxidation of NADH to NAD^+. The malate can cross the mitochondrial membranes and in the mitochondrion it is re-oxidized, transferring its electrons (originally from *cytosolic* NADH) to *mitochondrial* NAD^+. This produces NADH in the mitochondrion (as against $FADH_2$ in the glycerol phosphate shuttle) which, when re-oxidized to NAD^+ via the ETC, yields *three* ATPs per molecule. Thus the net effect of the malate–aspartate shuttle is the transfer of the two electrons from cytosolic NADH, via mitochondrial NADH, to the ETC, yielding three ATPs per molecule of cytosolic NADH oxidized.

In further contrast to the glycerol phosphate shuttle, the malate–aspartate shuttle is reversible, the electrons of cytosolic NADH *only* being brought into the mitochondrion if the $NADH/NAD^+$ ratio in the cytosol is higher than that in the mitochondrion. When operative the malate–aspartate shuttle allows a full 38 ATPs per glucose oxidized to carbon dioxide and water, unlike the 36 ATPs (Table 6.3) when only the glycerol phosphate shuttle is working. Unless stated otherwise, in this text we shall assume that only the glycerol phosphate shuttle is operating.

The overall reaction represented in Table 6.3 is

$$\text{glucose} + 36ADP + 36P_i + 36H^+ + 6O_2 \longrightarrow 6CO_2 + 36ATP + 42H_2O \qquad (6.24)$$

One way of expressing the overall energy yield is as a **P:O ratio**. This is the number of molecules of ATP formed from ADP (P) per *atom* of oxygen consumed (O). The P:O ratio here is 3 because 36 molecules of ATP are formed and 12 atoms of oxygen are consumed.

It is interesting to consider the overall energetics of glucose catabolism. The complete oxidation of glucose yields $2\,870\,kJ$ per mole under standard laboratory conditions. The energy stored in 36 molecules of ATP is $36 \times 30.5 = 1\,100\,kJ$ and so the efficiency of ATP formation from glucose is $(1\,100/2\,870) \times 100\% = 38\%$ under standard conditions. Although at first glance this percentage may not seem very impressive, by chemical standards it represents a highly efficient energy conversion system.

6.5.6 Respiratory control

Under most physiological conditions, electron transport is tightly coupled to ATP synthesis—electrons do not usually flow through the ETC to oxygen unless ADP is simultaneously phosphorylated to ATP. The most important factor that determines the rate of oxidative phosphorylation is the level of ADP in the cell. The rate of oxygen consumption by a tissue homogenate increases markedly when ADP is added and returns to its initial value when the added ADP had been converted into ATP (see Figure 6.17).

The regulation of the rate of oxidative phosphorylation by the level of ADP in the cell is called **respiratory control**.

◇ What is the physiological significance of this control mechanism?

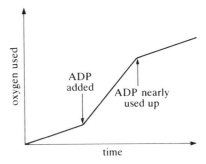

Figure 6.17 Respiratory control. The level of ADP (substrate for oxidative phosphorylation) controls the rate at which electrons are transferred along the ETC to oxygen and hence the rate of oxygen consumption.

◈ The level of ADP increases when ATP is consumed. This increased level of ADP stimulates oxidative phosphorylation and hence increases the rate of ATP synthesis. Thus ATP production is geared to ATP utilization.

This coupling of oxidative phosphorylation to the ETC means that electrons do not flow from fuel molecules to oxygen unless the cell requires ATP synthesis; in other words, unless the ratio of ATP/ADP in the cell falls, signalling a depletion in the level of ATP. At a high ATP/ADP ratio (where ATP is plentiful, and hence ADP relatively scarce) this acts to slow down further oxidative phosphorylation.

Thus, an extremely important mechanism whereby aerobic catabolism is regulated in accordance with the needs of the cell occurs through the coupling of ATP synthesis (by oxidative phosphorylation) to electron transport.

6.5.7 Uncouplers

Some compounds, such as 2,4-dinitrophenol (DNP), disrupt the tight coupling between electron transport and phosphorylation. These substances carry protons across the inner mitochondrial membrane. In the presence of these **uncouplers**, electron transport from NADH or $FADH_2$ to oxygen proceeds normally, but ATP is not formed by the mitochondrial ATPase complex because the electrochemical proton gradient and so the proton-motive force across the inner mitochondrial membrane is dissipated. This uncoupling thus leads to a loss of respiratory control by ADP. Loss of respiratory control leads to increased oxygen consumption and rate of oxidation of NADH and $FADH_2$. In contrast, uncouplers do not interfere with substrate level phosphorylation. Because uncouplers interfere specifically with the ETC, they have proved very useful in experiments on cell metabolism.

◇ What is the P:O ratio in uncoupled mitochondria?

◈ In completely uncoupled mitochondria, the P:O ratio is zero because no ATP is synthesized.

◇ When DNP is administered to rats, what would you expect to happen?

◈ Their oxygen consumption would increase and body temperature rise because the energy of oxidation, rather than being conserved as ATP, is lost as heat.

Mitochondrial oxidative phosphorylation can be uncoupled experimentally but uncoupling also occurs naturally and can be biologically useful. It is a means of generating heat to maintain body temperature, for example in hibernating animals, and in the newborn. Brown adipose (fat) tissue, which is very rich in mitochondria, is specialized for this process of *thermogenesis*. Hormones control the degree of uncoupling of oxidative phosphorylation in this tissue. Thus the mitochondria in brown adipose tissue can serve as generators of ATP or as generators of extra heat, depending on requirements.

In tightly coupled mitochondria neither electron flow nor ATP synthesis can operate independently of each other. If the cell has a build-up of ATP (a sign that its energy requirements are satisfied), its reserves of ADP will be low and the rate of synthesis of ATP will be low due to lack of substrate.

Of course, the rate of production of the 'substrates' of the ETC, i.e. the reduced coenzymes, NADH and $FADH_2$, (via the oxidative reactions of catabolism, notably the TCA cycle), is also subject to tight control. We have already seen how high ATP/ADP and NADH/NAD$^+$ ratios can 'turn down' catabolism, and hence the rate of formation of NADH and FADH$_2$, via their effect on key regulatory enzymes (Sections 6.2.5, 6.3.3 and 6.4.4). Such feedback control reinforces the direct effect of ATP/ADP balance on the ETC.

6.5.8 Energy transmission by proton gradients

We have discussed in some detail above the linking (coupling) of mitochondrial electron transport to ATP synthesis via a transmembrane electrochemical proton gradient, but ATP synthesis in bacteria and chloroplasts is also driven by proton gradients. In fact, gradients of protons (and other ions) can power a variety of energy-requiring processes, as you will see in the following chapter (Section 7.3).

Summary of Section 6.5

1 The electron transport chain (ETC), or respiratory chain, is the pathway via which reduced coenzymes (NADH and FADH$_2$) generated in the oxidative reactions of catabolism become oxidized by (i.e. reduce) molecular oxygen. This is achieved by the stepwise transfer of electrons from NADH and FADH$_2$ along a series of electron carriers. A proportion of the large amount of energy released during this process is used in the synthesis of ATP by oxidative phosphorylation. Oxidative phosphorylation generates 32 of the 36 molecules of ATP formed when a molecule of glucose is oxidized to carbon dioxide and water.

2 Oxidative phosphorylation is carried out by three kinds of respiratory complexes which are an integral part of the inner mitochondrial membrane.

3 The oxidation of one molecule of NADH yields three molecules of ATP, and oxidation of one molecule of FADH$_2$ yields two molecules of ATP.

4 The electron carriers of the ETC are mostly membrane-bound proteins i.e. flavoproteins, iron–sulphur (Fe–S) proteins and cytochromes, in association with the mobile carrier, coenzyme Q. One of the five cytochromes, cytochrome c, is also a loosely bound electron carrier.

5 Electrons from NADH are transferred to oxygen via the three respiratory complexes, which are linked by coenzyme Q (Q) and cytochrome c (cyt c):

NADH$\rightarrow$NADH–Q reductase$\rightarrow$Q$\rightarrow$QH$_2$–cyt c reductase$\rightarrow$cyt $c$$\rightarrow$cyt c oxidase$\rightarrow$O$_2$

6 With FADH$_2$ as electron donor, the electrons are transferred directly to the ETC because succinate dehydrogenase, of which FADH$_2$ is the coenzyme, is actually part of the ETC (as a component of the succinate–Q reductase complex):

succinate–Q reductase$\rightarrow$Q$\rightarrow$QH$_2$-cyt c reductase$\rightarrow$cyt $c$$\rightarrow$cyt c oxidase$\rightarrow$O$_2$
(contains the FADH$_2$)

7 One molecule of ATP is produced (per pair of electrons transported) at each of the three sites between NADH and oxygen, corresponding to the three respiratory complexes—thus accounting for the observed three ATPs per NADH.

8 Electrons from FADH$_2$ pass through only two of the three ATP synthesis sites—thus accounting for the observed two ATPs per FADH$_2$.

9 According to Mitchell's hypothesis, electron transport generates a proton gradient across the inner mitochondrial membrane, and it is the discharging of this gradient that leads to the synthesis of ATP.

10 The enzymes catalysing the synthesis of ATP are visible under the electron microscope as projections on the matrix side of the inner mitochondrial membrane, but when detached from the membrane they catalyse the reverse reaction—ATP hydrolysis.

11 A total of 36 molecules of ATP are produced via the complete oxidation of glucose: two ATPs via glycolysis, two ATPs 'directly' via the TCA cycle (substrate level phosphorylation via GTP) and 32 ATPs via the ETC (oxidative phosphorylation). In some tissues when the NADH/NAD$^+$ ratio is high the total energy yield can be 38 ATPs, 34 of them via oxidative phosphorylation.

12 The energy yield can be expressed as a P:O ratio—number of molecules of ATP formed (P) per atom of oxygen consumed (O).

13 The rate of oxidative phosphorylation is regulated by the ADP level in the cell.

14 Under normal conditions, electron transport is tightly coupled to oxidative phosphorylation (ATP synthesis), thus ensuring that fuel supplies are conserved when ATP is plentiful.

15 Uncouplers such as 2,4-dinitrophenol (DNP) disrupt this coupling so that electron transport to oxygen continues but no ATP is formed. Some degree of uncoupling is important in thermogenesis.

16 Ion gradients across biological membranes power a variety of energy-requiring processes in addition to ATP synthesis.

Question 9 (*Objectives 6.1, 6.6 and 6.8*) Consider the processes involved in the oxidation of cellular fuel molecules to carbon dioxide and water. Which of the following statements are true, and which are false?

(a) Proteins called cytochromes are involved.

(b) The rate of electron transport is increased when the ATP/ADP ratio in the cell is high.

(c) Oxidation of the two hydrogen atoms produced in the conversion of one molecule of succinate to fumarate can produce two molecules of ATP.

(d) If oxidative phosphorylation is uncoupled, electron transport can never occur.

(e) Oxidative phosphorylation takes place in the cytosol.

Question 10 (*Objectives 6.1, 6.2 and 6.10*) What is the yield of ATP when one molecule of each of the following substances is completely oxidized by a cell? Assume that, in (a)–(c), only the glycerol phosphate shuttle for cytosolic NADH is operating.

(a) Pyruvate

(b) Fructose 1,6-bisphosphate

(c) 2-Phosphoglycerate

(d) Glyceraldehyde 3-phosphate, in mammalian tissue when the malate–aspartate shuttle is operating.

Question 11 (*Objectives 6.1, 6.8 and 6.9*) What is the effect of 2,4-dinitrophenol (DNP) on electron transport and ATP formation and on the P:O ratio?

Question 12 (*Objectives 6.1 and 6.5*) When metabolic rates are low, the rate of production of ATP is low. In mammals, one of the main controls over ATP production is allosteric inhibition, by ATP and citrate, of which of the following regulatory enzymes?

(a) Phosphofructokinase

(b) Pyruvate dehydrogenase

(c) Isocitrate dehydrogenase

Question 13 (*Objectives 6.1 and 6.6*) In the electron transport chain, the final acceptor of electrons is which of the following?

(a) Cytochrome *b*

(b) Cytochrome a_3

(c) Oxygen

(d) Coenzyme Q

Question 14 (*Objectives 6.1 and 6.7*) The production of ATP by oxidative phosphorylation is *not* driven by energy from which of the following?

(a) An electrochemical potential across the inner mitochondrial membrane

(b) A pH gradient across the inner mitochondrial membrane

(c) The formation of NADH

(d) The passage of protons from the intermembrane space to the matrix of the mitochondrion.

6.6 AEROBIC AND ANAEROBIC METABOLISM

We have now completed our survey of the sequence of reactions by which glucose is totally oxidized to carbon dioxide and water, via glycolysis, link reaction, TCA cycle and ETC, as occurs during **aerobic** metabolism in a wide variety of cells and organisms. But some cells and organisms can exist in the absence of oxygen, under which circumstances metabolism must be **anaerobic**. In such cases the TCA cycle and ETC are inoperative. But glycolysis (i.e. from glucose to pyruvate) still occurs and allows the cells to derive some energy from glucose (see Section 6.6.3).

◇ How many molecules of ATP per molecule of glucose does glycolysis yield?

◈ A net yield of two (Table 6.1).

The subsequent fate of the pyruvate produced by glycolysis depends on the cells concerned and the conditions prevailing.

◇ What is the fate of pyruvate in aerobic metabolism?

◈ It is fed into the TCA cycle (Section 6.4) as acetyl CoA, via the link reaction (Section 6.3).

6.6.1 Alcoholic fermentation

In some micro-organisms, notably Brewer's yeast, where glucose metabolism is anaerobic the pyruvate produced by glycolysis is converted to ethanol. This is a two-step process:

$$\text{pyruvate} \longrightarrow \text{acetaldehyde} \longrightarrow \text{ethanol} \qquad (6.25)$$
$$+CO_2$$

There is no ATP production nor utilization in these two steps, yet they are indirectly vital to the organism's ability to gain any ATP at all from this, so-called, *alcoholic fermentation* of glucose. This is because the second step, from acetaldehyde to ethanol, catalysed by alcohol dehydrogenase, involves NADH as coenzyme:

$$\text{acetaldehyde} + \text{NADH} + \text{H}^+ \overset{\text{alcohol}}{\underset{}{\overset{\text{dehydrogenase}}{\rightleftharpoons}}} \text{ethanol} + \text{NAD}^+ \qquad (6.26)$$

◇ Can you suggest why this step should be vital to the production of ATP by metabolism of glucose (remember we are considering anaerobic metabolism)?

◈ In anaerobic metabolism pyruvate is *not* fed into the TCA cycle and thence the ETC. It is converted to ethanol. Glycolysis is the cell's sole source of ATP.

But glycolysis needs a constant supply of NAD^+, this being required for the reduction of glyceraldehyde 3-phosphate (reaction vi in Figures 6.2 and 6.4); the NAD^+ being reduced to NADH in the process. In aerobic cells, NAD^+ is regenerated from NADH via the ETC. In alcoholic fermentation, where no ETC operates, the alcohol dehydrogenase step (Equation 6.26) serves to regenerate enough NAD^+ to keep glycolysis and hence ATP synthesis going.

The overall reaction from glucose to ethanol is:

$$\text{glucose} + 2\text{H}^+ + 2\text{ADP} + 2\text{P}_i \longrightarrow 2\,\text{ethanol} + 2\text{CO}_2 + 2\text{H}_2\text{O} + 2\text{ATP} \qquad (6.27)$$

If you compare this with Table 6.1, you will see that, though the yield of ATP is the same, there is no net oxidation or reduction (no change in NAD^+/ NADH ratio) in this anaerobic fermentation.

6.6.2 Formation of lactate

Those who enjoy cheese with their wine may well have good cause to be grateful that ethanol is not the only possible product from anaerobic metabolism of pyruvate. In some micro-organisms, including ones important in the making of cheese, pyruvate is converted anaerobically (fermented) to lactate (lactic acid), imparting a 'sharp' taste to certain cheeses. This reaction, catalysed by lactate dehydrogenase (LDH), requires NADH and so, like that catalysed by alcohol dehydrogenase (Equation 6.26), regenerates NAD^+, permitting continued glycolysis:

$$\text{pyruvate} + \text{NADH} + \text{H}^+ \overset{\text{LDH}}{\rightleftharpoons} \text{lactate} + \text{NAD}^+ \qquad (6.28)$$

As for alcoholic fermentation, there is no net oxidation or reduction (no change in NAD^+/NADH ratio; see Equation 6.27) in the degradation of glucose to lactate and the ATP yield is the same as in glycolysis (i.e. as far as pyruvate; Table 6.1):

$$\text{glucose} + 2\text{ADP} + 2\text{P}_i \longrightarrow 2\,\text{lactate} + 2\text{H}_2\text{O} + 2\text{ATP} \qquad (6.29)$$

This anaerobic production of lactate is not the exclusive preserve of micro-organisms. The conversion of pyruvate to lactate can occur in some cells in higher organisms when supply of oxygen to those cells is inadequate and hence metabolism therein is effectively anaerobic. This may occur in skeletal muscle during intense activity. Formation of lactate allows some glycolysis, and hence ATP synthesis, to continue in these cells, as NAD^+ is regenerated via the LDH-catalysed reaction (Equation 6.28). Thus the cells 'buy time', accumulate a so-called **oxygen debt**, which can be repaid when muscular activity eases off and the oxygen supply can once again cope. Oxygen debt is not indefinite however and the time bought is limited, though it may be enough to successfully flee a predator and hence survive to repay the debt!

◇ The enzyme catalysing the conversion of pyruvate to lactate is lactate dehydrogenase (LDH). What do you recall about this enzyme in the various tissues of humans?

◆ LDH exists as five different isoenzymes; the extreme types being that predominant in skeletal muscle (type 5; M_4) and that predominant in heart (type 1; H_4), as detailed in Chapter 4, Section 4.8.

Though LDH isoenzymes have been widely studied, their function is far less certain than those of many other isoenzymes (Chapter 5, Section 5.3.1), though there has been no lack of speculation on the matter. Most suggestions for the roles of the different isoenzymes rest on the assumption that they are in some way connected with 'oxygen debt'. It is known that the H_4 isoenzyme is allosterically inhibited, *in vitro*, by high levels of pyruvate, while the M_4 isoenzyme is not. Pyruvate is of course also a substrate of the enzyme, or a product, depending in which direction you are considering that the enzyme is operating (Equation 6.28, above).

It may be that the insensitivity of M_4 to pyruvate inhibition helps it to operate effectively in the direction of pyruvate $\rightarrow$ lactate; this allowing NAD^+ regeneration, and hence continued glycolysis, in *skeletal muscle*. Lactate accumulated in skeletal muscle, during oxygen debt, can then pass into the bloodstream and be taken up by other organs, *such as the heart*, where it can serve as a fuel. To be a fuel it must be converted to pyruvate, meaning that the LDH must operate in the lactate $\rightarrow$ pyruvate direction; it may be that the predominant *heart* (H_4) isoenzyme is optimized to function in this direction. Thus, in summary, the role of M_4 may be to synthesize lactate from pyruvate, allowing oxygen debt in temporarily anaerobic skeletal muscle; that of H_4 may be to synthesize pyruvate from lactate, allowing lactate transported via the bloodstream to be used as a fuel in aerobic tissues like the heart.

This and other proposed schemes for the roles of LDH isoenzymes are not, however, uncontroversial. For example, there is the case reported of a man genetically without any H_4 in his heart. As he was in his sixties and apparently well, the need for H_4 seems, literally, somewhat less than vital. As we implied in Chapters 4 and 5, assigning function based on *in vitro* studies, to isoenzymes or anything else for that matter, is not always without its difficulties.

6.6.3 Energy yields from aerobic and anaerobic metabolism of glucose

Consider the amounts of energy that the anaerobic metabolism of glucose (as, say, in heavily exercising muscles) yields in comparison with the energy obtained when glucose is oxidized to carbon dioxide and water. Figure 6.18 shows the ATP yields from the anaerobic conversion of glucose to lactate, and from the complete oxidation of glucose to carbon dioxide and water, as occurs under aerobic conditions.

Under anaerobic conditions cells obtain energy from the conversion of glucose to lactate. The maximum energy available from this process is two ATPs per molecule of glucose, compared to the 36 ATPs available when glucose is oxidized to carbon dioxide and water. Thus, for the advantage of

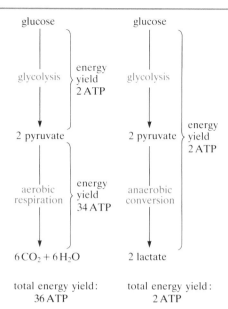

Figure 6.18 Comparison of ATP yields from the breakdown of one molecule of glucose under aerobic and anaerobic conditions. Less than 6% (i.e. $2/36 \times 100\%$) of the total ATP available can be obtained via the anaerobic route.

being able to extract energy from glucose in the absence of oxygen, an anaerobically functioning cell must waste more than 94% of the total ATP available via oxidation using molecular oxygen.

Because anaerobic degradation can extract only a small fraction of the total energy available from the glucose molecule, it might be expected that cells metabolizing anaerobically would use much more fuel per unit time per unit mass in order to do the same amount of cellular work as cells functioning aerobically. It has indeed been shown that under anaerobic conditions, cells may use more than ten times as much glucose as under aerobic conditions to do the same amount of work. This fact has some interesting consequences for so-called **facultative cells**. These are cells that can live under either aerobic or anaerobic conditions, and include many bacteria. In the absence of oxygen, they are able to extract energy from glucose by the same kind of anaerobic fermentation mechanisms as are used by *strict anaerobes*, but in the presence of oxygen they perferentially oxidize their foodstuffs completely (i.e. to carbon dioxide and water). Strict anaerobes are bacteria which grow *only* in the absence of molecular oxygen. Cancer cells are facultative and live aerobically or anaerobically. However, many types of cancer cell have a metabolic defect which causes them to use up very large amounts of glucose by anaerobic fermentation even when they are supplied with oxygen.

Summary of Section 6.6

1 Some cells and organisms can exist in the absence of oxygen, so their metabolism is anaerobic. Energy production is via glycolysis only, for the TCA cycle and the ETC are inoperative.

2 The pyruvate produced via anaerobic metabolism of glucose can be converted to ethanol (in micro-organisms) or lactate (in both micro-organisms and higher organisms). These chemical reductions of pyruvate use the NADH produced by glycolysis, thus regenerating the NAD^+ needed for glycolysis to continue.

3 In higher organisms the accumulation of lactate by skeletal muscle under anaerobic conditions is the price of a so-called 'oxygen debt'. This debt is 'repaid' when the lactate is reoxidized to pyruvate.

4 When glucose is catabolized anaerobically, only a small fraction (about 6%) of the ATP available via aerobic catabolism is extracted.

5 Facultative cells are those which can live under either aerobic or anaerobic conditions (e.g. many bacteria).

Question 15 (*Objectives 6.1 and 6.11*) (a) What is the net yield of ATP from glucose via:

(i) anaerobic metabolism to lactate?

(ii) aerobic metabolism to carbon dioxide and water?

(b) What is the source of the extra ATP generated via aerobic, as against anaerobic, catabolism of glucose?

SUMMARY OF CHAPTER 6

1 In aerobically metabolizing cells, glucose is completely oxidized to carbon dioxide and water. This occurs via glycolysis, the link reaction, the tricarboxylic acid (TCA) cycle and the electron transport chain (ETC).

2 Aerobic metabolism is compartmentalized: glycolysis occurs in the cytosol, while the link reaction, TCA cycle and electron transport take place in the mitochondria.

3 In glycolysis, which can also occur via anaerobic metabolism, glucose is degraded to pyruvate. Each molecule of glucose yields two molecules of pyruvate, and there is also a net gain of energy in the form of two molecules of ATP.

4 As well as providing energy (as ATP), glycolysis also helps provide building blocks for biosynthesis. Glycolysis is sensitive to these two 'outputs'—energy and building blocks—and is regulated by the cellular level of each of them.

5 In mammals, phosphofructokinase (PFK) is the central regulatory enzyme of glycolysis; hexokinase and pyruvate kinase are also involved in regulation.

6 NADH is produced in one of the reactions of the glycolytic pathway.

7 The link reaction links glycolysis to the TCA cycle and is catalysed by the pyruvate dehydrogenase (PDH) complex. The products of the link reaction are acetyl CoA, NADH and carbon dioxide.

8 The TCA cycle is a cyclic sequence of reactions in which the two carbon atoms of the acetyl group of acetyl CoA are converted to carbon dioxide with the concomitant reduction of the coenzymes NAD^+ and FAD. The energy thus 'trapped' as NADH and $FADH_2$ is subsequently used to produce ATP via the electron transport chain (oxidative phosphorylation).

9 The source of the acetyl CoA can be a variety of fuel molecules including pyruvate (via the link reaction), fatty acids (via β-oxidation) as well as certain amino acids.

10 The first step in the TCA cycle is the joining of acetyl CoA to oxaloacetate to form citrate.

11 In each turn of the cycle, citrate is converted, via a series of 6-carbon, 5-carbon and 4-carbon intermediates, back to oxaloacetate.

12 One reaction of the TCA cycle involves a substrate level phosphorylation; initially GTP is formed, and this can be then converted to ATP.

13 The TCA cycle is an important source of precursors for biosynthesis.

14   The TCA cycle is controlled directly at three points: the enzymes citrate synthetase, isocitrate dehydrogenase and α-oxoglutarate dehydrogenase are all subject to regulation by ATP/ADP and/or NADH/NAD$^+$ ratios. These effectors also control the link reaction enzyme, pyruvate dehydrogenase, and thus the rate of production of acetyl CoA—this serving as a further, indirect, control on the cycle.

15   The electron transport chain (ETC), or respiratory chain, is the pathway via which reduced coenzymes (NADH and FADH$_2$) generated in the oxidative reactions of catabolism become oxidized by (i.e. reduce) molecular oxygen. This is achieved by the stepwise transfer of electrons from NADH and FADH$_2$ along a series of electron carriers. A proportion of the large amount of energy released during this process is used in the synthesis of ATP by oxidative phosphorylation. Oxidative phosphorylation generates most of ATP formed when a molecule of glucose is oxidized to carbon dioxide and water.

16   Oxidative phosphorylation is carried out by three kinds of respiratory complexes which are an integral part of the inner mitochondrial membrane.

17   The oxidation of one molecule of NADH yields three molecules of ATP, and oxidation of one molecule of FADH$_2$ yields two molecules of ATP.

18   The electron carriers of the ETC are mostly membrane-bound proteins i.e. flavoproteins, iron–sulphur (Fe–S) proteins and cytochromes, in association with the mobile carrier, coenzyme Q (a quinone derivative). One of the five cytochromes, cytochrome c, is also a loosely bound electron carrier.

19   Electrons from NADH are transferred to oxygen via the three respiratory complexes, which are linked by coenzyme Q and cytochrome c.

20   One molecule of ATP is produced (per pair of electrons transported) at each of the three sites between NADH and oxygen, corresponding to the three respiratory complexes—thus accounting for the observed three ATPs per NADH. Electrons from FADH$_2$ pass through only two of the three ATP synthesis sites—thus accounting for the observed two ATPs per FADH$_2$.

21   According to Mitchell's hypothesis, electron transport generates a proton gradient across the inner mitochondrial membrane, and it is the discharging of this gradient that leads to the synthesis of ATP.

22   The enzymes catalysing the synthesis of ATP can be seen as projections on the inner mitochondrial membrane; when detached experimentally from the membrane they can catalyse the reverse reaction—ATP hydrolysis.

23   A total of 36 molecules of ATP are produced via the complete oxidation of glucose: two ATPs via glycolysis, two GTPs directly via the TCA cycle (substrate level phosphorylations) and 32 ATPs via the ETC (oxidative phosphorylation).

24   The energy yield can be expressed as a P:O ratio—number of molecules of ATP formed (P) per atom of oxygen consumed (O).

25   The rate of oxidative phosphorylation is regulated by the ADP level in the cell.

26   Under normal conditions, electron transport is tightly coupled to oxidative phosphorylation (ATP synthesis), thus ensuring that fuel supplies are conserved when ATP is plentiful.

27   Uncouplers such as 2,4-dinitrophenol (DNP) disrupt this coupling so that electron transport to oxygen continues but no ATP is formed. Some degree of, natural, uncoupling is important thermogenesis.

28   Ion gradients across biological membranes power a variety of energy-requiring processes in addition to ATP synthesis.

29 Some cells and organisms can exist in the absence of oxygen, so their metabolism is anaerobic. Energy production is via glycolysis only, for the TCA cycle and the ETC are inoperative.

30 The pyruvate produced via anaerobic metabolism of glucose can be converted to ethanol or lactate. These chemical reductions of pyruvate use the NADH produced by glycolysis, thus regenerating the NAD^+ needed for glycolysis to continue.

31 In higher organisms the accumulation of lactate by skeletal muscle under anaerobic conditions is the price of the so-called 'oxygen debt'. This debt is 'repaid' when this lactate is reoxidized to pyruvate.

32 When glucose is catabolized anaerobically, only a small fraction of the energy available via aerobic catabolism is extracted.

33 Facultative cells can live under either aerobic or anaerobic conditions.

OBJECTIVES FOR CHAPTER 6

Now that you have completed this chapter you should be able to:

6.1 Define and use, or recognize definitions and applications of, each of the terms printed in **bold** in the text.

6.2 Understand the outline of the catabolism of glucose to carbon dioxide and water, and recognize the intermediates. (*Questions 1, 2, 3, 5, 6, 8 and 10*)

6.3 Explain briefly the functions of each of the following coenzymes involved in metabolism: NADH, $FADH_2$ and coenzyme A (CoA). (*Questions 1 and 7*)

6.4 Give examples to illustrate the role of multi-enzyme complexes in metabolism. (*Question 4*)

6.5 Explain, in general terms, how metabolic pathways may be controlled, and describe how the aerobic catabolism of glucose is regulated in accordance with the needs of the cell. (*Questions 3 and 12*)

6.6 Describe how the arrangement of electron carriers within the inner mitochondrial membrane is important in linking electron transport to the formation of a proton gradient, and account for the observed ATP yields per molecule of NADH and $FADH_2$ oxidized. (*Question 9 and 13*)

6.7 Describe the structure and function of the mitochondrial ATP-synthesizing complex in relation to the discharge of the transmembrane proton gradient. (*Question 14*)

6.8 Describe the effects of uncouplers on the ETC and hence oxidative phosphorylation. (*Questions 9 and 11*)

6.9 Explain what is meant by the term P:O ratio. (*Question 11*)

6.10 Account for the 36 molecules of ATP produced in the complete catabolism of a molecule of glucose. (*Questions 2 and 10*)

6.11 Distinguish between aerobic and anaerobic catabolism of glucose in terms of net ATP yields and mechanism of regeneration of NADH. (*Question 15*)

MEMBRANES AND TRANSPORT ◆ CHAPTER 7 ◆

7.1 OVERVIEW

7.1.1 Fluidity and permeability of biological membranes

The early microscopists first encountered membranes as cell boundaries, staining on either side to give a three-layered appearance (alternating bands of dark-light-dark) as we saw in Chapter 2. Later on, molecular studies showed how this three-layered appearance relates to a two-layer reality, the **phospholipid bilayer** described in Chapter 3, where only the two outer faces take up the stain (see Figure 7.1). In Chapter 3 we also described the **fluid-mosaic model** of membrane structure. This visualizes the cell membrane as a *fluid* structure encircling the cell, quite distinct from the watery medium on either side of it, and with a *mosaic* of different proteins floating in it (see Figure 7.2, which we shall discuss in more detail later). Nowadays, it is recognized that this picture is too simple; proteins do not always float freely—some may be anchored to the cytoskeleton and there are also semi-rigid patches. Here several protein molecules stick together and move as a unit, at the same time constraining the movement of surrounding lipid molecules. This is particularly so in membranes with a high proportion of protein (e.g. the inner membranes of chloroplasts and mitochondria, with 80% protein). Nonetheless, even these membranes are fluid in comparison to other biological barriers like plant and bacterial cell walls.

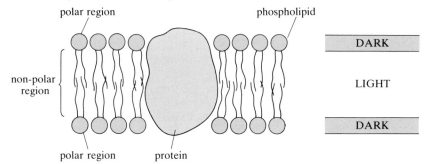

Figure 7.1 Relationship between fluid-mosaic model of membrane structure (left) and appearance under the microscope (right). (The terms *polar* and *non-polar* are described in Section 7.1.2.)

Fluidity is one of two key characteristics of biological membranes. As an example, it has been pointed out that nerve cells in the neck would crack every time a person nodded, were it not for the fluid nature of biological membranes! This fluidity can be traced directly to the chemical make-up of the lipid bilayer; as we shall see shortly, membrane lipids are held together not by rigid, covalent bonds, but by the non-covalent interactions (weak bonds) described in Chapter 3 (hydrophobic interactions, hydrogen and ionic bonding). Unlike the high-energy covalent bonds, these can be readily broken and remade with very little energy input.

The other key property of a biological membrane is its *selective permeability*; this is based on the *hydrophobicity* ('water hatred') of its component molecules. Because the phospholipid tails in the centre of the bilayer are

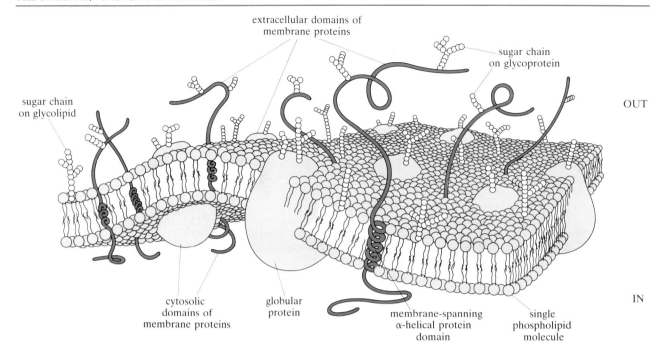

extracellular domains of
membrane proteins

sugar chain
on glycoprotein

OUT

sugar chain
on glycolipid

cytosolic
domains of
membrane proteins

globular
protein

membrane-spanning
α-helical protein
domain

single
phospholipid
molecule

IN

Figure 7.2 Fluid-mosaic model of membrane structure (here the *cell* membrane is depicted; intracellular membranes are similar but their components seldom carry sugars).

composed entirely of hydrophobic fatty acid chains, it is very difficult for water-soluble (hydrophilic) molecules to penetrate to the membrane interior. The result is a very effective permeability barrier. As we shall see, this barrier can be penetrated, but only via specific transport systems—an ideal way of controlling what goes in and out of the cell, and indeed what crosses from one subcellular compartment to another. Intracellular membranes link structure and function in just the same way as the outer cell membrane, whether they surround organelles (e.g. lysosomal and mitochondrial membranes) or form an intracellular network (nuclear and ER membranes).

In this chapter we concentrate particularly on membranes as permeability barriers. They control metabolism by restricting the flow of sugars and other water-soluble metabolites in and out of cells and between subcellular compartments—a phenomenon known as *compartmentation* (discussed in Chapter 5, Section 5.6). They store energy in the form of *transmembrane ion gradients* by allowing high concentrations of particular ions to accumulate on one side of the membrane. Controlled release of such a gradient can be used to extract nutrients from surrounding fluids, to pass electrical messages (nerve excitability) and to control cell volume and stop cells bursting with excess fluid (osmosis). These functions are discussed in Sections 7.2 to 7.4.

At the end of this chapter we turn to membrane functions based on fluidity.

Before discussing any membrane functions, we must look first at molecular structure, distinguishing *non-polar* (lipid-soluble or hydrophobic) groups of atoms of which the bilayer interior is composed, from *polar* (water-soluble or hydrophilic) groups to which it is largely impermeable.

7.1.2 Polar and non-polar interactions in membranes

In **polar** compounds, such as water or glucose, the electron cloud between certain sets of combining atoms is unsymmetrical; it is displaced towards the more electronegative of the pair, giving it a slight negative charge (Figure 7.3). Polar compounds therefore interact by mutual attraction between their positive and negative ends. Ions, like Na^+ or Cl^-, can be regarded as an extreme case, where one atomic nucleus has lost (or gained) complete control of the 'shared' electron, forming a positive (or negative) ion.

In contrast, the long fatty acid side chains of membrane phospholipids are composed entirely of **non-polar** groups of atoms including $-CH_2-$, $-CH=CH-$ and the terminal CH_3 (Figure 7.4). Here the electron cloud is symmetrically disposed over the group as a whole. However, it may be transiently polarized as constant fluctuations displace it to one or other end of the molecule (Figure 7.5). **Non-polar interactions** ('hydrophobic bonding') result when adjacent electron clouds fluctuate in synchrony, becoming trapped in the extreme position by mutual attraction of oppositely charged 'ends'. This is the essence of the hydrophobic interactions between phospholipid tails in the membrane bilayer. In eukaryotic membranes, similar interactions hold in place the flat hydrophobic steroid rings of cholesterol (see Chapter 3, Figure 3.53). Clearly much energy would be needed to insert polar molecules or ions into this intensely non-polar environment and this is why biological membranes present a *permeability barrier to polar compounds*.

Although phospholipid tails are non-polar, their heads are polar, and can enter into **polar interactions** (i.e. ionic or hydrogen bonding) at the cell surface.

◇ From the information in Figure 7.4a suggest which groups in the phospholipid head could take part in hydrogen or ionic bonding?

◆ Ionized groups like phosphate or $-\overset{+}{N}(CH_3)_3$, and polar groups like $-OH$ or $>C=O$ where the electron cloud is shifted towards the more electronegative atom of the pair (oxygen in both cases), as in Figure 7.3b.

These polar groups are important in the surface reactions of biological membranes, e.g. in binding to surface water, and to polar molecules dissolved in it.

The more specific and discriminatory of surface reactions involve polar amino acid groups on membrane proteins, or else the sugar chains that are covalently linked to many cell membrane proteins, and protrude from the cell surface. You may remember from Chapter 3 (Section 3.6.3) that proteins to which sugar chains are covalently attached are called *glycoproteins*. An enormous range of different sugar chains can be constructed, by using sugars that differ simply in the orientation of their $-OH$ groups, or by having an $-NHCOCH_3$, a $-COOH$ or other group in place of the $-OH$ (see Figure 3.45 in Chapter 3). Many lipids in the outer phospholipid layer of the cell membrane also carry covalently linked sugar chains; these are known as **glycolipids** (see Figure 7.2). Membrane glycoproteins and glycolipids are particularly important in discriminating between the myriads of molecules bumping up against the cell surface e.g. hormones, antibodies and other messenger molecules—including those bound to the surfaces of other cells. Indeed this last is often the basis of cell–cell recognition and communication.

Apart from sugars, membrane proteins carry polar groups on the side chains of amino acids such as Glu, Lys, Thr, Ser (look back at Table 3.2 in Chapter 3 for their formulae). The $>C=O$ and $>NH$ groups of the peptide bonds (Figure 3.14 in Chapter 3) are also polar. Non-polar groups are found in the side chains of Trp, Ala, Val, Phe etc.

Few membrane proteins are entirely embedded in the phospholipid bilayer; small portions (*domains*) may stick out into the cytosol or the cell exterior—or both. Sometimes these **cytosolic** and **extracellular domains** make up most of the protein, leaving only a small **membrane-spanning domain** to hold the

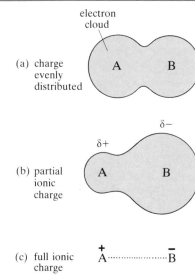

Figure 7.3 The type of compound formed by linking atoms A and B depends on the difference in their electronegativity (electron-attracting power). The symbol of $\delta+$ (or $\delta-$) indicates the loss (or gain) of only *part* of the charge associated with an electron. (a) equal electronegativity gives a *non-polar compound*; (b) slight difference in electronegativity give a *polar compound*; (c) large difference in electronegativity gives

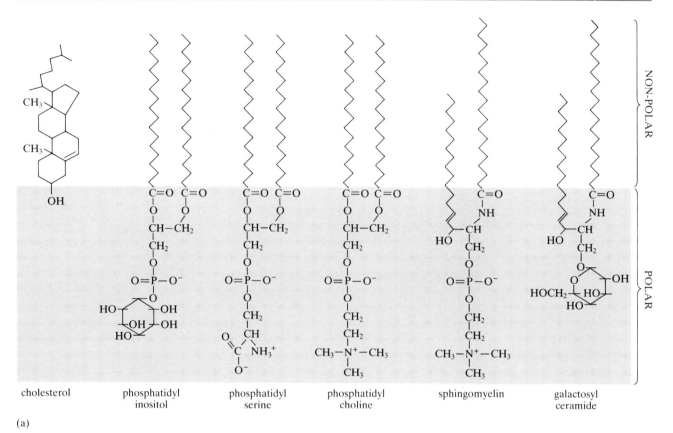

(a)

(b)

Figure 7.4 (a) Some membrane lipids. (b) Explanations of zigzag symbols representing fatty acid side chains.

protein in place. You can see in Figure 7.2 both 'fat' membrane-embedded domains (in globular proteins) and 'thin' membrane-embedded domains in what are sometimes called α-helical membrane proteins—for reasons that will become clear shortly. Obviously, the non-membrane-embedded domains will fold up in much the same way as other proteins in an aqueous medium by internalizing their non-polar residues (see Chapter 3, Section 3.4.3). However, because the interior of the membrane is non-polar, the reverse must be true for the membrane-spanning domains if they are to form a stable part of the membrane. Here the protein folds so as to externalize as many hydrophobic groups as possible, thereby maximizing opportunities for hydrophobic bonding with phospholipid hydrocarbon tails. Although entire globular proteins can be inserted into the membrane like this (as you can see in Figure 7.2) a common membrane-spanning device is a short length of α-*helical* polypeptide chain. You may remember from Chapter 3 that the amino acid side chains point away from the helix. So if this part of the polypeptide chain is composed of amino acids with non-polar side chains (R in Figure 3.14), it can readily be inserted into the bilayer. The structure labelled 'membrane-spanning α-helical protein domain' in Figure 7.2 has just such an arrangment.

The α-helical domain may form part of a *transmembrane channel* for the passage of polar **solutes** (ions or molecules in solution). The channel is composed of *several* α-helices, but here the only non-polar R groups are those on the outer edge of the channel, where hydrophobic interactions with phospholipid take place (Figure 7.6). The internal surface of each helix, i.e. the surface in contact with the channel, carries polar R groups; these are shielded from the non-polar tails of the surrounding phospholipids, and form a central polar core—very useful for the movement of ions and other polar solutes. Membrane proteins involved in transporting specific polar com-

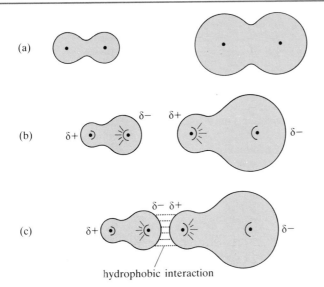

Figure 7.5 Hydrophobic interaction between two non-polar compounds. The dots or 'eyes' mark the centres of fluctuating electron clouds. The electron clouds may start off symmetrical (a) and then become transiently polarized (b); in (c) this transient polarization has become stabilized.

pounds across membranes are variously known as transporters, **transport proteins** or channel proteins. Strictly this last term should be reserved for those known to use α-helical structures as channelling devices. Some transport proteins are thought to act more like 'mobile carriers', picking up the transported solute at one side of the membrane, and moving it through the membrane, e.g. by a conformational change that alters the position of the solute binding site (Figure 7.7).

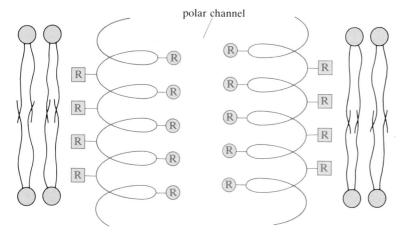

Figure 7.6 Transmembrane channel lined by lengths of protein (shown in blue) folded into regular α-helices. Polar R groups (side chains) are shown as blue circles, non-polar ones as squares.

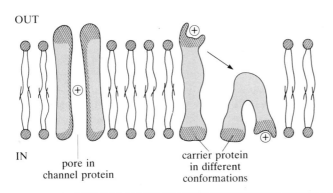

Figure 7.7 Transport of a polar solute across a biological membrane via transport protein acting as a *channel* (left) or a *mobile carrier* (right). Hatched areas indicate polar surfaces. (The conformational change shown for the mobile carrier is highly diagrammatic.)

If not involved in *transport*, most of the membrane proteins studied so far act as **receptors**. Receptor proteins are usually glycoproteins, and have binding sites for many different messenger molecules including hormones and *neurotransmitters* (hormone-like molecules released from nerve endings); they may also bind recognition molecules that protrude from the surfaces of other cells, acting as address labels or *identity tags* (Figure 7.8). In all cases the membrane-bound receptor protein has a binding site complementary to the molecule it recognizes. This molecule is known as a *ligand* (see also Chapter 3, Section 3.2.4). Receptor and ligand form a non-covalently bound complex, just like the ES complex you met with enzymes (Chapter 4).

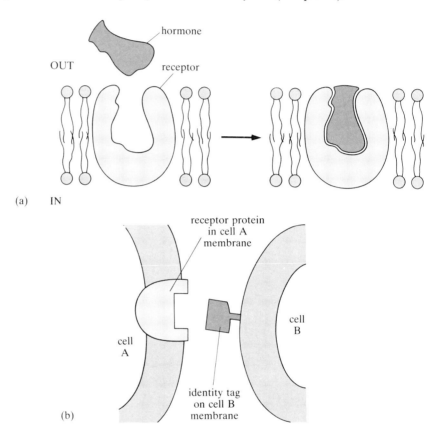

Figure 7.8 Receptor protein in the cell membrane with a binding site complementary to (a) a hormone molecule and (b) cell-bound identity tag.

Summary of Section 7.1

1 Biological membranes are non-covalently bound aggregates of protein and lipid molecules. This structural device underpins their unique characteristics of fluidity and low permeability to polar solutes.

2 All membrane protein and lipid molecules have both polar (hydrophilic or water-soluble) groups and non-polar (hydrophobic or lipid-soluble) groups. In biological membranes, these are aligned to minimize contact between non-polar groups and polar water molecules, and to maximize hydrophobic interactions with each other. The result is a fluid lipid bilayer with a non-polar core, containing a 'mosaic' of proteins.

3 All membrane proteins have a hydrophobic, membrane-spanning domain. Many also have large, hydrophilic domains that are cytosolic and/or extracellular.

4 Many membrane proteins act as transporters (transport proteins). They have binding sites specific for the solute being transported. Some have an α-helical membrane-spanning channel lined with polar groups (channel proteins).

5 Some membrane proteins, particularly those in the outer surface of cell membranes, act as receptors. They have binding sites for specific messenger molecules in the surrounding medium e.g. hormones, and for recognition molecules on other cell surfaces.

6 Cell surface glycolipids and glycoproteins often form part of receptor binding sites.

Question 1 (*Objectives 7.1 and 7.2*) Which of the following compounds would be able to penetrate the cell membrane without the assistance of special transport proteins: glucose; glucose 6-phosphate; a triglyceride (see Chapter 3, Figure 3.49); a steroid hormone (formula similar to cholesterol in Figure 3.53); vitamin D (another cholesterol analogue); the amino acid serine (see Table 3.2)?

Question 2 (*Objectives 7.1 and 7.2*) Using the formulae given in Table 3.2 of Chapter 3, pick out the non-polar groups of atoms in the side chains of the following amino acids: Ala, Met, Tyr, Trp, Ile, Leu.

Question 3 (*Objectives 7.1 and 7.3*) Are the following molecules in organelles more likely to form specific interactions with the extracellular, membrane-spanning or cytosolic domain of a membrane protein: cytoskeleton (see Chapter 2); hormone secreted into the bloodstream; surface identity tag of nearby cell; polar molecule entering the cell via a channel protein?

7.2 MEMBRANE TRANSPORT BY DIFFUSION

Having described the structure of membrane components, we can now ask how they operate. How do they block the passage of some compounds while allowing others to move freely in and out of the cell? Before answering this, we need to describe how molecules move about in free solution.

In this section we begin by describing *simple* or free *diffusion* and then show how the same driving force is used to drive polar solutes across the membrane by *facilitated diffusion*.

7.2.1 Simple diffusion

Small polar substances like water and ions can slowly 'leak' through biological membranes, by mechanisms that are not quite clear. One suggestion is shown in Figure 7.9 where they appear to slip between fluid phospholipid side chains as these wave about in random motion. In general, however, biological membranes present a formidable permeability barrier to polar compounds. Indeed, the cell membrane is best viewed as an outer barrier, surrounding and closely linked to the cytosol, with its highly organized collection of macromolecules, ions and small-molecule metabolites. The intracellular concentration of each of these molecules may be strikingly different from that of the outside medium. If the membrane were freely permeable to all molecules—large or small—no such concentration differences would be seen.

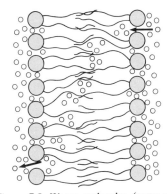

Figure 7.9 Water molecules (open circles) 'leaking' across a phospholipid bilayer.

To demonstrate concentration differences, we need first to describe a basic physical phenomenon called *diffusion*. Imagine a blob of, say, blue ink dropped into a large glass of water and left for several hours. Instead of remaining as a distinct blob, the ink particles become distributed throughout the fluid in the beaker. The energy for the redistribution process comes from the exchange of kinetic energy when water molecules and ink particles collide with each other. Equilibrium is reached when the ink particles are evenly distributed among the water molecules in the beaker, that is, when there is no gradient (difference) in ink particle concentration between any two areas within the beaker. The process of redistribution until an equilibrium is reached is called **diffusion.** It can be defined as the spontaneous movement of particles down a concentration gradient.

Here the ink is a *suspension* of particles rather than a true solution. You can see the diffusion of molecules in *solution* by putting a spoonful of treacle at the bottom of a glass of water (try it!). Sugar molecules from the highly concentrated solution (treacle) will slowly spread throughout the water.

The *rate of diffusion* of molecules can be predicted from **Fick's law**, which relates diffusion rate to the difference in concentration or *concentration gradient*, $C_H - C_L$, where C_H is higher than C_L. This relationship was derived by the German physiologist Adolph Fick in 1885, for a situation like that in Figure 7.10a. Here a nylon net has been suspended across the top of the glass. The net is permeable to the blue ink and, according to Fick's law, the rate at which ink particles diffuse through it is given by the equation:

$$\text{diffusion rate} = -\frac{DA(C_H - C_L)}{x} \tag{7.1}$$

where x is the thickness of net (or membrane), A the area of net exposed to the solution, and D the diffusion coefficient. The negative sign is a convention to show that diffusion always proceeds from high to low concentration. If rates are expressed in cm per second (cm s^{-1}), then so is D; this assumes of course that A and x are in cm^2 and cm, respectively. The term **flux** is often used to describe movement of solute through a solution. For comparative purposes, fluxes are usually expressed as quantity of solute moved per unit time *per unit area*. For movement across a membrane—which of course is what interests us here—*area* refers to the area of membrane exposed to solution.

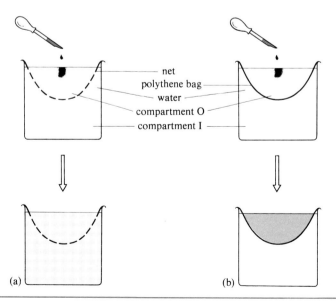

Figure 7.10 Free diffusion across barriers of different composition. (a) The *net* is not a barrier to free diffusion so the solute molecules (blue) distribute themselves evenly throughout the solution. (b) The *polythene bag* is a barrier to free diffusion, so the solute molecules are confined to compartment O.

Because the net (in Figure 7.10a) is totally permeable to the ink, it presents no barrier to free diffusion. Mixing continues until ink particles are evenly distributed throughout the solution. The situation is quite different, however, if a small polythene bag is put in place of the nylon net (Figure 7.10b). The difference between the two situations should be obvious: the net is permeable to ink particles whereas the polythene is not. In just the same way, biological membranes can act like the polythene bag, forming a barrier to free diffusion.

Imagine first what would happen if membranes were composed only of phospholipid. Experiments show that that non-polar substances cross lipid bilayers fairly rapidly, though the rate of movement is not the same for all non-polar solutes because some dissolve in lipids more easily than others. The diffusion coefficient D depends on the size and hydrophobicity (non-polar nature) of the diffusing solute, and also on the composition of the bilayer it is diffusing across. An unfortunate example of this comes from chemotherapy, using slightly polar drugs that cross the cell membrane by free diffusion—the only route open to drugs that have no specific transport mechanisms. Many drugs that act intracellularly must be able to cross the cell membrane by free diffusion, and this is very important in drug design, since the *effective* dosage for a drug that acts intracellularly will be influenced by its hydrophobicity *or* by any change in membrane permeability. This is thought to explain why some tumours become resistant to chemotherapy. Patches of highly non-polar triglyceride and cholesterol derivatives accumulate in the tumour cell membrane, impeding the entry of (slightly polar) anti-cancer drugs that would otherwise have crossed by free diffusion.

For polar solutes like ions, sugars and amino acids that the cell encounters regularly, the situation is quite different. Biological membranes are *not*, after all, made up entirely of lipid. Typically, some 50% of the membrane is composed of protein. So the passage of polar compounds can be *mediated* by specific transport proteins inserted into the lipid bilayer.

7.2.2 Facilitated diffusion of uncharged molecules

A molecule can, of course, be polar but uncharged (like glucose) i.e. it can have no *net* charge, but still be sufficiently polar for its transport across the lipid bilayer to need 'mediating' by special transport systems. These work by effectively masking either the *polarity* of the molecule being transported (by combining it with a membrane-soluble transport protein) or the *non-polarity* of the membrane (by forming a polar channel in the middle of a membrane-spanning protein). These two possibilities were shown diagrammatically in Figure 7.7.

In this section we concentrate on **facilitated diffusion**, which is mediated transport that does not need metabolic energy. Movement is by diffusion, down a concentration gradient (or down an *electrochemical gradient* for charged solutes like ions, but we shall come on to this later). A good example is the *glucose transporter*, a membrane-spanning protein found in many mammalian cell membranes; it allows them to extract glucose from surrounding nutrient fluids. The protein has a binding site specific for glucose, though chemically similar molecules may also bind, as we shall see later. Binding of glucose triggers a still obscure conformational change in the protein; this moves the glucose molecule across to the other side of the membrane, from where it diffuses off into the cytosol. This leaves the transporter free to return to its original state, ready to pick up another glucose molecule from the cell exterior. Transport continues until the glucose concentration gradient no longer exists. It then stops, because the transporter can only facilitate or mediate movement *down* a concentration gradient.

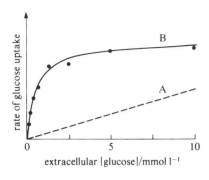

Figure 7.11 Influx of glucose into human red blood cells plotted against the glucose concentration in the bathing medium. A, Theoretical data predicted from Fick's law; B, experimental data.

Although the mechanism of facilitated diffusion is not fully understood at the molecular level, we can learn quite a lot about the process at the physiological level by studying the kinetics of glucose transport in red blood cells. Figure 7.11 shows the results of an experiment in which human red cells were incubated in radioactively labelled glucose solutions of different concentrations. If glucose had entered the cell simply by free diffusion (obeying Fick's law), the rate would have been proportional to concentration—as in curve A. However the observed influx (uptake rate) of glucose was rather different from the theoretical influx predicted by Fick's law, suggesting that glucose had not crossed the cell membrane simply by free diffusion. At low glucose concentration (below 2 mmol l^{-1}), influx was linear, but much greater than expected from diffusion alone; however, increasing glucose concentration beyond this point gave little further increase in glucose influx (curve B).

◇ Compare the two curves in Figure 7.11 with Figures 4.6 and 4.7 in Chapter 4 and then suggest a reason for the shape of curve B.

◈ Glucose could be interacting with a *limited number* of transport protein molecules. These enhance transport at low glucose levels, but are saturated at high levels. Glucose uptake would therefore obey Michaelis–Menten kinetics—hence the similarity with plots of enzyme reaction rate against substrate concentration (e.g. Figure 4.7 in Chapter 4).

Experimentally determined plots of flux against concentration gradient for many biologically important polar substances are very similar to curve B in Figure 7.11. Since permeability can be explained by membrane-bound proteins providing an alternative route across the otherwise impermeable lipid bilayer, this phenomenon is generally called *mediated permeability*. It includes both active transport (which we discuss later) and facilitated diffusion (the subject of this section).

◇ Suppose that in a similar experiment to that shown in Figure 7.11, white blood cells were used and the same sort of graph was obtained. What two quantities could be used to compare the transport proteins involved in the movement of glucose across the two different cell membranes? (Recall how the properties of enzymes are often quantified—if you cannot remember, look up Chapter 4, Section 4.4)

◈ Because these systems can be described mathematically in a similar manner to the interaction of enzymes and their substrates, it is usual to apply the same conventions. The terms v_{max} and K_M can thus be used to describe and compare different transport proteins, as well as different enzymes.

Most biologists use slightly different symbols for v_{max} and K_M in this context. Maximum rate of reaction (v_{max}) becomes J_{max} (*maximum rate of transport*), and is related to the number of transport protein molecules per unit area of membrane. K_M becomes K_t (where the 't' stands for transport); the constant K_t is used as an indicator of *affinity* i.e. the ease with which the transport protein binds to the molecule being transported. Here J_{max} and K_t are obtained from plots of J (rate of uptake) against the concentration of glucose, in exactly the same manner as is used for v_{max} and K_M. With transport proteins, K_t is defined as the solute concentration at which transport is half maximum. This K_t is always a true *binding constant*, measuring the affinity between transport protein and the molecule being transported. (In contrast, K_M is only a measure of the enzyme–substrate affinity for *some* enzymes, because its value is affected by later changes inflicted on the substrate by the enzyme.) High K_t indicates low affinity and low K_t indicates high affinity.

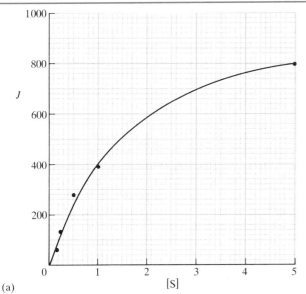

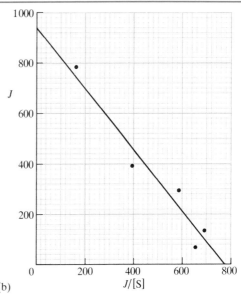

(a)   (b)

Figure 7.12 Kinetics of transport of the amino acid phenylalanine across isolated segments of rat intestine. (a) Plot of flux, J, in nmol (mg protein)$^{-1}$ h^{-1} against phenylalanine concentration, [S], in mmol l^{-1}. (b) Another way of plotting the data in (a). *Note*: (i) So that the preparation remained as fresh as possible, measurements were made over a 15 minute period—even though they are expressed here as uptake rate per hour. (ii) So that different preparations of rat intestine can be compared, uptake rates are expressed *per mg of protein*.

◇ Figure 7.12a shows a plot of J, against outside (extracellular) concentration of the amino acid phenylalanine, for epithelial cells from rat intestine, and Figure 7.12b shows a linear plot of the same data. What name is given to this latter type of plot? (cf. Figure 4.12 in Chapter 4.) Calculate J_{max} and K_t for phenylalanine uptake from Figure 7.12b.

◆ Figure 7.12b is a Hofstee–Eadie plot, so J_{max} is the intercept on the vertical axis. This is about 940 nmol (mg protein)$^{-1}$ h^{-1}. K_t can be estimated from the slope of the graph, which equals $-K_t$, or from the intercept on the horizontal axis. The intercept, which equals J_{max}/K_t, is at about 780.

The units of J_{max}/K_t are the same as for $J/[S]$

Therefore $J_{max}/K_t = 780$ nmol (mg protein)$^{-1}$ h^{-1} mmol^{-1} l

Substituting for $J_{max} = 940$ nmol (mg protein)$^{-1}$ h^{-1} and rearranging, we get

$$K_t = \frac{940 \text{ nmol (mg protein)}^{-1} \text{ h}^{-1}}{780 \text{ nmol (mg protein)}^{-1} \text{ h}^{-1} \text{ mmol}^{-1} \text{ l}}$$

Cancelling similar units, we get

$$K_t = \frac{940}{780} \text{ mmol l}^{-1}$$

$$= 1.20 \text{ mmol l}^{-1}$$

Notice that just as K_M always comes out in units of substrate concentration, so K_t always comes out in units of solute concentration, irrespective of the units used for J.

Enzymes and transport proteins share the typical characteristics of globular proteins described in Chapter 3. For example, they both exhibit *specificity*. The data in Table 7.1 emphasize this in two ways. First, the transport protein in this particular membrane has a higher affinity, i.e. lower K_t, for D-glucose

Table 7.1 Uptake of different sugars by the glucose transporter of the human red blood cell membrane

Sugar	K_t/mmol l^{-1}
D-glucose	4
D-galactose	30
D-xylose	60
L-galactose	>3 000
L-glucose	>3 000

than D-xylose. (Recall from Chapter 4 that for some enzymes low K_M values indicate a *high* affinity for substrate.) Secondly, you can see from Table 7.1 that transport is stereospecific, because there is a vast difference in affinity for different stereoisomers (cf. D-glucose and L-glucose). Although mediated permeability systems often show discrimination between different types of solute molecule, they are frequently able to transport several structurally related compounds, but usually to different extents. Like enzymes, transport proteins display *competitive inhibition*.

Quite a number of mediated permeability processes can be stopped by substances that at first sight appear to be unrelated to the compound being transported. These inhibitors bind covalently to the transport molecule and are equivalent to the irreversible inhibitors described for enzymes in Chapter 4 (Section 4.4.3). Many are highly toxic in minute amounts; this emphasizes the importance of membrane transport in living organisms. For example, the explorer Captain Cook recorded in his ship's log for September 1774 that two of the ship's naturalists were given puffer fish by the 'friendly' natives of a Pacific island. The ship's cook served the livers and roes to the captain and his friends and threw the remains of the fish to a pig kept on board ship. The following morning, Cook and the naturalists had numb mouths and tongues and partly paralysed limbs. Fortunately they recovered, but the pig died! The substance responsible for these effects has since been isolated and is known as *tetrodotoxin*. The pure compound is extremely toxic (1 mg would kill you), yet its mode of action is quite simple. It binds covalently to the *sodium channel protein* that makes nerve membranes permeable to Na^+. (We shall describe this protein in detail later.) *Transport inhibitors* such as tetrodotoxin are valuable research tools and enable the experimenter to 'switch off' specific transport systems. Radioactively labelled forms of these compounds can be used to estimate the number of transport molecules in a particular cell by measuring the number of radioactive molecules that bind to a known quantity of membrane. These techniques have shown that the membrane concentration of a particular transport protein can vary throughout the life of the cell, depending on the nutritional and developmental state.

We have now outlined the basic features that characterize mediated permeability. The features that distinguish it from free diffusion through the lipid bilayer are:

1 Transport rate and membrane permeability are greater than predicted from the molecular size and lipid solubility of the molecule being transported (at low concentrations of this molecule).

2 The transport rate saturates at high solute concentrations (Michaelis–Menten type of kinetics).

3 Transport is highly specific, and the properties of a transport protein molecule can be expressed in quantitative terms (i.e. K_t and J_{max}) that are analogous to those used for enzymes.

4 The transport rate can be altered by substances closely related to the solute being transported, or by other molecules that are able to bind covalently to the transport proteins. This is equivalent to competitive and irreversible inhibition respectively.

5 In mediated permeability processes, the gradient can be a concentration gradient as in the examples described so far, or an electrochemical gradient if charged solutes are involved (see later).

6 Facilitated diffusion is a form of mediated permeability that does not need metabolic energy, because transport is down a concentration gradient. But however great the concentration gradient, the transported molecule can move through the membrane only by combining with a transport protein.

7.2.3 Facilitated diffusion of ions: electrochemical gradients and membrane potential

For polar molecules with no net charge (e.g. glucose) facilitated diffusion across a biological membrane is essentially the same as the free diffusion of non-polar (lipid-soluble) molecules across a lipid bilayer. Until the transport protein is saturated, the driving force is the difference in concentration across the membrane. Transport continues until an equilibrium position is eventually reached, i.e. until the concentrations on either side of the membrane are equal. (That is not to say movement across the membrane ceases at this point—just that the *net* flux is zero because transport rates are the same in both directions.)

Transport of charged solutes (i.e ions) is rather different. Although ions tend to move spontaneously down their concentration gradient, their transport is inevitably affected by the electrical gradient produced when positive and negative charges are unevenly distributed across the membrane. The cell interior tends to be negative, largely because ionizable groups on intracellular macromolecules are predominantly negative at cellular pH. These so-called 'fixed charges' enhance the entry of positive ions, and repel negative ones. There comes a point when the tendency of an ion to move one way down its *concentration* gradient is exactly balanced by its tendency to move the other way, down its *electrical* gradient. This is the point of *electrochemical equilibrium*. Until this point is reached, an ion—in combination with its transport protein—will spontaneously cross the membrane, driven by a combination of concentration and electrical gradients i.e. the **electrochemical gradient**.

To visualize in more detail how the electrochemical gradient works, look at the theoretical membrane which in Figure 7.13a separates two compartments (O and I) each containing an ionic substance such as potassium chloride (KCl) at a concentration of 10 mmol l^{-1}. If the membrane is permeable only to K^+ ions, these will move across the membrane at equal rates from O to I or back again, from I to O. (The flux of chloride ions will of course be zero, because the membrane is impermeable to Cl^-.) Suppose we now increase the concentration of KCl in compartment O to 100 mmol l^{-1} (see Figure 7.13b). There is now a concentration difference between compartments O and I, and hence K^+ will tend to diffuse from O to I. However, as soon as the first K^+ crosses from compartment O to I, another driving force comes into operation. Originally, the positive charge on every K^+ in both compartments was balanced by a negatively charged Cl^-. When a K^+ migrates from O to I, a

Figure 7.13 Voltage produced by facilitated diffusion of ions across a membrane that is permeable to K^+ and impermeable to Cl^-; the situation at electrochemical equilibrium when KCl concentrations in compartments O and I are (a) equal and (b) unequal. In (b), the electrical and concentration driving forces work in opposite directions, and electrochemical equilibrium is reached only by having an imbalance of charges across the membrane. (The terms J^K_{OI} and J^K_{IO} indicate K^+ flux from O to I, and from I to O respectively.)

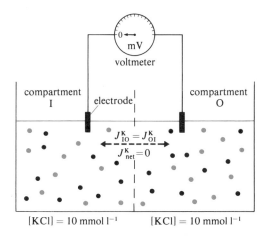

(a)

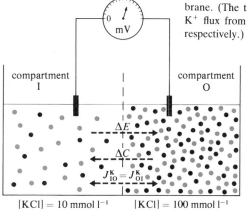

KEY
- potassium ions
- chloride ions

(b)

difference in electrical potential will develop between the two compartments; compartment O will have a net negative charge, and compartment I a net positive charge. This difference in electrical potential or voltage can be measured with a voltmeter. It can be visualized as the work needed to move an electrostatic charge against the repulsive force that exists between similarly charged particles.

Since like charges repel each other, the next K^+ that tries to move across the membrane will have two driving forces acting on it: one due to the difference in concentration between compartments O and I, which will tend to drive the K^+ in the direction O→I; and a second one due to the difference in electrical potential, which will tend to drive the K^+ in the opposite direction I→O. At first, the electrical potential force is weaker than that due to the concentration difference, but as more K^+ cross from O to I the electrical potential difference (ΔE) becomes larger (Δ means *difference in*), and it is more difficult for K^+ to move from O to I. Eventually, these two oppositely directed forces will come into balance and electrochemical equilibrium is reached.

By putting a figure on these charge and concentration differences, we can measure the electrical force (i.e. the voltage) that exactly counterbalances the concentration difference at equilibrium i.e. we can find out what the voltmeter would read. This reading, of potential difference across a biological membrane, is known as the **membrane potential**, E_m.

Some closely reasoned mathematical argument is needed to calculate membrane potential. For both ions the concentration difference (ΔC) is known $(100 - 10 \text{ mmol l}^{-1} = 90 \text{ mmol l}^{-1})$, so the question being asked is, in theory, quite simple: what voltage has the same driving force as a concentration difference of 90 mmol l^{-1}? Both ΔE, the voltage difference, and ΔC are measures of force but unfortunately they are not expressed in identical units. In order to compare them, we need to convert these two driving forces into the same energy units, normally joules. Electrical potential differences (voltages) are converted to joules by multiplying ΔE by a suitable conversion factor. The chemist Faraday was able to show that the movement of one mole of a monovalent ion is equivalent in electrical terms to the transfer of 96 500 coulombs of electrical charge. (For divalent ions, the quantity transferred is $2 \times 96\,500$ coulombs.) The figure of 96 500 coulombs mol^{-1} is known as the *Faraday constant* (F). Ion valency and charge are usually embodied in the symbol Z, e.g. Z for K^+ is $+1$, for Cl$^-$ it is -1, and for Ca^{2+} it is $+2$. The necessary conversion factor for ΔE is $Z \times F$, or ZF. Similarly, chemists have shown that the driving force caused by a concentration difference, ΔC, can be converted to joules by using as a conversion factor the gas constant (R) multiplied by the absolute temperature of the system (diffusion is dependent on temperature). So the conversion factor is $R \times T$, or RT. When equilibrium is reached, the two driving forces are equal and opposite. So, in symbols

electrical driving force = $ZF\Delta E$ (7.2)

concentration driving force = $RT\Delta C$ (7.3)

where $\Delta E = E_I - E_O$, the electrical potential difference between compartment O and compartment I (it is written $E_I - E_O$ because the driving force due to electrical potential difference acts in the direction I→O). Therefore Equation 7.2 becomes

electrical driving force = $ZF(E_I - E_O)$ (7.4)

The concentration difference, $\Delta C = C_O - C_I$, continually changes as ions move from compartment O to I until equilibrium is reached. Therefore a *better* expression for the concentration difference turns out to be $2.303\,(\log_{10}C_O - \log_{10}C_I)$. This means that Equation 7.3 becomes

$$\text{concentration driving force} = 2.303RT\log_{10}C_O/C_I \tag{7.5}$$

At equilibrium, the electrical and concentration driving forces are exactly balanced. So by substituting for the two driving forces from Equations 7.4 and 7.5, we arrive finally at a mathematical description of the situation at equilibrium:

$$ZF(E_I - E_O) = 2.303RT\log_{10}\frac{C_O}{C_I} \tag{7.6}$$

By manipulating this equation, we can now answer the question: what would the voltmeter read at equilibrium, i.e. what is the membrane potential, E_m? At equilibrium $E_I - E_O = E_m$, so

$$E_m = \frac{2.303RT}{ZF}\log_{10}\frac{C_O}{C_I} \tag{7.7}$$

This equation was originally derived in 1889 by the chemist Walter Nernst, so it is usually called the **Nernst equation**.

◇ Use the Nernst equation to calculate the value of E_m for the situation shown in Figure 7.13b. Assume that the gas constant (R) is 8.314, and the temperature is 18 °C. (Remember the Faraday constant for a monovalent ion is 96 500 coulombs mol^{-1}, Z is +1 for positively charged, monovalent ion, and T is the absolute temperature, i.e. here $(273 + 18)$ K or 291 K.

◈ Substituting in Equation 7.7 we get

$$E_m = \frac{2.303 \times 8.314 \times 291}{+1 \times 96\,500} \times \log_{10}\frac{100}{10}\ \text{volts}$$

$$= 0.058 \times \log_{10}10\ \text{volts}$$

$$= +0.058\ \text{volts} \qquad (\text{remember } \log_{10}10 = 1)$$

Rather than talk in terms of fractions of a volt, it is usual to express E_m values in millivolts (mV). Thus at equilibrium, compartment I in Figure 7.13b is 58 mV more electropositive than compartment O (recall $E_m = E_I - E_O$). This is due to the migration of K^+ into compartment I.

Summary of Section 7.2

1 In free diffusion, particles, molecules and ions all move spontaneously down their concentration (or electrochemical) gradients. Rate of movement (flux) is proportional to this gradient; it can be calculated from Fick's law.

2 Facilitated diffusion is mediated by specific transport proteins. Diffusion proceeds spontaneously down a concentration gradient (for uncharged solutes) and down an electrochemical gradient (for ions and other charged solutes). Metabolic energy is not required.

3 Transport kinetics show that each transport protein has a binding site specific for the transported solute, which becomes saturated at high solute concentrations, and can be competed for by chemically similar molecules.

Maximum flux (J_{max}) and transporter–solute affinity (K_t) are equivalent to v_{max} and (less rigorously) to K_M, in the interaction of enzyme and substrate to form ES.

4 Facilitated diffusion of ions produces a membrane potential, E_m, whose magnitude depends on the position of electrochemical equilibrium for each particular ion. This is influenced by the overall charge on either side of the membrane, and can be calculated from the Nernst equation.

Question 4 (*Objectives 7.1 and 7.4*) Patients with diseased kidneys tend to accumulate toxic nitrogen compounds (e.g. urea) in the bloodstream. These can be removed by dialysing the blood in an artificial kidney i.e. by allowing it to equilibrate with a urea-free solution, from which it is separated only by a thin urea-permeable cellophane membrane. Using Fick's law, explain how dialysis can be speeded up by (a) replacing the urea-free dialysis fluid at regular intervals and (b) changing the length of cellophane membrane in the dialysis machine.

Question 5 (*Objectives 7.1 and 7.6*) Uptake of the amino acid serine from the gut is much impaired if the amino acid threonine is also present in the gut lumen. If a small quantity of a small molecule X that binds irreversibly to amino acid transport protein is inadvertently swallowed, uptake of both serine and threonine is drastically reduced. Explain both observations in terms of the competitive and irreversible inhibition of transport proteins.

Question 6 (*Objectives 7.1 and 7.6*) Are the following statements true or false?

(a) Once J_{max} for a particular solute X has been reached, the rate J at which X is transported across a cell membrane cannot be increased, however much the concentration of X is raised.

(b) If K_t for X is higher than for another solute Y, the transport protein binds X less strongly than Y.

Question 7 (*Objectives 7.1 and 7.7*) A squid nerve cell membrane is passively permeable to chloride ions. The intracellular and extracellular concentrations of chloride ions are 60 mmol l^{-1} and 540 mmol l^{-1}, respectively. What would you expect the transmembrane potential (E_m) of this membrane to be (in millivolts), assuming that it was measured at 18 °C? (Remember that chloride ions have a negative charge and the Faraday constant for a monovalent ion is 96 500 coulombs mol^{-1}. Note also that the gas constant, R, is 8.314 and $\log_{10}9$ is 0.95.)

7.3 MEMBRANE TRANSPORT REQUIRING METABOLIC ENERGY

7.3.1 Introducing active transport

Although facilitated diffusion (sometimes called *passive transport*) is the commonest form of protein-mediated transport across membranes, it tends to be overshadowed by **active transport**. Here the driving force is not diffusion; solutes do not move down their concentration gradients, but are actively 'pumped' up a gradient using energy from another source. This apparently perverse idea has two advantages over passive transport. It allows desirable solutes to be accumulated and undesirable ones to be removed. Furthermore, much of the energy used in the uphill pumping is conserved. So active transport can provide a very neat way of *storing energy*. By moving back *down* the concentration gradient the stored solute can release the energy needed for its accumulation, putting it to a variety of ingenious uses. We shall return to this shortly in more detail.

At the physiological level, active transport was first discovered from observations like the following:

◇ Certain kidney cells are responsible for reabsorbing valuable solutes such as sugars from the liquid filtered from the blood, immediately before this filtrate passes to the bladder for excretion as urine. Suppose that the concentration of glucose in the cytosol of the kidney cells and in the filtrate are 40 mmol l^{-1} and 5 mmol l^{-1}, respectively. In terms of glucose transport, what would you expect to happen if the membrane of these cells were made permeable to glucose?

◆ Glucose should move *out* of the cell *down* the prevailing concentration gradient until the concentration of glucose outside the cell is the same as that inside the cell. However, this does not happen. These cells *take up* further glucose under these conditions! They must therefore be capable of a different sort of transport process.

There are several ways of testing whether a substance is moving by active transport or not. The most obvious is to see whether concentrations of the substance in question are the same in the tissue of interest, as in the medium surrounding it.

◇ A higher concentration in the tissue would indicate glucose uptake by active transport. What might a higher concentration in the medium suggest?

◆ That glucose is being *secreted* into the medium, again by active transport.

Another way of distinguishing between active and passive transport is to add to the incubation medium a metabolic inhibitor that will block the major pathways of energy production. Most active transport systems are driven by metabolic energy in the form of ATP—sometimes directly, as in the *sodium pump*, and sometimes indirectly through an ion gradient (see below). Whatever the mechanism, metabolic inhibitors deprive active transport of its driving force and are therefore useful experimental tools.

7.3.2 Molecular mechanisms that link metabolic energy and active transport

Linking metabolic energy to ion transport is possible only because of the remarkable capabilities of certain transport proteins. Most important is the **sodium pump**, a membrane-spanning protein found in virtually all animal cell membranes, and consuming up to 30% of the cell's energy—70% in the case of nerve cells. Blocking the pump with the specific inhibitor ouabain (pronounced w-ar-bane) has disastrous consequences for the uptake of nutrients, for nerve and muscle excitability and for the osmotic balance of all cells—as you will see by the end of this chapter.

The sodium pump has been intensively studied, though even now a few of its molecular mechanisms remain obscure. It had been known for a long time that most animal cells can pump Na^+ ions out of the cell against an electrochemical gradient, if K^+ ions are simultaneously pumped in. The energy source for this active transport became much clearer when it was discovered (in 1957) that red cell membranes have an ATPase that needs Na^+ and K^+ ions for maximum activity. (Hence the sodium pump is often called the *Na^+/K^+ ATPase*.) Red cell membranes are a very convenient tool for transport studies, because empty 'ghosts' of cells can be made by gently lysing (bursting) the cell membrane, allowing the cytoplasmic contents to leak out, and replacing them with known solutions—e.g. one containing ATP—before the membranes reseal. These preparations are known as *resealed ghost cells*. By varying the composition of intracellular and extracellular solutions in this way, it was shown that the sodium pump binds ATP and Na^+ at its inner, cytosolic face and K^+ at its outer face, and that ouabain inhibits both ion transport and ATP hydrolysis by competing for the K^+ binding site. But the key to how these two activities are coupled had to wait till this complex, multimeric protein had been purified.

The protein has two subunits, one of unknown function and one whose function is well characterized. This is known as the catalytic subunit. It has an M_r of 100 000 and during ATP hydrolysis a single aspartate residue gets phosphorylated by the inorganic phosphate released in converting ATP to ADP. This simple chemical modification appears to trigger a conformational change across the entire, membrane-spanning protein (Figure 7.14), such that Na^+ already sitting in its cytosolic binding site is transported to the other side of the membrane and released (i.e. pumped out of the cell). Loss of Na^+ is thought to trigger a series of further conformational changes that open up the K^+ binding site so that extracellular K^+ becomes bound. This enables phosphate to be cleaved off the aspartate residue, K^+ is released to the cell interior, and the protein returns to its original conformation, ready to bind Na^+ and ATP again. The proposed sequence is shown in Figure 7.15. Note the general point that proteins can convert one form of energy to another by changing from a relaxed, low-energy conformation to a strained high-energy one. By relaxing the constraints, energy is made available for an energy-requiring process—like ion pumping. In the sodium pump, the relaxed conformation is the Na^+-binding, unphosphorylated form; phosphorylation produces the constrained conformation.

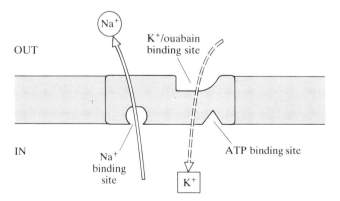

Figure 7.14 Diagram linking the functions of the sodium pump to specific binding sites on the protein surface.

Any system that converts one form of energy to another is known as an *energy transducer*. Biological membranes, as well as individual proteins, can act as energy transducers but only if they are impermeable to polar solutes like ions. Ions pumped from one side of the membrane to the other (as by the sodium pump) will then be unable to move back again. The result is a *transmembrane ion gradient* where the concentration of ions is much greater on one side of the membrane than the other. The gradient is a form of stored energy. With the sodium pump example, the *energy source* was the energy of ATP hydrolysis used to expel sodium ions *against* their electrochemical gradient. If the membrane then becomes permeable to sodium ions, they will flood back into the cell, *down* their electrochemical gradient, allowing the stored energy to be used to drive a variety of *energy-requiring* processes, or so-called *energy sinks*.

This sort of analysis leads to an interesting question: can such energy exchange processes work backwards? Can energy be transferred from the downhill flow of ions to an energy-requiring chemical reaction? This question has been tested experimentally by preparing resealed red cell ghosts with a *low* K^+ concentration. This *reverses* the normal cellular ion gradient. In most cells, sodium ion concentration is low, and K^+ concentration high. Under these normal conditions, Na^+ is pumped out of the cell (and K^+ is pumped in) via the ATPase, but in these specially prepared cells ATP is *synthesized*. So a 'downhill' flow of ions can be used to drive an energy-consuming chemical reaction (ATP synthesis).

Ion gradients provide the link between energy source and sink in a wide variety of membrane-based energy exchange systems (Figure 7.16). The familiar concept of active transport, which has been known to physiologists for a long time, can now be seen as just one of these systems, part of a wider phenomenon that we can now look at in more detail.

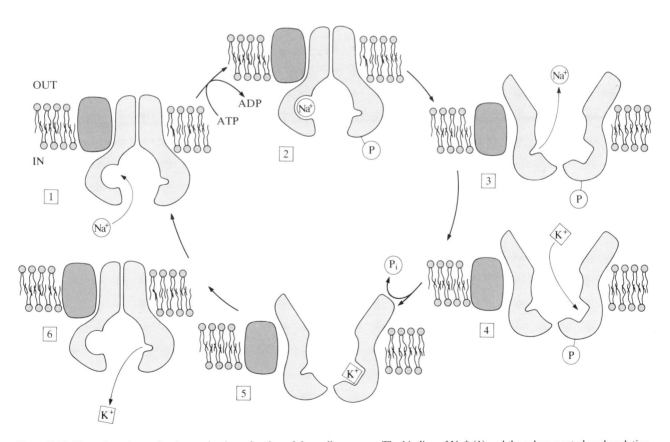

Figure 7.15 Tentative scheme for the mechanism of action of the sodium pump. The binding of Na^+ (1) and the subsequent phosphorylation (2) on the cytosolic face of the ATPase (light blue) induce the protein to undergo a conformational change that transfers the Na^+ across the cell membrane and releases it on the outside (3). Then the binding of K^+ on the external surface (4) and the subsequent dephosphorylation, releasing P_i, (5) return the protein to its original conformation, and transfers the K^+ across the membrane and releases it into the cytosol (6).

7.3.3 Ion gradients and energy exchange: membranes as energy transducers

The role of ion gradients in energy exchange was first put forward in the 1960s as part of Peter Mitchell's *chemiosmotic hypothesis*, for which Mitchell was awarded the Nobel Prize in 1978. You met this hypothesis in Chapter 6 (Section 6.5.4) where it was described how a proton gradient (built-up using the energy of electron transport) can be used to drive ATP synthesis. When he first introduced this idea, Mitchell was working on transport across bacterial membranes, where the transported ion was Na^+ and the key protein was a sodium pump similar to the one just described. Mitchell realized that the energy in *any* electrochemical gradient can be used to perform biological work, provided that the downhill driving force for ion movement can be *coupled* to other cellular processes. When the ion gradient is dissipated and its energy used to transport a second substance against an electrochemical gradient, stage II in Figure 7.16 is sometimes called *secondary active transport. Primary active transport* then refers to the process of building up the ion gradient in the first place (stage I in Figure 7.16).

A good example of the link between primary and secondary active transport comes from the gut. Sugars and amino acids from the digestion of food have to be transported into the intestinal cells lining the gut lumen, even though—particularly after a meal—their intracellular concentrations may already be quite high. This means transporting the sugars and amino acids against a concentration gradient into the intestinal cells. Various transport proteins in the intestinal cell membrane can bring about the *co-transport* of Na^+ ions and glucose (or other sugars, or amino acids). One such protein is the **Na^+-powered glucose transporter**. It is also found in cells lining kidney tubules, as we saw at the beginning of this section; here it re-absorbs glucose in the formation of urine. In both cases, glucose molecules enter the cell

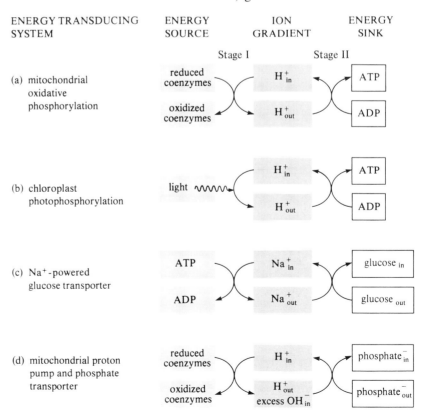

Figure 7.16 Energy exchange systems using transmembrane ion gradients. (Note that in (b) 'out' refers to the thylakoid space and 'in' refers to the stroma—Chapter 2, Section 2.9.)

against a concentration gradient, by allowing Na^+ ions to move simultaneously back *down* their electrochemical gradient. The transporter has binding sites for both Na^+ and glucose, both of which must be occupied before transport can take place. The Na^+ gradient exists because the energy of ATP hydrolysis has previously been used by the sodium pump to eject Na^+ ions against an electrochemical gradient (Figure 7.16c).

◇ What would happen to glucose transport if the cells were bathed in a solution containing ouabain?

◆ Glucose transport would cease as soon as intra- and extracellular concentrations were equal. By inhibiting the sodium pump, ouabain deprives the Na^+-powered glucose transporter of its energy source.

Like the sodium pump, the Na^+-powered glucose transporter links two separate activities (an energy-providing *source* and an energy-requiring *sink*) so that one cannot proceed without the other.

◇ What are these two linked activities in (a) the Na^+-powered glucose transporter and (b) the sodium pump?

◆ In the glucose transporter, the two linked activities are Na^+-transport (energy source) and glucose transport (energy sink). In the sodium pump they are ATP hydrolysis (energy source) and Na^+ transport (energy sink).

Table 7.2 Ion gradients and energy exchange systems

Energy exchange system	Energy source for formation of ion gradient	Ion gradient	Energy sink for dissipation of ion gradient
Inner mitochondrial membrane	Reduced coenzymes ($6\,H^+/e^-$ pair)	H^+	*Chemical work*: ATP synthesis
			Active transport: Ca^{2+} uptake, ATP export or ADP uptake, phosphate uptake, Na^+ export
Bacterial cell membrane e.g. in *E. coli*	Reduced coenzymes ($4\,H^+/e^-$ pair)	H^+	*Chemical work*: ATP synthesis
			Active transport: uptake of Na^+, Ca^{2+}, sugars, amino acids
			Mechanical work: rotation of flagellum (1 rotation per $256\,H^+$)
Chloroplast membrane	Light	H^+	*Chemical work*: ATP synthesis
Purple patches on cell membrane of the bacterium *H. halobium*	Light	H^+	*Chemical work*: ATP synthesis
			Active transport: ion uptake
Na^+ pump in cell membrane	ATP hydrolysis	Na^+	*Active transport*: uptake of glucose, amino acids, Cl^-

The energy sink does not always involve secondary active transport. Nor does the ion gradient always need Na^+ ions. Table 7.2 lists a number of quite different energy exchange systems, where energy sources and sinks are linked through gradients of Na^+ or H^+, generated by either the sodium or proton pump. Figure 7.16 shows the same thing diagrammatically. In all cases, membranes form a key component of the energy-exchange system. Some **proton pumps** are driven by chemical energy and others by light energy. You have already met pumps driven by chemical or metabolic energy in oxidative phosphorylation (Chapter 6, Section 6.5). Here, energy released from the re-oxidation of reduced coenzymes via the electron transport chain, is used to pump protons across the inner mitochondrial membrane against an electrochemical gradient. An excess of protons builds up in the space between inner and outer mitochondrial membranes. The gradient can be dissipated only at certain points in the inner membrane, where a complex transmembrane protein (the F_0–F_1 ATPase) links the inflow of protons to the synthesis of ATP (Section 6.5.4 in Chapter 6).

Several proton pumps are driven by light energy. The simplest of these is the protein *bacteriorhodopsin*, a key component of the purple patches found in the membrane of *Halobacterium halobium* (a bacterium that lives in very salty water).

Bacteriorhodopsin consists of a light-absorbing pigment, retinal, covalently bound to a single polypeptide chain that winds back and forth across the membrane. The protein contains a proton channel. Light of about 560 nm wavelength is absorbed by the retinal, changing its colour from purple to white. As this activated form of the pigment returns to its initial state, protons are pumped through the protein channel, setting up a gradient that can be made use of in a variety of ways. These include both ATP synthesis and active transport. A similar light-driven energy transducer has been found in the rod cells of vertebrate eyes and in photoreceptors of invertebrates. In vertebrates, rod cells lining the retina appear able to convert the energy of absorbed light into a nerve impulse. Although the molecular mechanism for this is still unknown, it is surely no coincidence that 80% of the rod cell membrane is composed of rhodopsin, a retinal-linked protein very similar in secondary structure to bacteriorhodopsin. The light-controlled channel that it forms across the membrane may however transmit sodium ions, rather than protons.

Light is also the energy source in the more complex energy-transducing membrane of the chloroplast. Here retinal is replaced by chlorophyll and other pigments, and light energy is used to promote the flow of electrons along a series of membrane-bound carriers, from water to $NADP^+$. The reduced form of the coenzyme (NADPH) represents the end-point of electron flow, unlike its counterpart (NADH) which is the starting-point in mitochondria. Structurally, electron carriers of chloroplast and inner mitochondrial membranes have much in common. However, in chloroplasts, protons are pumped into the closed membrane-bound sacs (thylakoids) inside the chloroplast (see Plate 2.7), rather than being expelled to the outside (i.e. into the stroma).

So proton pumps can use a variety of energy sources to create the key transmembrane ion gradient. This stored energy can spill out into a variety of energy *sinks* (Figure 7.16). In photophosphorylation, the chloroplast uses the proton gradient mainly for driving ATP synthesis. This is catalysed by a membrane-bound ATP synthetase which closely resembles the mitochondrial F_0–F_1 complex. Bacteria like *E. coli* may use the proton gradient not only for ATP synthesis and active transport, but also for *mechanical work*. The motility of some bacteria depends on the propeller-like action of a flagellum

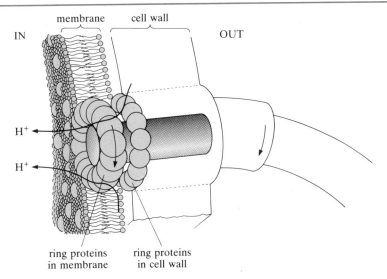

Figure 7.17 Tentative model for flagellar rotation in *E. coli* driven by proton influx.

(Chapter 2, Section 2.11). This slots into a ring of membrane proteins to form what must be the smallest electric motor in existence (see Figure 7.17). Ring proteins seem to provide an alternative to the F_0 component as a re-entry channel for protons. Somehow *rotation of the flagellum* in this ring is powered by proton influx, about 256 protons being required for every full flagellar revolution.

Finally, we come back to active transport. In both mitochondria and bacteria, dissipation of the proton gradient may be linked to the uphill flow of some other substance against its electrochemical gradient. In this respect, it is important to note that the proton pump is *electrogenic* i.e. it produces a *net transfer of charge* across the membrane. In mitochondria, for example, it leaves one OH^- behind for every H^+ expelled. (Since both ions, H^+ and OH^-, come largely from dissociation of H_2O, they are originally present in equal amounts.) Many positively charged substances exploit this internal abundance of OH^- to gain entry to the mitochondria against an electrochemical gradient. For example, Ca^{2+} ions can be accumulated in mitochondria so long as the proton pump is operating to maintain an internal excess of OH^- ions.

Summary of Section 7.3

1 Active transport is the movement of solutes across a membrane, against their concentration or electrochemical gradient.

2 The energy for this uphill movement may come directly from ATP hydrolysis (e.g. sodium pump); more often it comes from the simultaneous dissipation of an ion gradient.

3 Many different energy-producing systems (energy sources) can store energy in the form of a transmembrane ion gradient (Mitchell's chemiosmotic hypothesis). These can be tapped by many different energy-requiring systems (energy sinks).

4 Examples of energy sources are light (chloroplast membrane, bacteriorhodopsin), electron transport for coenzyme re-oxidation (inner mitochondrial membrane), ATP hydrolysis (sodium pump).

5 Examples of ion gradients used as intermediaries between energy sources and sinks are: Na^+ ions (as produced by the sodium pump and used by the Na^+-powered glucose transporter), and H^+ ions (proton pump).

6 Examples of energy sinks are: Na^+-powered glucose transporter, ATP synthesis (F_0–F_1 ATPase), flagellar rotation, and active transport of mitochondrial Ca^{2+} ions against an electrochemical gradient.

7 Formation of an ion gradient can be described as primary active transport; dissipation of the gradient to move other solutes 'uphill' can be described as secondary active transport.

Question 8 (*Objectives 7.1 and 7.5*) Which of the following three criteria must be satisfied if substance X is to be transported across a cell membrane by (a) free diffusion, (b) facilitated diffusion or (c) active transport?

Criterion A The concentration of X must be lower on the far side of the membrane.

Criterion B X can combine with a membrane-embedded carrier protein.

Criterion C The concentration of X can be higher on the far side of the membrane, providing transport is linked to the utilization of metabolic energy.

Question 9 (*Objectives 7.1, 7.6 and 7.10*) Figure 7.18 shows a series of Lineweaver–Burk plots of deoxyglucose uptake. (Deoxyglucose is a non-metabolized form of glucose with a similar K_t.) The plots were obtained by bathing sheets of hamster intestine in solutions with different concentrations of sodium ions. Each line represents data obtained at a particular fixed extracellular $[Na^+]$ but variable [deoxyglucose]. (You may find it helpful to revise Chapter 4, Section 4.4.2 on the Lineweaver–Burk plot before attempting the question.)

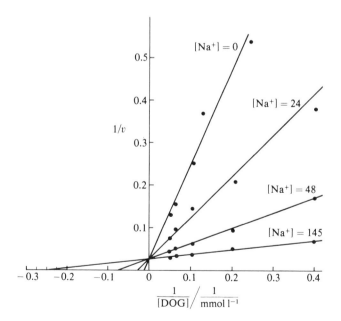

Figure 7.18 The effect of different sodium concentrations (in mmol 1^{-1}) on the kinetics of deoxyglucose (DOG) transport across the hamster intestinal wall.

(a) When the extracellular $[Na^+]$ is increased, does the K_t for deoxyglucose (i) increase, (ii) decrease or (iii) remain the same? (Na^+ concentrations are in mmol 1^{-1}.)

(b) What does this indicate about the affinity of the transport molecule for deoxyglucose (i.e. the ease with which it binds deoxyglucose)?

(c) How do these findings support the theory that Na^+ and glucose have to be transported simultaneously?

Question 10 (*Objectives 7.1 and 7.8*) An extraterrestrial bacterium brought back from Mars is found able to accumulate intracellularly the rare sugar L-fucose, when this is added to the bacterial suspension medium. Transport is dependent on hydrolysis of the nucleotide GTP. When the bacteria are transferred to fresh medium (after a few days on Earth) fucose transport ceases, even in the presence of ample GTP. It can be restored by adding small quantities of lanthanum ions (La^{2+}) to the medium.

(a) Draw a simple diagram illustrating the possible role of an La^{2+} gradient in an energy exchange system linking GTPase and fucose transporter proteins.

(b) A compound that makes the bacterial cell membrane permeable to La^{2+} is added to the medium. How would you expect this to affect the ability of the bacterium to hydrolyse GTP and to transport fucose?

7.4 CELL FUNCTIONS DEPENDENT ON ION MOVEMENT

In this section we bring together the different molecular mechanisms so far described for the movement of ions across membranes, and show how they relate to cell function.

By now, you should be well aware that ions can be transported either passively or actively:

1 Passive movement *down* an electrochemical gradient is by facilitated diffusion. The transport proteins that mediate this movement are often known as ion channels, because their membrane-spanning device is thought to be an α-helix-lined pore studded with polar groups as described in Section 7.1.2 (Figure 7.6).

2 Conversely, ion pumps move ions *up* electrochemical gradients by transport proteins that are able to trap and use energy from another source—light, ATP etc.

In this section we describe how cells coordinate the activity of these two devices, producing some remarkable cell properties.

7.4.1 Excitability

In any cell, ion pumping can produce a membrane potential, as described in Section 7.2.3. For the resting state the membrane potential is negative, since there are more negative ions inside the cell than out. The unique **excitability** of nerve and muscle cells—their ability to pass electrical signals—comes from altering this balance of charges at a very rapid rate. This is done by combining the actions of ion pumps and particularly voltage-sensitive ion channels known as **voltage-gated ion channels**.

Nerve cells have thousands of *voltage-gated sodium channels*. When these open, Na^+ ions expelled by the sodium pump flood back into the cell down their electrochemical gradient. So there is no longer an intracellular excess of negative ions and the membrane potential collapses almost instantaneously to around zero (in fact it overshoots to about $+25\,mV$ i.e. point B in Figure 7.19). It rapidly builds up again to $-70\,mV$ as the sodium channels close, and the sodium pump expels the extra positive ions, leaving the usual excess of negative charges within the cell. So within milliseconds the membrane potential is restored to normal, but meanwhile sodium channels in the adjacent stretch of membrane have opened, and the sequence is repeated.

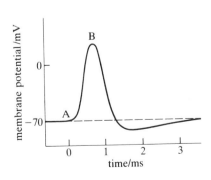

Figure 7.19 The action potential in a nerve cell. The sodium channels open at point A and close at point B.

These rapid changes in membrane potential result in a wave of electrical activity—the so-called *action potential*—passing rapidly along the nerve at over 100 metres per second.

The key to this rapid propagation of nerve impulses is the opening of the sodium channels. These channel proteins are extremely sensitive to small movements of surrounding ions—such as you would expect if the adjacent stretch of membrane is experiencing the dramatic voltage changes of the action potential. Presumably these small ionic movements change the distribution of charge across the protein surface, slightly altering the non-covalent interactions that maintain its higher-order structure. At a critical point, the protein flips from closed to open conformation (see Figure 7.20). Hence the term *voltage-gated* to describe an ion channel that is usually closed (gated) and opened only in response to changes in membrane potential (voltage).

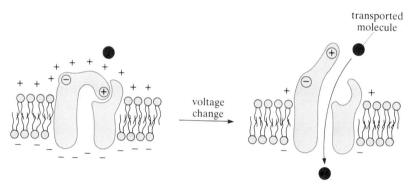

Figure 7.20 A voltage-gated ion channel.

Sodium channels are found along the length of the nerve, but at the extreme end of the nerve axon (see Figure 7.21a) there is a preponderance of *voltage-gated calcium channels*; these are instrumental in transmitting the nerve impulse across the gap or *synapse* between cells. At the axon terminal, the electrical message is converted to a chemical signal, by a *voltage-induced* influx of calcium ions through the calcium channel. These Ca^{2+} ions stimulate the cell to release small hormone-like molecules such as acetylcholine. These are known as **neurotransmitters**. They diffuse across the synapse and are picked up by specific membrane-bound proteins or *receptors* on the far side (Figure 7.21b). As we said in Section 7.1.2, receptors resemble enzymes, in having a binding site complementary to the particular molecule, the *ligand*, with which they interact. Just as enzymes recognize their substrates, receptors recognize their particular hormone or neurotransmitter. A non-covalently-bound hormone–receptor complex is formed (HR; or NR, where N is a neurotransmitter). These complexes are exactly equivalent to the ES complex. However, receptors, unlike enzymes, do not catalyse the chemical conversion of their ligands (H or N); they simply undergo a conformational change indicating 'message received', and allow H or N to diffuse off again into the extracellular medium.

With many acetylcholine receptors, the conformational change induced in the receptor has the effect of opening a nearby ion channel. This changes the membrane potential of the second, post-synaptic cell, and a wave of electrical activity passes along *its* membrane. In this way, the nerve impulse crosses the gap between cells.

The acetylcholine receptor is an example of a **ligand-gated ion channel** (Figure 7.22). In general, the effects of ligand-gated ion channels on membrane potential are less abrupt than those of voltage-gated ion channels, and allow

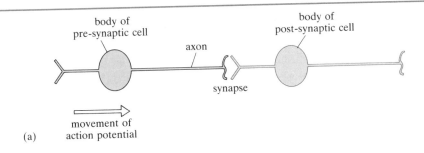

(a)

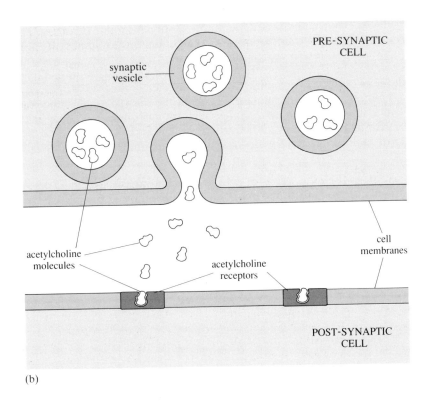

(b)

Figure 7.21 The role of acetylcholine in transmitting nerve impulses across a synapse. (a) Diagram showing the position of a synapse between two nerve cells; (b) details of synapse.

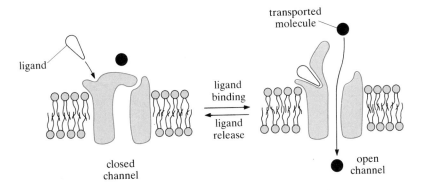

Figure 7.22 A ligand-gated ion channel.

more scope for subtle distinctions and nuances to be built into the electrical message. Ligand-gated channels are found in a variety of cells, and are particularly important in transmembrane signalling by calcium ions. Other ligand-gated channels release K^+ and Cl^- ions.

7.4.2 Membranes in polarized cells: polarizability and cell polarity

The typical cell is often presented as roughly spherical, suspended in a homogeneous medium and surrounded by a membrane of uniform composition. This is an oversimplification. Many cells face different environments on different sides, and their cell membranes may be divided into sections with specialized components—both lipid and protein—that cope with the particular functions appropriate to each cell surface. These sections are known as *membrane domains*. (They are not to be confused with the different domains within a single protein molecule! It is unfortunate that the different 'parts' of a membrane or of a protein are both called 'domains'. However, the concepts are similar in that structure determines function, whether you are talking about the specialized parts of a membrane or of a protein.) Figure 7.23 clearly shows the two membrane domains of a cell lining the gut lumen. Because the domains have different compositions and therefore different properties, the cell is said to be **polarized**. Such cells have been extensively studied in animal epithelia, where cells are joined into a thin sheet, often only one cell thick. Such epithelia are found lining the gut and the kidney tubules and form the innermost layers of skin. The *apical* 'end' or *pole* of an intestinal cell is the one in contact with the gut lumen contents (the top end in Figure 7.23). The opposite pole, in contact with the extracellular fluid, is known as the *basal pole*.

In the gut epithelium, adjacent cells are packed tightly together, preventing unwanted components in the gut contents from entering the rest of the body. Adjacent cells are held together by tight junctions (see Figure 2.4 in Chapter 2) that encircle each cell (like a zip fastener) and interlock with similar structures on neighbouring cells (Figure 7.23). The important point is that tight junctions prevent the free diffusion of protein and lipid molecules *in the plane of the membrane*. So the composition of the *apical domain* remains

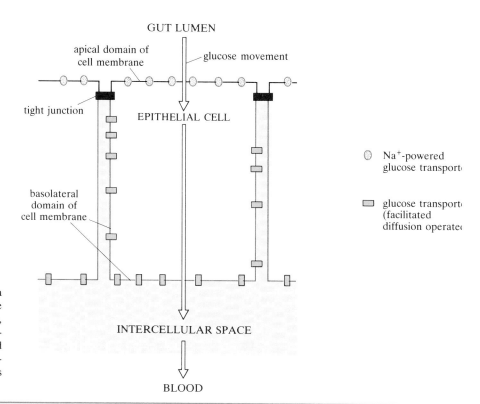

Figure 7.23 Glucose transport across a polarized epithelial cell. Only parts of the tight junction are seen here. In reality, the tight junction encircles the cell, preventing the movement of protein or lipid molecules between basolateral (blue outline) and apical (black outline) domains of the membrane.

distinct from that of the *basolateral* ('base-sides') *domain*. In gut epithelium, these structural differences clearly tie in with the different functions of the two different membrane domains.

◇ The apical surface is bathed by a solution of nutrients (digested food) in the gut lumen. These have to be taken into epithelial cells that may already have absorbed quite a lot of food. What type of transport protein might you expect to find here?

◆ One capable of transporting nutrients into the cell against a concentration gradient.

An example of such a protein is of course the Na^+-powered glucose transporter described earlier (Section 7.3.3). In gut epithelial cells, this protein is confined to the apical surface. At the basolateral surface is the other type of glucose transporter, capable only of facilitated diffusion (Section 7.2.2). This allows glucose to move out of the cell, down its concentration gradient, and so into the blood.

So by locating different types of transport protein in different membrane domains, the cell can create a *gradient* in the concentration of a particular solute, starting high at one end of the cell and dropping continuously towards the opposite end. So far we have considered only uncharged molecules like glucose. But if ion channels and ion pumps are also asymmetrically distributed in the cell membrane, it is not hard to see how the cell as a whole can have a *polarity*—with an excess of positive charge at one end and an excess of negative charge at the other (Figure 7.24). Indeed in the egg of the brown seaweed *Fucus*—a single rather large cell—there is a measurable flow of current from apical to basal pole. This polarizability with respect to ions could have exciting consequences which are only now being evaluated. It could influence embryonic development for example, since in theory at least, cell polarity would determine the way cells pack together to form a tissue. These ideas are still tentative however.

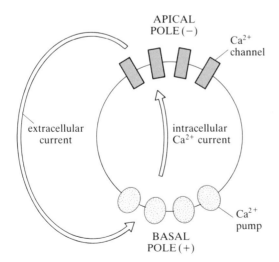

Figure 7.24 Diagram to show how asymmetrical distribution of ion pumps (blue symbols) and ion channels (grey symbols) for Ca^{2+} could generate current.

7.4.3 Osmosis: the movement of water

By now you should be well aware that the permeability properties of cell membranes introduce numerous opportunities for controlling the entry of polar solutes, with spin-offs for energy exchange, excitability and cell polarizability.

The one molecule whose entry we have almost ignored is water. As we showed in Figure 7.9 earlier, biological membranes are 'leaky' to water. But there is no *net* movement of water in or out of the cell, unless it is drawn in a particular direction by **osmosis**. This is a phenomenon in which the movement of water is linked to the movement of substances dissolved in it, i.e. its solutes. So osmosis can be demonstrated only when aqueous solutions of different concentration are separated by a **semipermeable membrane**, i.e. one permeable to water but impermeable to solute (except via special transport proteins). For aqueous solutions, the phospholipid bilayer provides just such a membrane.

So too does the cellophane wrapping round say, a box of chocolates. Imagine this made into a bag and filled with treacle. When immersed in water (Figure 7.25a) the bag will gradually fill up as water is drawn in to dilute the solute (here sugar) in the highly concentrated solution that makes up treacle (Figure 7.25b). This urge for water to follow solute can be quantified. By exerting pressure—like the head of water, *h*, in Figure 7.25c—the entry of water can be prevented; otherwise it will continue till the sugar concentration is the same on either side of the membrane. The pressure needed to prevent water being drawn in to dilute a solution is known as the **osmotic pressure** of that

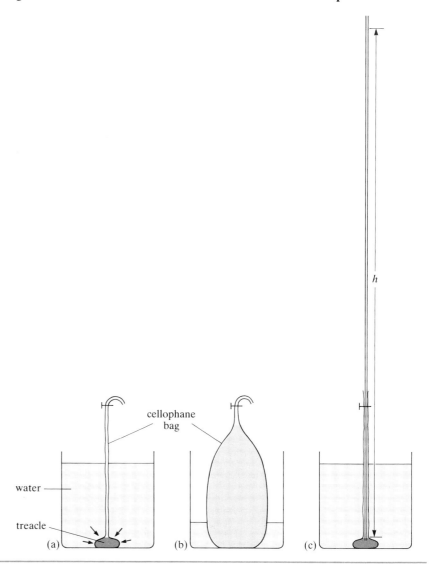

Figure 7.25 Osmosis and osmotic pressure. (a) Water moves across the semipermeable membrane (cellophane) by osmosis to dilute the concentrated sugar solution (treacle). (b) The cellophane bag has filled up with water. In (c) a column of water of height *h prevents* entry of water into the concentrated solution in (a). The pressure exerted by this head of water is the osmotic pressure, *Π*, of the solution.

solution. Not surprisingly it is proportional to the difference in solute concentration across the membrane (ΔC) and also to temperature.

The exact relationship is:

$$\text{osmotic pressure, } \Pi \text{ (pronounced 'pi')} = RT\Delta C \qquad (7.8)$$

where R is the gas constant and T absolute temperature.

Death from cholera is a good example of uncontrolled osmosis. Cholera toxin binds to the sodium pump in gut epithelial cells, leaving it permanently jammed on. So Na^+ ions are continuously ejected into the gut lumen. Water is then drawn out of the cells, following the Na^+ ions by osmosis. Excessive diarrhoea and eventually death from dehydration, are the result. The opposite (though still fatal) sequence of events can be seen when the sodium pump is inhibited, either directly, or through a metabolic inhibitor that deprives it of ATP energy. Excess Na^+ ions accumulate in the cells, so that they draw in water by osmosis until they burst. (This is how 'slug-death' pellets work!)

Higher plants can use osmosis to control movement. Among the best studied examples are the guard cells that surround gas-exchange pores in leaf surfaces. In the light, they allow entry of CO_2 for photosynthesis, but at night they minimize water loss by closing. Similar are the 'sleep' movements of leaves that close up at night, and the movements of touch-sensitive plants like *Mimosa*. The molecular mechanism is thought to be similar in all cases. The initial stimulus—be it light, pressure or metabolite—is thought to open an *ion channel*. (Remember the ligand-gated ion channels in Section 7.4.1.) As ions pour into the cell down their electrochemical gradient, water follows by osmosis. This increases the volume and 'rigidity' (turgor) of the cell. An increase in turgor of the guard cells causes the pores to open.

Summary of Section 7.4

1 Many cell functions depend on the coordinated activity of ion pumps and ion channels.

2 In excitable cells, the ion channels may be voltage- or ligand-gated.

3 In polarized cells, channel and pump proteins for a particular solute are located in different membrane domains. Proteins and lipids from different domains are prevented from intermingling by tight junctions. For uncharged solutes, the 'channel' may be a transport protein like the glucose transporter. For charged solutes, the combined activities of pump and channel may produce a polarity (a gradient of charge across the cell), with apical and basal poles being oppositely charged.

4 In osmosis, the movement of water across a semipermeable membrane is linked to the movement of solute to which the membrane is impermeable except via channels. So cell volume and related phenomena can be controlled by transport proteins e.g. ion channels.

Question 11 (*Objectives 7.1, 7.9 and 7.14*) A bacterial toxin binds irreversibly to the voltage-gated sodium channel, leaving it permanently open. Explain how infection by the bacterium would lead to paralysis. (Muscles are normally stimulated to contract by electrical impulses passing to them from a nerve.)

Question 12 (*Objectives 7.1 and 7.9*) A hypothetical tranquillizer drug T acts by binding reversibly to the acetylcholine receptor in nerve synapses. Explain how this tranquillizes the patient by reducing nerve excitability. You may assume that T *competitively inhibits* the binding of acetylcholine to the receptor protein, just as xylose and glucose compete for the glucose transport protein (Table 7.1).

Question 13 (*Objectives 7.1, 7.10 and 7.14*) A less depressing example of osmosis in animal cells than that cited in Section 7.4.3, concerns the mechanism that allows cells of the eye lens to combat detrimental osmotic changes during diabetes. The cells remain in osmotic balance, despite increasingly high levels of extracellular glucose found in diabetic lens. They are thought to do this by raising intracellular concentrations of a metabolically inert but osmotically active compound known as sorbitol.

Use Equation 7.8 to explain why, in the absence of sorbitol, the diabetic lens cells would shrink.

7.5 MEMBRANE PROPERTIES THAT DEPEND ON FLUIDITY

7.5.1 Overview

Having described many functions that biological membranes can achieve only because of their *low permeability to polar compounds*, we now turn to a second key property that distinguishes them from rigid, covalently-linked barriers; this is *fluidity*. Movement across the membrane may need special transport systems, but movement sideways can be very fast, simply by lateral diffusion in the plane of the membrane. Although membrane lipids rarely move between inner and outer monolayers (leaflets) of the bilayer, they can change places with neighbours in the same leaflet up to 10^7 times a second. This means a lipid molecule can travel the entire cell length in a few seconds.

There are some very basic cell phenomena that rely on membrane fluidity. One is **protein sorting**. Newly synthesized proteins may be targeted towards any one of a variety of sites, e.g. an organelle like the lysosome or chloroplast, or the cell membrane or the extracellular medium (Figures 7.26 and 7.27a). But in eukaryotes, most proteins are synthesized somewhere quite different—on ribosomes, either free in the cytosol or bound to the endoplasmic reticulum (ER), as described in Chapter 2. So how do proteins get from their site of synthesis to their final destination? Very simply, the answer is that many proteins are inserted directly from the ribosome into the ER; this forms a major part of the internal membranous network of the cell. From the ER, proteins move to the Golgi apparatus, and thence to their final destination. The journey is not that simple however, because the intracellular network is not continuous—and anyway the final destination may be an isolated organelle. This problem is solved by **transport vesicles**, small, membrane-enclosed sacs that bud off from one membrane, travel through the cytosol carrying the protein with them, and *fuse* with a second membrane. Figure

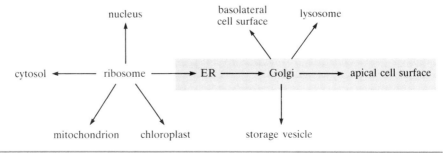

Figure 7.26 Simplified diagram to show the routes followed by newly synthesized protein from the ribosome to its final destinations. The secretory pathway (see Section 7.5.5 later) is shown within the blue tone area.

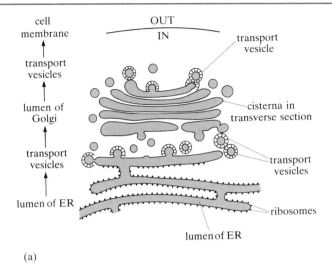

(a)

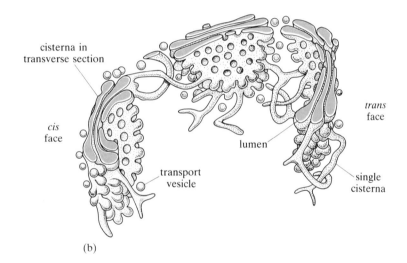

(b)

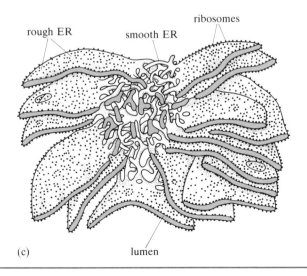

(c)

Figure 7.27 Organelles involved in the secretory pathway. (a) Two-dimensional view of the whole pathway; (b) three-dimensional picture of the Golgi apparatus; (c) three-dimensional picture of the ER.

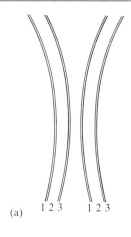

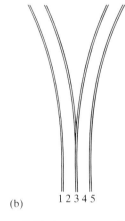

(a) 1 2 3   1 2 3

(b) 1 2 3 4 5

Figure 7.28 Appearance of biological membranes under the electron microscope: (a) before fusion; (b) during fusion. (For explanation of numbers, see text. The outer edges of each phospholipid bilayer appear dark, as shown in Figure 7.1.)

7.27, to which we return in Section 7.5.5, shows vesicles in transit from ER to Golgi membranes. (This was also illustrated in Chapter 2, Figure 2.5.) Clearly transport vesicles could neither *bud off* nor *fuse* with other membranes, were it not for the fluidity of the lipid bilayer.

Membrane fusion is needed not only for protein sorting, but for numerous very basic cell activities; meiosis, fusion of egg and sperm, mitosis and sometimes even differentiation (e.g. to form multinucleate muscle cells). Just occasionally, membranes caught in the act of fusion can be seen in the electron microscope (Figure 7.28). Instead of the usual three-layered picture (dark-light-dark), there are short five-layer stretches where the membrane components from the two outer leaflets of combining membranes intermingle and redistribute themselves into a single bilayer. Clearly this needs fluidity.

We shall now describe how the *molecular* properties of bilayer components promote fluidity. We can then look in more detail at some *cellular* properties that derive from it—bulk flow, endocytosis and protein sorting.

7.5.2 Fluidity and membrane composition

Fatty acid side chains of membrane phospholipids can be saturated or unsaturated, as we saw in Chapter 3. The term *unsaturated* implies that there is room for additional H atoms i.e. the structure includes one or more sequences of $-CH=CH-$ rather than $-CH_2-CH_2-$. Atoms can be arranged about the double bonds in two different ways, *cis* or *trans* (Figure 7.29a). In ***cis*-unsaturated** chains, the double bonds cause kinks, preventing adjacent chains packing closely together (Figure 7.29b).

◇ So which membrane will be more fluid: one whose phospholipids have a high proportion of *cis*-unsaturated fatty acids, or one with predominantly saturated or *trans*-unsaturated fatty acids?

◆ *Cis*-unsaturated fatty acids can pack less tightly and so tend to *raise fluidity*. So membranes with a high proportion of *cis*-unsaturated fatty acids will be more fluid.

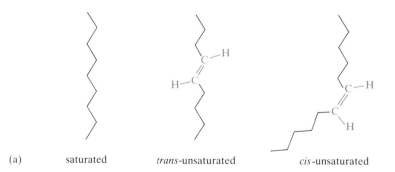

(a) saturated   *trans*-unsaturated   *cis*-unsaturated

Figure 7.29 (a) The three types of fatty acid side chain; (b) *cis*-unsaturated side chains prevent close-packing in a lipid bilayer.

(b)

Highly fluid cooking oils (liquid at room temperature) are converted to less fluid and more convenient (solid) margarines simply by hydrogenating the double bonds. Conversely, many animals can adapt to cold temperatures by raising the proportion of unsaturated fatty acids in their membrane phospholipids, and hence maintaining fluidity. Cholesterol also affects fluidity. At the concentration found in most normal eukaryotic cell membranes it promotes fluidity by inserting itself between phospholipid side chains (Figure 7.30). But at higher concentrations, the rigidity introduced by the fused cholesterol ring system tends to stiffen side chains and so reduce fluidity.

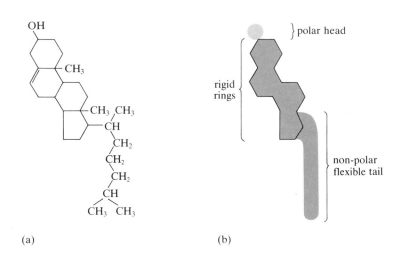

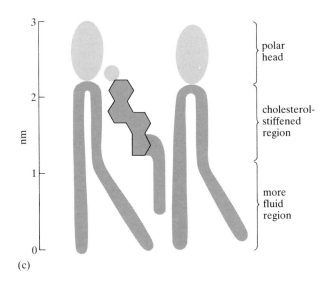

Figure 7.30 Effect of cholesterol on bilayer fluidity. (a) Chemical formula; (b) schematic formula showing rigid and flexible regions of cholesterol molecule; (c) interaction of cholesterol and phospholipid in single leaflet of bilayer.

7.5.3 Bulk membrane flow

Imagine an elongated membrane-enclosed organelle, where vesicles are always pinched off at one end of the organelle, and oncoming vesicles always fuse in near the opposite end (Figure 7.31). The result will be that some proteins are pushed apart (A and B in Figure 7.31) and that proteins like A are moved along this stretch of membrane.

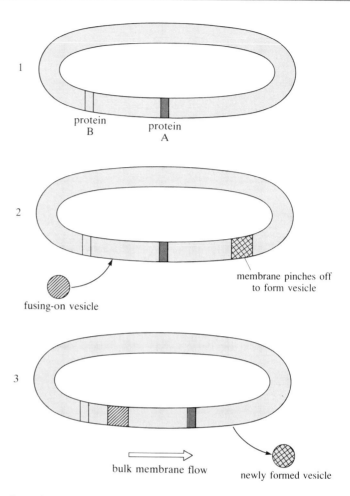

Figure 7.31 Diagram to show how budding and fusion of vesicles at different membrane sites can produce bulk membrane flow.

So *directionality* of vesicle budding and fusion can produce **bulk membrane flow**. This is particularly useful in protein sorting e.g. where movement along both ER and Golgi networks is always in the same direction.

Both protein and lipid components of the membrane are moved along in bulk flow; so too are soluble proteins in the lumen (interior) of the organelle. There does however seem to be a mechanism for some membrane-bound proteins to 'stand still' presumably by binding firmly to cytoskeletal or other proteins in the adjoining cytosol. This is particularly important in the Golgi for instance, since it allows for enzyme proteins that are not in transit, but have a particular job to do at that location.

Bulk flow of the *cell* membrane takes place on a surprisingly large scale. The entire membrane may be recycled in 30 minutes, though rates vary with cell type.

◇ Can you suggest how bulk membrane flow could push individual isolated subunits of a membrane protein closer together, so forming a functional aggregate from inactive subunits?

◈ If vesicles bud off from *between* subunits and reinsert on their opposite sides, the subunits will be pushed together. (First imagine the budding process as shown in Figure 7.32b. Then imagine vesicles budding off, rather than fusing on to, the membrane between proteins A and B in Figure 7.31).

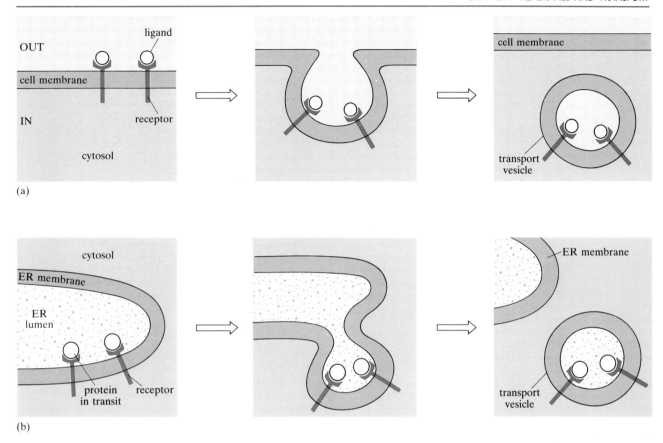

Figure 7.32 (a) Budding of transport vesicles from the cell membrane in *endocytosis*. (b) A similar mechanism postulated for budding of transport vesicles from ER membranes in the *sorting* of proteins in the secretory pathway (Section 7.5.5).

Bulk flow of the cell membrane could also explain cell locomotion, e.g. the formation of 'cell feet' (pseudopodia) and related activities where the cell as a whole crawls across the underlying surface. This last idea is somewhat controversial, but what is not disputed is the role of bulk membrane flow in *endocytosis* (Chapter 2, Section 2.2.2), where it was first discovered.

7.5.4 Receptor-mediated endocytosis

It has been known for a long time that cells can take up very large nutrient molecules from their surroundings—too large to use the transport mechanisms described earlier in this chapter. Such molecules enter by *endocytosis* where large molecules or particles, together with small amounts of surrounding fluid, are internalized in *transport vesicles* formed by the invagination and pinching off of cell membranes (see Figure 7.32a). As discussed in Chapter 2 (Section 2.2.2), endocytosis is a general phenomenon that includes *pinocytosis* ('cell drinking') where the cell can be seen continuously taking in minute droplets of its surrounding fluid. It also includes **receptor-mediated endocytosis**, in which this continual drinking has been adapted so that certain *specific* molecules are taken in with the imbibed fluid, by binding to receptors specific for them in the cell membrane (Figure 7.33a). (Remember from Sections 7.1.2 and 7.4.1 that receptors have binding sites for specific ligands; so this entry mechanism helps ensure that only desirable molecules are internalized.)

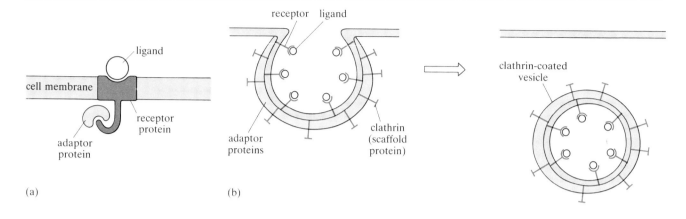

(a)

(b)

Figure 7.33 Diagram to show a suggested sequence for the formation of a clathrin-coated vesicle by interaction of receptor, adaptor and scaffold proteins. (a) Ligand binds to receptor which itself binds to adaptor protein—detail for a single receptor. (b) Invagination of the cell membrane to form a clathrin-coated vesicle; clathrin binds to the adaptor protein on the outside of the vesicle membrane.

The whole receptor–ligand complex is then drawn into the cell as part of a transport vesicle (Figure 7.33b). The essential role of the receptors, then, is to concentrate their specific ligands from the extracellular fluid and thereby bring them into the cell by endocytosis.

Many transport vesicles are stabilized by a surrounding scaffold of protein molecules. So far only one such protein has been identified and this is **clathrin**, which has been extensively studied. It is found throughout the cytosol and when isolated under the right conditions it spontaneously aggregates to form *clathrin cages* (Figure 7.34). It is not hard to see how these could enclose a roughly spherical transport vesicle. It is less clear how they come to surround the vesicle, but the clathrin molecules are known to interact with *adaptor proteins* which lie between the membrane of the vesicle and the clathrin coat and may bind to the cytoplasmic 'tail' domain of membrane-bound receptor proteins (Figure 7.33). Thus they act as 'adaptors' between clathrin and receptors.

There is also a great deal of research into what happens to the molecules internalized by receptor-mediated endocytosis. It is a complex process that helps ensure that the various internalized ligands are delivered to their correct intracellular sites (e.g. cytosol, lysosome etc.) and details vary depending on the particular ligand and receptor. Transport vesicles lose their clathrin coats shortly after budding from the cell membrane, and may then fuse with other specialized vesicles known as *endosomes*. The endosome interior is so acidic that internalized ligands dissociate from their receptors. Just how they escape from the endosome interior is still unclear, but in some cases the receptor is recycled to the cell surface.

Whatever their precise mechanisms, these three processes of vesicle budding, vesicle fusion and dissociation of ligand from receptor, are all part of a remarkably efficient intracellular transport system. So far we have mentioned only movement from outside in, i.e. the uptake of extracellular molecules by receptor-mediated endocytosis and their delivery to specific intracellular destinations. But exactly the same basic mechanisms must apply to transport in the opposite direction. Molecules synthesized at specific sites *within* the cell can be targeted to the cell surface, or else to specific organelles like mitochondria or lysosomes.

In the following section we discuss one such process, the delivery of proteins from their site of synthesis—the ribosomes—to the various destinations shown in Figure 7.26. This is the process known as *protein sorting* (Section 7.5.1).

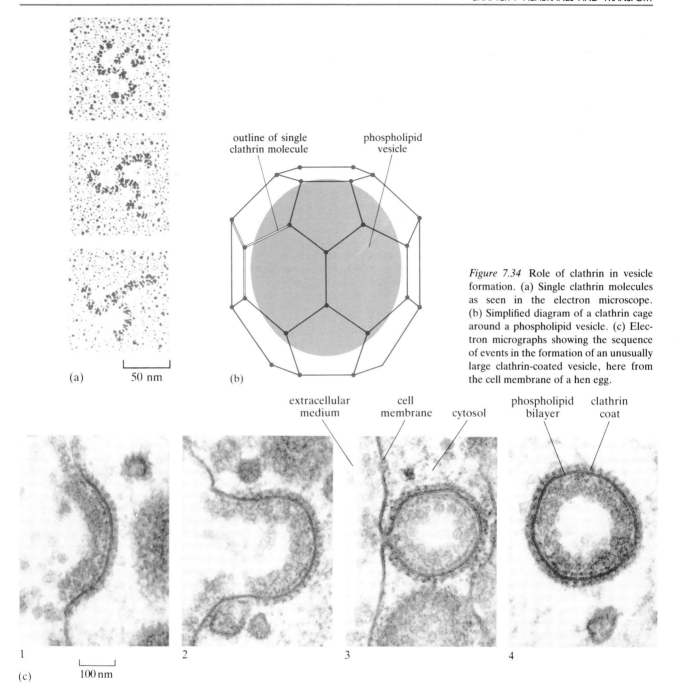

outline of single clathrin molecule   phospholipid vesicle

(a)   50 nm   (b)

Figure 7.34 Role of clathrin in vesicle formation. (a) Single clathrin molecules as seen in the electron microscope. (b) Simplified diagram of a clathrin cage around a phospholipid vesicle. (c) Electron micrographs showing the sequence of events in the formation of an unusually large clathrin-coated vesicle, here from the cell membrane of a hen egg.

extracellular medium   cell membrane   cytosol   phospholipid bilayer   clathrin coat

1   2   3   4

(c)   100 nm

7.5.5 Protein sorting

In our discussion of the sequence of events in protein sorting we shall concentrate on the major route shown in Figure 7.26, from ribosome to ER to Golgi apparatus to cell exterior—the so-called **secretory pathway**. This was discovered by George Palade in the 1960s, while working on the secretion of pancreatic enzymes. It has since given numerous insights into the basic mechanisms of protein sorting. It also gives us two examples of **post-translational processing**, i.e. changes in the primary structure of a protein that are not under direct genetic control. Most such changes are catalysed by enzymes e.g. partial hydrolysis of the polypeptide backbone to liberate active protein from a precursor—as in the conversion of pro-insulin to insulin (Chapter 3, Section 3.4.3).

The ER can be regarded as a very elongated organelle, with a narrow, fluid-filled interior—the ER lumen—bounded on either side by a phospholipid bilayer. The big question is: how does a water-soluble protein like one of the pancreatic enzymes move from the ribosome, across the lipid bilayer, and into the lumen? You may remember from Section 7.1.2 that water-soluble proteins fold to maximize surface polar groups. Yet to cross the bilayer they must come into intimate contact with *non-polar* phospholipid side chains. This enigma was solved by the finding that newly synthesized proteins insert into the membrane *before* folding up—indeed before the polypeptide chain is complete (Figure 7.35).

Proteins are always synthesized from N- to C-terminus, and the N-terminal part of the new polypeptide chain acts as a **signal peptide**, binding to a specific receptor or *docking protein* found only in the ER membrane. (Similar proteins in the membrane of chloroplasts, mitochondria and nuclei are thought to interact specifically with proteins destined for these organelles). Once safely inserted into the ER lumen, the protein loses its signal peptide, this being cleaved off by a membrane-bound enzyme (an example of post-translational processing). Then the protein makes its way along the ER by bulk membrane flow, driven by the budding off of vesicles at one 'end' of the membrane and the fusion of (different) vesicles at the other (see Figures 7.27 and 7.32b earlier). The protein then enters one such vesicle which travels—probably by interaction with the cytoskeleton—to the *cis* face of the **Golgi apparatus**.

This organelle was identified by microscopy in cell sections as early as 1898; it is now known to be a series of flat, interconnecting membrane-bound sacs or

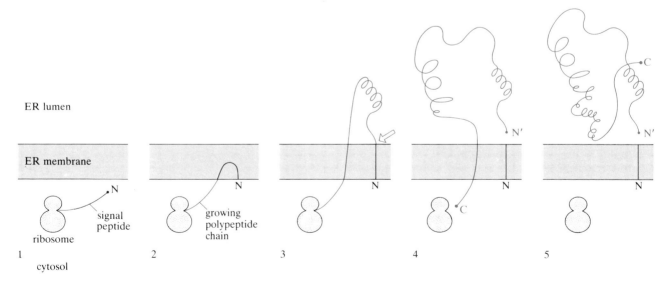

Figure 7.35 Diagram to show how a newly synthesized protein may cross a membrane. The numbers indicate the sequential steps in the process. In (1) synthesis has begun (from the N-terminus) and the signal peptide (black) is emerging from the ribosome. As the polypeptide chain grows the signal peptide enters the membrane (2). In (3) the signal peptide has been completely inserted into the membrane and the growing polypeptide chain is being pushed into the lumen. The signal peptide is usually removed around this stage by a membrane-bound enzyme (signal peptidase, marked with an arrow), producing a new N-terminus (N′) on the still growing polypeptide chain (4). In (5) the protein folds to its final water-soluble conformation, drawing the rest of the chain through the membrane.

cisternae, stacked parallel to one another as in Figure 7.27b. It is a site of *glycosylation*, another example of post-translational processing. In glycosylation, short sugar chains are added to specific amino acids on the protein, converting it to a glycoprotein (described in Section 7.1.2 and Figure 7.2). Specific glycosylation takes place in the ER and in the Golgi. So taking the correct route through the cell ensures that the protein not only finds the right destination, but is structurally complete.

The newly synthesized protein arrives at the *cis* face of the Golgi in a transport vesicle pinched off from the ER. It is transferred to the Golgi lumen when vesicle and Golgi membranes fuse, allowing the vesicle contents to mingle with those of the Golgi lumen. The protein continues its journey towards the *trans* Golgi, moving along each cisterna by bulk membrane flow, and *between* cisternae by transport vesicle (see Figure 7.27a). There is particular interest in what happens as the protein enters the last vesicle, in which it leaves the Golgi apparatus *en route* for its final desination—be it organelle, apical or basolateral cell membrane. These post-Golgi vesicles are often clathrin-coated and it is assumed that adaptor proteins are again involved in their formation, and that as with receptor-mediated endocytosis, the crucial job of selecting the right protein for the right vesicle lies with membrane-bound receptors.

It is a reasonable assumption that the protein cannot enter the transport vesicle unless it has *first* bound to its specific receptor. Although many different *cell surface* receptors are known, no equivalent *intracellular* receptor has yet been isolated from the membranes of secretory pathway organelles (ER or Golgi) nor from their transport vesicles. Nonetheless, all transport vesicles may well turn out to target their proteins most efficiently when budding follows the same basic mechanism, and involves the interaction of receptor, adaptor and scaffold proteins as postulated in Figure 7.33. Although budding can take place without scaffold proteins, the resulting vesicles seem to be much less efficient at targeting specific proteins to specific destinations. This is an area of intense research. The best characterized vesicle proteins are those involved in receptor-mediated endocytosis but many intriguing questions are still unanswered.

Despite these uncertainties, research into protein sorting mechanisms has helped to dispel the original static view of cell structure. This was exceedingly useful in its time, but it is now clear that the intracellular membrane network behaves more like a three-dimensional motorway, where not the cars but the roadways themselves are in constant, unidirectional flow. Sections of roadway may bud off into 'airborne' cabins, transporting traffic to different levels—a traffic controller's nightmare but not apparently for the cell.

Summary of Section 7.5

1 Membrane fluidity is enhanced by phospholipids with *cis*-unsaturated fatty acid side chains; too much cholesterol may reduce fluidity.

2 Bulk membrane flow is caused by coordinating the relative location of vesicle budding and fusion sites. It can alter the distance between membrane proteins, can move proteins in one direction through the intracellular membrane network, and may also be involved in cell locomotion.

3 In receptor-mediated endocytosis, extracellular ligands are internalized by binding to specific receptors; ligands are thereby concentrated inside transport vesicles that bud off from the cell membrane.

4 It is suggested that all transport vesicles with an efficient delivery or targeting mechanism bud off through the combined action of receptor,

adaptor and scaffold proteins (e.g. clathrin). This applies to ER receptors (for recently synthesized proteins) and to cell membrane receptors (for receptor-mediated endocytosis).

5 In the secretory pathway, protein sorting (for proteins that use the secretory pathway) begins by interaction between signal peptide and docking protein (a receptor in the ER membrane). Proteins move through ER and Golgi to their final destination by bulk membrane flow and transport vesicle. Important post-translational processing is catalysed by membrane-bound enzymes along the way.

Question 14 (*Objectives 7.1, 7.11 and 7.12*) Are the following statements true or false?

(a) A crucial stage in the constant recycling of cell membrane components is the fusion of transport vesicle and endosome.

(b) The acid pH of the endosome interior encourages ligand–receptor dissociation.

(c) A protein P in the centre of a stretch of membrane AB will be moved towards A, if a phospholipid vesicle is pinched off from the membrane somewhere between P and A, and another one fuses with the membrane somewhere between P and B.

Question 15 (*Objectives 7.1 and 7.12*) Which of the following statements are true for receptor-mediated endocytosis and which for protein sorting?

(a) The phospholipid membrane of the vesicle is stabilized, during budding, by an outer frame of scaffold protein.

(b) Adaptor proteins embedded in the vesicle membrane are thought to interact with both receptor and scaffold protein.

Question 16 (*Objectives 7.1 and 7.13*) Mutations in the signal peptide of pre-pro-insulin (the original precursor protein for the hormone insulin) prevent the secretion of active insulin from the cell surface of pancreatic cells. (a) From what you know of protein sorting mechanisms, explain this finding. (b) Mutations could also occur in a different part of the insulin precursor, namely in the area that binds to membrane-bound receptor in the post-Golgi transport vesicle. Would you also expect a failure of insulin secretion with this mutation?

SUMMARY OF CHAPTER 7

1 Biological membranes are non-covalently bound aggregates of protein and lipid molecules. This structural device underpins their unique characteristics of fluidity and low permeability to polar solutes.

2 All protein and lipid molecules have both polar (hydrophilic or water-soluble) groups and non-polar (hydrophobic or lipid-soluble) groups. In biological membranes, these are aligned to minimize contact between non-polar groups and polar water molecules, and to maximize hydrophobic interactions with each other. The result is a fluid lipid bilayer with a non-polar core, containing a 'mosaic' of proteins.

3 All membrane proteins have a hydrophobic, membrane-spanning domain. Many also have large, hydrophilic domains that are cytosolic and/or extracellular.

4 Many membrane proteins act as transporters (transport proteins). They have binding sites specific for the solute being transported. Some have an α-helical membrane-spanning channel lined with polar groups (channel proteins).

5 Some membrane proteins, particularly those in the outer surface of cell membranes, act as receptors. They have binding sites for specific messenger molecules in the surrounding medium e.g. hormones, and for recognition molecules on other cell surfaces.

6 Cell surface glycolipids and glycoproteins often form part of receptor binding sites.

7 In free diffusion, particles, molecules and ions all move spontaneously down their concentration (or electrochemical) gradients. Rate of movement (flux) is proportional to this gradient; it can be calculated from Fick's law.

8 Facilitated diffusion is mediated by specific transport proteins. Diffusion proceeds spontaneously down a concentration gradient (for uncharged solutes) and down an electrochemical gradient (for ions and other charged solutes). Metabolic energy is not required.

9 Transport kinetics show that each transport protein has a binding site specific for the transported solute, which becomes saturated at high solute concentrations, and can be competed for by chemically similar molecules. Maximum flux (J_{max}) and transporter–solute affinity (K_t) are equivalent to v_{max} and (less rigorously) to K_M, in the interaction of enzyme and substrate to form ES.

10 Facilitated diffusion of ions produces a membrane potential, E_m, whose magnitude depends on the position of electrochemical equilibrium for each particular ion. This is influenced by the overall charge on either side of the membrane, and can be calculated from the Nernst equation.

11 Active transport is the movement of solutes across a membrane, against their concentration or electrochemical gradient.

12 The energy for this uphill movement may come directly from ATP hydrolysis (e.g. sodium pump); more often it comes from the dissipation of an ion gradient.

13 Many different energy-producing systems (energy sources) can store energy in the form of a transmembrane ion gradient (Mitchell's chemiosmotic hypothesis). These can be tapped by many different energy-requiring systems (energy sinks).

14 Examples of energy sources are light (chloroplast membrane, bacteriorhodopsin), electron transport for coenzyme re-oxidation (inner mitochondrial membrane), ATP hydrolysis (sodium pump).

15 Examples of ion gradients used as intermediaries between energy sources and sinks are: Na^+ ions (as produced by the sodium pump and used by the Na^+-powered glucose transporter), and H^+ ions (proton pump).

16 Examples of energy sinks are: Na^+-powered glucose transporter, ATP synthesis (F_0–F_1 ATPase), flagellar rotation and active transport of mitochondrial Ca^{2+} ions against an electrochemical gradient.

17 Formation of an ion gradient can be described as primary active transport; dissipation of the gradient to move other solutes 'uphill' can be described as secondary active transport.

18 Many cell functions depend on the coordinated activity of ion pumps and ion channels.

19 In excitable cells, the ion channels may be voltage- or ligand-gated.

20 In polarized cells, channel and pump proteins for a particular solute are located in different membrane domains. Proteins and lipids from different

domains are prevented from intermingling by tight junctions. For uncharged solutes, the 'channel' may be a transport protein like the glucose transporter. For charged solutes, the combined activities of pump and channel may produce a polarity (a gradient of charge across the cell), with apical and basal poles being oppositely charged.

21 In osmosis, the movement of water across a semipermeable membrane is linked to the movement of solute to which the membrane is impermeable except via channels. So cell volume and related phenomena can be controlled by transport proteins e.g. ion channels.

22 Membrane fluidity is enhanced by phospholipids with *cis*-unsaturated fatty acid side chains; too much cholesterol may reduce fluidity.

23 Bulk membrane flow is caused by coordinating the relative location of vesicle budding and fusion sites. It can alter the distance between membrane proteins, can move proteins in one direction through the intracellular membrane network, and may also be involved in cell locomotion.

24 In receptor-mediated endocytosis, extracellular ligands are internalized by binding to specific receptors; ligands are thereby concentrated inside transport vesicles that bud off from the cell membrane.

25 It is suggested that all transport vesicles with an efficient delivery or targeting mechanism bud off through the combined action of receptor, adaptor and scaffold proteins (e.g. clathrin). This applies to ER receptors (for recently synthesized proteins) and to cell membrane receptors (for receptor-mediated endocytosis).

26 In the secretory pathway, protein sorting (for proteins that use the secretory pathway) begins by interaction between signal peptide and docking protein (a receptor in the ER membrane). Proteins move through ER and Golgi to their final destinations by bulk membrane flow and transport vesicles. Important post-translational processing is catalysed by membrane-bound enzymes along the way.

OBJECTIVES FOR CHAPTER 7

Now that you have completed this chapter, you should be able to:

7.1 Define and use, or recognize definitions and applications of, each of the terms printed in **bold** in the text.

7.2 Given structural formulae of membrane components (i.e. phospholipids, cholesterol, amino acid side chains), pick out the polar and non-polar groups of atoms. (*Questions 1 and 2*)

7.3 (a) Describe, in terms of polar and non-polar interactions, how membrane proteins are inserted into the phospholipid bilayer. (b) Describe how the domain structure of membrane proteins relates to their function. (*Question 3*)

7.4 Use Fick's law to explain the effect of solute concentration on flux. (*Question 4*)

7.5 Distinguish between free diffusion, facilitated diffusion and active transport, giving examples. (*Question 8*)

7.6 Compare the following terms and concepts in membrane transport and enzyme kinetics: specificity, K_t, K_M, v_{max}, J_{max}, competitive inhibition, irreversible inhibition. (*Questions 5, 6 and 9*)

7.7 (a) Describe quantitatively how transmembrane differences in electrochemical potential produce a membrane potential. (b) Given the Nernst equation, calculate membrane potential from experimental data. (*Question 7*)

7.8 Describe, with the aid of simple diagrams, how transmembrane ion gradients can be used to link light and chemical energy with a variety of energy-requiring processes. Give examples. (*Question 10*)

7.9 Describe how the following proteins contribute to the excitability of nerve cells: sodium pump, voltage-gated and ligand-gated ion channels. (*Questions 11 and 12*)

7.10 Give examples of cell functions dependent on (a) transmembrane ion gradients and (b) transmembrane differences in the concentration of uncharged solutes. (*Questions 9 and 13*)

7.11 Explain how bulk membrane flow can give rise to the transport of membrane-bound proteins. (*Question 14*)

7.12 Describe the role of receptor, adaptor and scaffold proteins in vesicle budding during receptor-mediated endocytosis and protein sorting. (*Questions 14 and 15*)

7.13 Describe how signal peptides and docking proteins draw newly synthesized proteins into the ER, and how receptor specificity could control both receptor-mediated endocytosis and the specific targeting of proteins from the secretory pathway. (*Question 16*)

7.14 Give examples where membrane fluidity and impermeability are important for biological activity. (*Questions 11 and 13*)

ANSWERS TO QUESTIONS

CHAPTER I

Question 1 Resolving power is used correctly in (a) and (d); (b) is wrong—if this happened, it would be very inconvenient; in (c), magnification and resolving power are confused.

Question 2 (a) The correct order is: fixation; dehydration; embedding; sectioning; rehydration; staining; mounting.

(b) The main differences in embedding, staining and specimen mounting.

For embedding, the medium is related to the thinness of the sections—much thinner for the electron microscope (50–100 nm) than for the light microscope (1–10 μm)—so resin is normally used for electron microscopy and paraffin wax for light microscopy.

For staining: for the light microscope, stains are dyes that absorb different wavelengths of transmitted light; but for the electron microscope, stains are metal compounds that impart differences in electron density when taken up by cell components.

For specimen mounting: for the light microscope, the stained specimen is mounted in a transparent medium on a transparent (glass) slide; for the electron microscope the stained specimen is mounted on a fine-mesh metal grid (usually copper) which supports it.

For fixation, dehydration and sectioning; no differences in principle, but differences in substances used.

Question 3 For circumstances (b) and (c), you would choose a TEM because a high resolving power is needed and the preparative processes will not destroy the part you want to see.

For circumstances (a) and (d), you do not need high resolving power and as you need to scan fairly large objects for fairly large structures, you would choose a light microscope with appropriate magnification.

Question 4 The only object from which the shape could *not* have been obtained is A in Figure 1.6.

CHAPTER 2

Question 1 A: mitochondrion. This organelle is surrounded by a double membrane, which means that it could not be either a lysosome or a peroxisome as these are surrounded by only one membrane. Double membranes surround the nucleus, chloroplast and mitochondrion, and should be visible at high magnification. The question states the cell is an animal cell, so chloroplast is eliminated as a possibility.

The internal structure of parallel membranes is typical of mitochondria. Chloroplasts can also be eliminated because, although they contain internal membranes, these are arranged in stacks (Plate 2.7).

B: rough endoplasmic reticulum. The obvious characteristics of B are that it is not fibrous (so eliminating cytoskeleton) and not surrounded by a membrane (so eliminating mitochondrion, chloroplast, nucleus, lysosome, peroxisome or vacuole of any kind). It is obviously made of parallel membranes, which are covered with small electron-dense granules (the ribosomes) on one side only. This eliminates *smooth* endoplasmic reticulum.

C: cell membrane (between facing arrows). The best clue that this is a cell membrane (probably a lateral cell membrane), is that it runs as a boundary around the stacks of RER and the two apposing membranes diverge at the upper right-hand corner of the micrograph. There is also a difference in appearance between the cell areas on either side of the membrane, and both areas contain RER so it cannot be a nuclear envelope.

D: nuclear membranes/envelope (between facing arrows). The most obvious point of recognition is that there is RER on one side and denser, nuclear-like material on the other. Most importantly, ribosomes are visible only on one side (the cytosolic side) but not the other.

E: nucleus (or you might have said nucleoplasm). From the rationale for D, the choice of nucleus is clear. Even if these arguments for D had not been made, the absence of structures such as mitochondria and RER would lead one to identify E as a nucleus.

F: nucleolus. This is an electron-dense, apparently structureless body inside the nucleus. No other body has these two characteristics.

Note: No Golgi apparatus is visible in this electron micrograph.

Question 2 (a) C: the fact that the mitochondrion has burst indicates that the ionic concentration of the fixative was too low—enough water has entered the mitochondrion to make it swell and burst.

(b) B: this mitochondrion is shrunken—the fixative used had too high an ionic concentration.

(c) A: a satisfactorily fixed mitochondrion.

Question 3 The processes would be: (a) synthesis and export of proteins (Section 2.4); (b) secretion (Section 2.5); (c) absorption (Section 2.2.1); (d) secretion of steroids or detoxification of drugs (Section 2.4).

CHAPTER 3

Question 1 (a) You might try using differences in solubility, such as ammonium sulphate fractionation, although this is a somewhat crude method. Electrophoresis (without SDS) could be used to separate the proteins on the basis of charge. A commonly used procedure is SDS–PAGE, which separates on the basis of size. Gel filtration could be used: a series of columns packed with different gels would give increasing degrees of purification. If the protein binds to a known ligand, then an affinity chromatography column might be produced with the ligand bound to it, which *might* give precise separation.

(b) An essential requirement during purification is an *assay* for the particular protein of interest; without this you cannot know what is happening to this protein. Binding to a specific ligand can be used as a 'label' to detect our protein during purification.

Question 2 (a) There are four peaks, which means that there are at least four different proteins in the mixture. Two or more proteins of the same size might appear as a single peak, so each peak might represent a mixture of proteins.

(b) A given antibody reacts very specifically with a single macromolecule (antigen). Binding of the antibody to the protein from peak 3 suggests strongly that this peak contains avidin (see (a), (d)).

(c) Larger molecules are eluted from a gel filtration column before smaller ones, so avidin is smaller than the molecules in peak 1.

(d) Samples from peak 3 could be treated to gel electrophoresis (PAGE). If only one band appears on the gel, then this suggests that peak 3 contains a single protein, i.e. pure avidin. Several bands on the gel would indicate that peak 3 contains more than one protein.

Question 3 There are two possible explanations for this result. Firstly, you have managed to isolate the particular DNA molecule, but the preparation is contaminated with two other molecules, so you still have a mixture that requires further purification. An alternative explanation is that during the purification the DNA molecule has been damaged i.e. broken into three fragments (these large molecules are especially susceptible to breakage).

Question 4 (a) 24.

(b) Yes, because there are more than 20 amino acids in the sequence. Below 20 the polypeptide would be described as a peptide.

(c) Alanine (Ala) with six representatives.

(d) Alanine is at the C-terminus. By convention, the sequence is read from the N-terminus (on the left) to the C-terminus (on the right).

(e) Methionine is at the N-terminus.

(f) This polypeptide can form tertiary structure by folding of the chain. Since there is only one chain it cannot form quaternary structure.

(g) A disulphide bridge could not form because there are no cysteine (Cys) residues present.

(h) A hydrophobic core would form from amino acids with non-polar side chains. The core would therefore consist of the sequence -Val-Phe-Leu-Ile-Val-Leu-.

(i) An ionic bond would form between amino acids whose side chains or terminal charged groups have opposite charges, i.e. one positively charged and one negatively charged. Lys, Arg and the N-terminal Met are positively charged, whereas Glu and the C-terminal Ala are negatively charged. Ionic bonds could therefore form from different combinations of the charged groups of these amino acids.

Question 5 Hair resists stretching by intrachain hydrogen bonds (i.e. hydrogen bonds formed within a single polypeptide chain) running along the 'stairwell' of each of its keratin α-helices, and also by occasional disulphide bridges between the chains. (Of course, stretching of the backbone of covalent peptide bonds cannot occur.)

Question 6 Since all three—collagen, fibroin and RNAase—are proteins, they all have primary structure which consists of a sequence of amino acids linked together by *covalent bonds*.

Collagen is a fibrous protein and has an unusual primary structure which is a repeat of Gly-X-Pro. Its secondary structure consists of an extended helix (not an α-helix), which is hydrogen-bonded. Three helices are intertwined to form a superhelix, which again is stabilized by hydrogen bonds. Superhelices associate to form collagen fibres, which are covalently linked. Collagen therefore has only one type of regular secondary structure.

Fibroin is another fibrous protein, again with a repeat sequence in its primary structure. Most of the secondary structure consists of β-pleated sheets which interact firstly through hydrogen bonds. Many sheets lie against one another, and are further stabilized by hydrophobic interactions.

RNAase, in contrast to collagen and fibroin, is a globular protein. Its secondary structure consists of a mixture of α-helices and β-structure, stabilized by hydrogen bonds, interspersed by irregularly folded regions. Tertiary structure is stabilized by a mixture of disulphide bridges, hydrophobic interactions, ionic bonds, and hydrogen bonds.

Question 7 (a) Blood clots formed at wounds prevent blood loss. It would be dangerous if they formed in other parts of the blood system, where clot formation would retard or prevent the flow of blood. Indeed, thrombosis results when clots form where they should not. The synthesis of fibrin as the inactive precursor fibrinogen ensures that clots form usually where they are required and nowhere else.

(b) Conversion of fibrinogen to fibrin occurs at a wound by removal of part of the polypeptide chain of fibrinogen. This is analogous to the conversion of pro-insulin to insulin. Fibrinogen can be reversibly denatured by treatment with urea and mercaptoethanol because it consists of a single polypeptide chain and the information required for folding and refolding is contained in the linear sequence of amino acids. Some of this information is removed when fibrinogen is converted to fibrin, so fibrin cannot be reversibly denatured.

(c) The conversion of fibrinogen to fibrin enables the monomer to aggregate to form the multimeric complex which is the basis of the clot, because fibrin has a different shape (conformation) compared with fibrinogen. Probably amino acids in the polypeptide chain, buried in the centre of fibrinogen, are exposed in fibrin and enable interactions to occur between the monomers.

(d) Each fibrinogen molecule consists of a single polypeptide chain and therefore cannot have a quaternary structure. Fibrin, in contrast, is multimeric, i.e. many identical monomers interact to give quaternary structure.

Question 8 Treatment of the protein with SDS has separated it into two components, i.e. the protein has quaternary structure consisting of subunits. Weak bonds holding these together are disrupted by the SDS treatment. The subunits are non-identical because they have different M_r values. The quaternary structure appears to consist of one of each of the two different subunits, since the M_r values of the separated subunits add up to give the M_r value of the protein.

Question 9 Polypeptides are polymers of amino acids; if there are less than 20 amino acids in the chain then the molecule is referred to as a peptide. The statement is therefore only partially true. A protein may have quaternary structure, in which two or more polypeptide chains associate to form the active molecule. In this case the individual polypeptide does *not* constitute a protein, but rather represents a subunit of it. All polypeptides are therefore not strictly proteins. The terms 'peptide', 'polypeptide' and 'protein' are often used interchangeably and the dividing line between them is often arbitrary. The statement is therefore not strictly true and might even be considered to be ambiguous.

Question 10 Presumably gentle heating of the DNA solution disrupts secondary structure, which is maintained by weak bonds: hydrogen bonding and hydrophobic interactions (base-stacking). Once these are disrupted the two polynucleotide strands separate and form separate random coils, i.e. base-pairing is lost. On cooling the base-pairs reform and so secondary structure is regained. This indicates that the information required to form secondary structure is contained within the primary structure, i.e. the linear sequence of nucleotides (or more precisely, the bases). So, as with proteins, primary structure determines higher-order structure which in turn determines biological function.

Question 11 Altogether, 52 hydrogen bonds are formed. Each G—C pair forms 3 bonds; 3×12 pairs = 36 bonds. Each A—U pair forms 2 bonds; 2×8 pairs = 16. So the total number of hydrogen bonds formed is $36 + 16 = 52$. Note that you were asked to assume that the unusual bases form the same number of hydrogen bonds as do their 'normal' counterparts, and that uracil forms hydrogen bonds with adenine in the same way as thymine does (see Figure 3.39).

Question 12 Your structure should look like this:

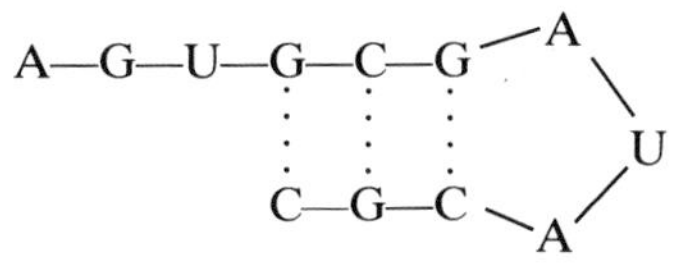

Nine hydrogen bonds are possible in this structure ($3 \times$ G—C), which is the maximum possible. No other loop can form which allows more than 9 bonds.

Question 13 The model for the RNA structure seems unlikely based on current knowledge. Some RNA molecules have tertiary structure, for example the tRNAs, where the hydrogen-bonded secondary structure is folded to give precise tertiary structure. Nucleic acids do not usually form quaternary structures in the absence of interactions with proteins. Indeed some nucleic acid tertiary structures appear to require proteins to maintain stability.

Question 14 Cellulose is prevented from stretching by (1) intrachain hydrogen bonds (i.e. those within a single chain), (2) interchain hydrogen bonds (i.e. those between different chains) between adjacent chains, and (3) covalent bonds with non-cellulose polymers. It is also, of course, prevented from stretching by the backbone of covalent bonds between the sugar residues (glycosidic bonds).

Question 15 Fatty acid chains of neutral fats (triglycerides) containing two or more double bonds are polyunsaturated. Plant oils tend to be rich in polyunsaturated fatty acids, whereas animal fats tend to be richer in saturated fatty acids. A diet 'high in polyunsaturates' would therefore consist of a high percentage of plant oils and fats, relative to animal ones, e.g. vegetable oils such as sunflower seed oil and soft margarine, rather than butter and lard.

Question 16 The structure is simply a rearrangement of a lipid like those shown in Figure 3.52. The correct answers are (c) and (e).

(a) Double bonds are present, so it is not a saturated lipid.

(b) Sphingosine is not present so it cannot be a sphingolipid.

(c) The phosphate group (and the presence of glycerol) should tell you that it is a phospholipid.

(d) It is not a triglyceride as there are only two fatty acid chains; triglycerides have three.

(e) There is a polar head group and two long hydrophobic tails, so it is amphipathic.

(f) It contains fatty acid chains so it cannot be a simple fat such as cholesterol. The ring structure of this lipid is also absent.

Question 17 None of the treatments gives direct information about possible covalent linkages. Treatment (a) suggests that the complex is not stabilized by ionic interactions, which would be disrupted by NaCl. Treatment (b) implies that hydrophobic interactions are also not involved. Treatment (c) is independent of the nature of the linkage, since all interactions between DNA and protein are destroyed by removal of the protein.

CHAPTER 4

Question 1 The correct options are (c), (d) and (f). (c) relates to the high specificity of enzymes; (d) to their massive catalytic power; (f) where systematic names have been given these end in -*ase*.

(a) is usually not true; though enzymes are 'natural resources' and hence do not have to be made synthetically, they are often (but not always) difficult and expensive to isolate and purify (Chapter 3). (b) is not true; enzymes, being proteins, are usually of high M_r, while non-biological catalysts are usually much simpler and of low M_r. (e) is neither true for enzymes nor for non-biological catalysts; both are affected by temperature, enzymes often dramatically because of the denaturation of the protein (Chapter 3).

Question 2 (i) If + IIc; B absorbs u.v. light, T does not. As B is converted to T absorption at 340 nm will *decrease*.

(ii) Ic + IIb; L reacts with dye to give a red colour. Note that u.v. absorption cannot be used because both B and G absorb it at the same wavelength (340 nm).

(iii) Id + IIb (in solution N is yellow, L is not; as L is formed the colour will *decrease*. A decrease in a colour can be followed just as can an increase; Section 4.3.1). *Or* Ic + IIb (L reacts with dye to give red colour).

(iv) Ih + IIa *or* Ih + IIf (*Note*: measuring the amount of alkali added gives a measure of the H^+ formed). u.v. absorption cannot be used because both B and G absorb at 340 nm.

(v) Ih + IIf (measures H^+ produced as amount of alkali needed; see (iv) above—must keep pH steady as enzyme is acid-sensitive) *or* If + IIc (B absorbs u.v. light) *or* Ic + IIb (L reacts with dye to give red colour).

(vi) If + IIc; G absorbs u.v. light; both L and Z react with dye to give red colour, therefore this cannot be used.

(vii) Ia. There is nothing to distinguish Q from T among the factors listed (Table 4.2). But production of T can be followed by coupling reaction (vii) (Q→T) to reaction (i) (B→T) performed *in reverse* (i.e. T→B). The reaction T→B can be followed by an *increase* in absorption at 340 nm (as against a decrease in (i); see answer to (i)). As here, B is the product, this employs principle Ig and apparatus IIc. The complete answer is therefore Ia + Ig + IIc.

Question 3 (b) is correct.

(a) is incorrect. Observations made on an enzyme cannot be 'because of an equation'. The equation is merely consistent with the behaviour of the enzyme.

(c) is a possibility for some enzymes but unlikely to be true for all; it does not explain the initial increase in rate.

(d) is untrue. As the rate is increasing with increasing substrate concentration, it cannot be due to the substrate being depleted.

Question 4 Graphs (a) and (d).

Remember that competitive inhibitors do *not* alter v_{max} but *do* alter K_M. In (a) and (b), which are Lineweaver–Burk plots, v_{max} is found from $1/v_{max}$, the intercept on the y-axis (and K_M from $-1/K_M$, the intercept on the x-axis). In (a), but not in (b), the vertical intercept is the same in the presence and absence of inhibitor; hence (a) represents competitive inhibition (confirmed by $-1/K_M$ being altered, as, hence, is K_M).

(c) and (d) are Hofstee–Eadie plots; here v_{max} is given by the intercept on the y-axis, and K_M from $-K_M$, the slope of the line. In (d), but not in (c), v_{max} is unaltered (intercept on the y-axis) while the slope, $-K_M$, is altered. Hence (d) represents competitive inhibition.

Question 5 As a Hofstee–Eadie plot is v against $v/[S]$, you must first calculate the values of $v/[S]$ from the data (i.e. v and $[S]$) given on page 124:

[S]	v	$v/[S]$
0.20×10^{-5}	16.7	83.5×10^5
0.22×10^{-5}	17.5	79.5×10^5
0.25×10^{-5}	19.6	78.4×10^5
0.27×10^{-5}	20.0	74.1×10^5
0.50×10^{-5}	28.6	57.2×10^5
0.77×10^{-5}	37.0	48.1×10^5
1.00×10^{-5}	38.5	38.5×10^5

The graph obtained on plotting v against $v/[S]$ should be *something like* that shown in Figure 4.36.

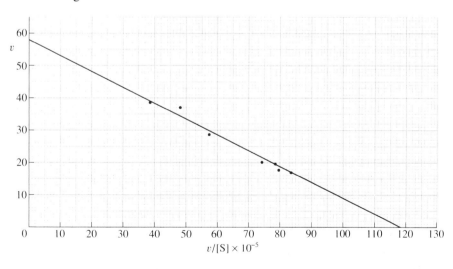

Figure 4.36 A Hofstee–Eadie plot of the data given on page 124.

From this v_{max} is given by the intercept on the y-axis, that is 58; hence v_{max} is 58 μmol min^{-1}.

K_M can be obtained from the intercept on the x-axis (which equals v_{max}/K_M) or the slope (which equals $-K_M$). Using the intercept on the x-axis we find that $v_{max}/K_M = 118 \times 10^5$. As v_{max} is $58\,\mu\text{mol min}^{-1}$, K_M must be $58/(118 \times 10^5)\,\text{mol l}^{-1}$, that is $4.9 \times 10^{-6}\,\text{mol l}^{-1}$.

If you look back to pages 124–5, you will find that these values are not both quite the same as those obtained from the Lineweaver–Burk plot of the same data ($v_{max} = 57.1\,\mu\text{mol min}^{-1}$ and $K_M = 4.9 \times 10^{-6}\,\text{mol l}^{-1}$). Indeed they are probably also not precisely the numbers that you obtained. Differences between your numbers and ours and between those obtained from the Hofstee–Eadie plot and the Lineweaver–Burk one reflect a number of factors.

The basic problem is that there is a scattering of the experimental points such that they do not all lie clearly on the same straight line. This means that you must decide where to place the line. If you are deciding by eye alone, as we were for both plots (Figure 4.11 and Figure 4.36), it is unlikely that any two people will choose exactly the same straight line. Then there is the problem of reading off the intercepts; how accurately can you do this? Add to these problems errors introduced by rounding off numbers, and it is hardly surprising that some differences might occur between the values obtained by different individuals or the same individual using one or other plot (Lineweaver–Burk or Hofstee–Eadie), or, as we discovered, even constructing the same type of plot more than once! Fortunately mathematical procedures exist that remove the need to guess where to draw the best straight line and these would be employed if more accurate values were needed.

Question 6 (a) False—some coenzymes are probably *derived* from vitamins.

(b) False—what is a vitamin for one organism may not be for another (for example, the other organism may be able to synthesize a required coenzyme from simple starting materials; Section 4.5.1).

(c) True—but they are a particular class of irreversible inhibitors designed to resemble substrates chemically and thereby bind (irreversibly) to active sites. Thus while affinity labels are irreversible inhibitors *not all* irreversible inhibitors are affinity labels.

(d) True—for example, by dehydrogenases (Sections 4.3.1 and 4.3.3).

(e) False—some, at least, are known to be metabolized, for example, to produce coenzymes.

(f) False—many enzymes have an optimum pH at around 7.4, but some operate in 'atypical' environments, (for example, pepsin in the acid environment of the stomach), and have optimum pHs far removed from pH 7.4.

(g) False, or at least partially so—their sensitivity to acid *and* alkaline conditions is due to ionization of amino acid side chains; however, strong acids present over prolonged periods can cause hydrolysis of peptide bonds.

(h) False—the photographs are far removed from actual 'pictures' of the three-dimensional structures (Chapter 3). Each photograph is a series of 'spots' related to the occurrence and size of particular atoms in the molecule. The 'translation' of the photographs into 'pictures' of the actual structures is a complex mathematical process.

Furthermore the structures thus derived are 'stills'. Therefore catalysis cannot be seen actually occurring; its progression can, to some extent, be *inferred* from structures derived from enzymes crystallized with and without substrates bound. (Note, however, the promise of Laue diffraction; Box 3.3.)

Question 7 (a) False—if anything, this supports the hypothesis Section 4.6).

(b) False—activity can be restored by zinc or cobalt.

(c) Possibly true, but data insufficient to allow conclusion.

(d) Possibly true, but data insufficient to allow conclusion—it could inhibit for a variety of reasons.

(e) Possibly true, but data insufficient to allow conclusion.

(f) True—as it is needed for activity; it is a coenzyme by definition (Section 4.5.1).

Question 8 (a) Possibly true, but data insufficient to allow conclusion. It seems unlikely in view of other possible roles of the tyrosine side chain (see below).

(b) Probably true—no large change in conformation occurs on removal of zinc, but activity is lost. Therefore zinc is more likely to be involved in catalysis than in maintaining overall enzyme structure.

(c) Possibly true, but data insufficient to allow conclusion. Movement of the tyrosine side chain over to the bond to be broken is not evidence that it helps break the bond. Similarly inhibition by acetylation of the tyrosine side chain is also not 'proof' of its catalytic role; it is, however, quite suggestive.

(d) Possibly true, but data insufficient to allow conclusion—we have no information given on any change in *substrate* conformation after movement of the tyrosine side chain; hence we do not know whether the substrate appears strained. Perhaps, however, the *active site* is strained (tyrosine moves).

You will have noted the frequent 'possibly true, but data insufficient to allow conclusion' in the answers to Questions 7 and 8. This obviously depends on the data (which are real) and questions that we chose, but it also reflects the actual situation in studying enzymes, where it is very difficult to ascertain what chemical groups are involved in active sites and what their precise role is (binding of substrates or coenzymes, breaking bonds in the substrate, or whatever; Figure 4.19). You will also see how some techniques can contradict what otherwise seems good evidence. For example, note how the hitherto reasonable *suggestion* (see (c)) that the tyrosine was involved in catalysis (presumably involving its hydroxyl group) has been rendered very unlikely by the recent application of protein engineering discussed in Section 4.5.5.

Question 9 (a) True, but only in a trivial sense. The existence of isoenzymes argues for *natural selection*, in that particular refinements of biological role, as afforded by isoenzymes, are assumed to have been selected for during evolution.

(b) True—e.g. LDH (Section 4.9.1).

(c) False—only used in diagnosis.

(d) True—is a pure enzyme one enzyme or a mixture of several isoenzymes? Or is one pure enzyme just to be considered as one isoenzyme of several?

Question 10 (a) False—though a heart attack will likely lead to an increased plasma level of LDH, an increased level is not in itself sufficient indication of a heart attack. There can be other causes.

(b) True—α-amylase is synthesized in the pancreas; pancreatitis could well lead to damage of pancreatic cells and a subsequent increased leakage of pancreatic enzymes into the plasma (Section 4.9.1).

(c) False—though the increased presence of type 1 LDH isoenzyme is consistent with a suspected heart attack, the presence of increased isocitrate dehydrogenase is consistent with liver damage and cannot be said to corroborate the LDH data (Section 4.9.1).

Question 11 (a) Disadvantageous, *generally*—they will not last long and hence will require frequent replacement. Though there might be circumstances where the enzyme having 'done its stuff' we wish to inactivate it; in which case, its instability should facilitate this and be *advantageous*.

(b) Disadvantageous—coenzymes are often expensive to produce and some may be consumed during the reaction, or like enzymes be unstable, and need replacement.

(c) Advantageous, *generally*—specificity, notably product specificity, permits the processes to be carried out with high yield; i.e. little of the substrate is wasted by its conversion to the unwanted by-products of side reactions, as is often the case with non-enzymic reactions. There might, however, be some circumstances where a less specific enzyme, able to utilize a wider variety of substrates, might be more advantageous than a highly specific one. For example, a protease able to hydrolyse many of the proteins found in food would be considered more useful in a detergent than one with a narrower, more specific range of protein substrates.

(d) Advantageous—micro-organisms are often easy and cheap to grow in large bulk, hence enzymes that can be produced from micro-organisms are relatively easy to obtain in quantity.

CHAPTER 5

Question 1 Statements (c) and (d) take a 'designer' approach.

(a) is merely a statement of fact about the observed behaviour of threonine deaminase; it applies no argument about the supposed functional reason (the purpose).

(b) is, like (a), an observation without actually stating a supposed purpose.

(c) definitely implies a purpose for the shape of an active site, something that we assume is a reason for the shape. A 'non-designer' re-statement of the facts could be along the lines of 'A substrate fits precisely the active site of an enzyme, whereas non-substrates do not'.

(d), unlike (a), imputes a definite functional 'reason', a 'role' for threonine deaminase.

As this question shows, a teleological/designer type of statement often contains phrases such as 'so as to . . .' or 'have the function of . . .' or 'the role of . . .'. To avoid teleology, such statements can often be re-cast omitting such phrases or re-stating them in terms of natural selection; e.g. 'The advantage gained by bacteria that can control their production of isoleucine according to how much isoleucine is present has led to selection of threonine deaminase that is inhibited by high levels of isoleucine'.

Question 2 (a) Possibly true.

(b) Probably false, *in fact almost certainly so*—X *increases* the production of Y; a feedback inhibitor would not be expected to increase production of Y. Moreover, X is the *starting substance* in the pathway from X to Y, so could hardly be called a *feedback* inhibitor.

(c) Possibly true—this, too, could lead to an increase in the production of Y.

An additional possibility is that an enzyme of which X is a substrate was operating below its v_{max} (as the level of X in the cells is low) and hence addition of X to the cells increases v and thus ultimately the production of Y.

Question 3 (a) Probably false—the inhibitor of protein synthesis should not affect the activation of the enzyme by X, because this does not depend on an increased amount of enzyme.

(b) Still probably false—the additional data on using the inhibitor are not relevant.

(c) Probably true—this would be expected to be affected by an inhibitor of protein synthesis, because enzymes are proteins.

Question 4 The most likely answer is that shown in Figure 5.19.

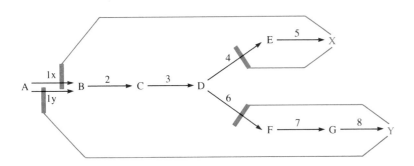

Figure 5.19 The likely feedback loops in the pathway from A to X and Y, in order to control production of X and Y. (For answer to Question 4)

Question 5 Look at Figure 5.19. (a) It will inhibit steps 6 and 1 (via 1y). If we assume the production of B by 1y is just enough to produce the amount required for the pathway leading to Y, then the synthesis of only B, C, D, F, G and Y will decrease. Hence B, C, D, F and G will be decreased in amount (Y is high as it is added to the cells).

(b) D gives rise to E, X, F, G and Y. Therefore a large amount of D will lead to an excess of X and Y. This will in turn lead to inhibition of step 1 (via isoenzymes 1x and 1y) and steps 4 and 6. Therefore, ultimately, synthesis of all the substances leading from A will be reduced.

Question 6 (a) False (Sections 5.2 and 5.4); e.g. CTP is quite unlike either of the two substrates of ATCase (Figure 5.9).

(b) True—cooperativity can only work where there is more than one active site. A single active site cannot be cooperative with itself.

(c) False—though Michaelis–Menten kinetics indicate than an ES complex is formed, their absence does not necessarily indicate that such a complex is *not* formed.

(d) False—though in some pathways more than one enzyme is involved, it need not be the case. ATCase is a case in point; the same enzyme is sensitive to both allosteric activation (by ATP) and allosteric inhibition (by CTP) and here the two effectors compete for the same allosteric site.

Question 7 (a) True—assuming that the feedback inhibition observed here is due to X being an allosteric inhibitor of enzyme 1. The active sites and allosteric sites of enzyme 1 being separate entities (perhaps even on separate subunits as is the case for ATCase), it is, in principle, possible to destroy one and leave the other intact.

(b) False—irreversible inhibition would not permit the pathway to rapidly switch off *and* switch on the production of end-product as conditions (i.e. the level of end-product) altered.

(c) Might be true; *might* is critical here—really we have no information to decide one way or the other; but pathways, or even individual enzymes within pathways (e.g. ATCase), can be subject to both inhibition and activation.

(d) Might be true—assuming X is an allosteric inhibitor of enzyme 1. Allosteric inhibitors may, via long-range interactions, affect binding of *substrate* to the active site (e.g. effect on affinity of CTP on ATCase). As Z is a competitive inhibitor of the enzyme it too can bind at the active site and therefore its binding to this site may also be affected by the same long-range effects caused by X.

Question 8 (a) False, not always—e.g. the reaction by which AMP is added to or subtracted from glutamine synthetase is catalysed by the same enzyme (adenylyl transferase; Section 5.5.2).

(b) False—e.g. glutamine synthetase is sensitive to allosteric control by up to eight end-products *and* subject to reversible covalent modification (Section 5.5.2).

(c) True—e.g. glycogen phosphorylase (Section 5.5.1) and glutamine synthetase (Section 5.5.2).

Question 9 (a) True—e.g. gluconeogenesis is split between three compartments (lumen of endoplasmic reticulum, cytosol and mitochondrion; Section 5.6).

(b) False—though the boundary (membrane) may be the site of some regulation and control (i.e. a control point), it need not be the only control point for the pathway.

(c) False. For example, say that the starting substance for a pathway in compartment A is the end-product of another pathway that is present in compartment B. Then any controls that operate to alter either the production of the substance in compartment B, or to affect its rate of entering compartment A will also affect the rate of operation of the pathway in compartment A.

CHAPTER 6

Question 1 Glyceraldehyde 3-phosphate dehydrogenase is an oxidoreductase (reaction vi in Figure 6.2). In this dehydrogenation reaction NAD^+ is reduced to NADH.

Question 2 One molecule of ATP is used in each of the two kinase reactions early on in the glycolytic pathway (reactions i and iii in Figure 6.2). Subtraction of this two ATP outlay from the gross ATP yield, i.e. four ATPs, gives the net value of two ATPs observed.

The result of the two kinase reactions is that two phosphate groups are added per molecule of glucose substrate, giving fructose 1,6-bisphosphate. This (6-carbon) intermediate is subsequently cleaved to give two 3-carbon intermediates, each now containing one of the phosphate groups. A second phosphate is added (in reaction vi). These phosphate groups are used in the two ATP syntheses—substrate level phosphorylations—later on in glycolysis (reactions vii and x).

Question 3 The three regulatory enzymes of glycolysis are phosphofructokinase (PFK), hexokinase and pyruvate kinase. These enzymes respond (either directly, or indirectly in the case of hexokinase) to the energy level in the cell: high ATP/ADP or ATP/AMP ratios—indicative of high energy levels—reduce their activities while low ATP/ADP or ATP/AMP ratios lead to increased activities. Both PFK and pyruvate kinase are also sensitive to the level of precursors: the former is controlled by the level of the TCA cycle intermediate, citrate; the latter by the level of the amino acid, alanine.

Question 4 Examples are the pyruvate dehydrogenase (PDH) complex in the link reaction and the α-oxoglutarate dehydrogenase complex in the TCA cycle (Figure 6.6, step iv). Separate components of these complexes catalyse sequential steps in a series of chemical transformations. This arrangement has the advantage of enabling the product of one reaction to pass immediately to the active site of the enzyme catalysing the next reaction.

Question 5 The TCA cycle takes part in the oxidation of sugars, fatty acids and other respiratory fuels that can be catabolized to acetyl CoA. In this cyclic sequence of reactions, the two carbon atoms of the acetyl group of acetyl CoA are converted to carbon dioxide and the hydrogens are transferred to the coenzymes NAD^+ and FAD. The cycle is important to the cell's economy in terms of biosynthesis (it produces a variety of metabolic intermediates) and energy (GTP directly and reduced coenzymes which are linked to ATP formation via oxidative phosphorylation).

Question 6 (a) True (Equation 6.8; and reaction i in Figure 6.6).

(b) True—it is a *cyclic* series of reactions.

(c) False. The carbon dioxide comes from two of the carbon atoms of oxaloacetate. The two carbons of the acetyl group from acetyl CoA do *not* emerge as carbon dioxide in their *first* turn of the cycle (see Figure 6.6).

Question 7 All of them. In all, except one, of the dehydrogenation reactions involved in the complete oxidation of glucose, NAD^+ is reduced to NADH, and in the remaining reaction FAD is reduced to $FADH_2$. All three coenzymes are involved in the oxidative decarboxylation reactions of pyruvate and α-oxoglutarate (see Figure 6.5).

Question 8 No, because, although the acetyl groups (of acetyl CoA) contain two carbon atoms and these are fed into the TCA cycle, two carbon atoms are lost in the two decarboxylation steps of the cycle (Figure 6.6, steps iii and iv). Hence, there is no *net* synthesis of any of the cycle intermediates. Oxaloacetate can, however, be synthesized from pyruvate and carbon dioxide in the reaction catalysed by pyruvate carboxylase (Section 6.4.3).

Question 9 (a) True (see Section 6.5.3).

(b) False. The opposite is true in a coupled system (see Section 6.5.6).

(c) True. In the oxidation of succinate to fumarate by succinate dehydrogenase its coenzyme, FAD, is reduced to $FADH_2$. There are only *two* ATP synthesis sites along the ETC between $FADH_2$ and oxygen, accounting for the two ATPs/$FADH_2$ oxidized. (In contrast, there are *three* ATP synthesis sites between NADH and oxygen, hence three ATPs/NADH oxidized.) Figure 6.13 shows the sequence of the electron-transfer complexes involved.

(d) False. However the reverse is true i.e. oxidative phosphorylation cannot occur without a functioning, coupled, electron transport chain.

(e) False. Oxidative phosphorylation takes place in the mitochondria (Figure 6.12).

Question 10 (a) 15. Three molecules of ATP are formed from the molecule of NADH produced in the oxidative decarboxylation of pyruvate and a further one molecule of ATP from the GTP generated by one turn of the TCA cycle. Two ATPs result from the $FADH_2$ and nine ATPs from the three NADH molecules formed during one turn of the TCA cycle, when they are fed into the ETC.

(b) 38. One molecule of fructose 1,6-bisphosphate generates *two* molecules of pyruvate which then yield 30 (i.e. 2×15) molecules of ATP (see answer to part (a)). In addition, there are eight more ATPs generated between fructose 1,6-diphosphate and pyruvate: two molecules of glyceraldehyde 3-phosphate give two molecules of (cytosolic) NADH (reaction vi in Figure 6.2) which yield four ATPs via the ETC, and a further four ATPs are generated by the two substrate level phosphorylations (reactions vii and x in Figure 6.2).

(c) 16. 2-Phosphoglycerate is converted to phosphoenolpyruvate (reaction ix in Figure 6.2); one ATP is formed by the dephosphorylation of this (reaction x) and a further 15 from the complete oxidation of pyruvate (see answer to part (a)).

(d) 20. When the malate–aspartate shuttle is operative, the cytosolic NADH formed during the conversion of glyceraldehyde 3-phosphate to 1,3-bisphosphoglycerate produces *three* ATPs via the ETC. The dephosphorylation of 1,3-bisphosphoglycerate to 3-phosphoglycerate generates another ATP. 3-Phosphoglycerate is converted to 2-phosphoglycerate (reaction viii in Figure 6.2) and 16 more ATPs are formed by oxidizing this (see answer to part (c)).

Question 11 DNP abolishes ATP synthesis by dissipating the electrochemical proton gradient across the inner mitochondrial membrane. It relieves respiratory control and the rate of electron transport in practice increases. As no ATP is synthesized, the P:O ratio must be zero.

Question 12 (a). (b) is inhibited by GTP and (c) is inhibited by ATP, but neither is inhibited by citrate.

Question 13 (c). (d), (a) and (b) are (in this order) intermediate electron carriers in the ETC.

Question 14 (c) is the only one which does not drive ATP production via oxidative phosphorylation. It is the *re-oxidation* of NADH via the ETC that generates the energy for ATP synthesis.

(a) This provides the proton-motive force.

(b) A proton gradient inevitably means there is a pH gradient—i.e. proton gradient and pH gradient are essentially synonymous terms, as $pH = -\log_{10}[H^+]$.

(d) It is the flow of protons *down* the gradient from intermembrane space to matrix via the F_0 channels that results in the synthesis of ATP.

Question 15 (a) (i) Two ATPs (see Table 6.1 and Equation 6.29); (ii) 36 ATPs (see Table 6.3).

(b) The majority (32 molecules) of the extra 34 ATPs are produced via oxidation of reduced coenzymes (NADH and FADH$_2$) in the mitochondria. The other two are generated via a substrate level phosphorylation in the TCA cycle (Table 6.3).

CHAPTER 7

Question 1 Unable to penetrate the cell membrane without the assistance of transport proteins are: glucose (because of its polar −OH groups); glucose 6-phosphate (even more polar than glucose, because of its charged phosphate group); and serine (charged −NH$_3^+$ and −COO$^-$ groups, plus polar −OH in side chain). However the high hydrophobicity (non-polar nature) of fatty acid side chains in the triglyceride, and of the non-polar rings in the steroid hormone and vitamin D, mean that all three readily pass through the cell membrane by dissolving in the non-polar phospholipid bilayer. (Small polar groups like $>$C=O in triglyceride, and −OH in cholesterol analogues, are 'swamped' by the other, highly hydrophobic components.)

Question 2 Non-polar groups are −CH$_3$ in Ala, −CH$_3$ and −(CH$_2$)$_2$− in Met, −CH$_2$− and the aromatic rings of Tyr and Trp, and the entire side chain for each of Ile and Leu—since these are composed entirely of −CH$_3$, −CH$_2$− and $>$CH− groups.

Question 3 The cytoskeleton is thought to anchor itself to the membrane by specific interaction with the *cytosolic* domain of membrane proteins. Both hormone and cell surface identity tag interact specifically with the *extracellular* domain of receptor proteins. The polar molecule may enter the cell after first binding specifically to the *extracellular* domain of a transport protein of the channel type; but it will then certainly interact with the *membrane-spanning* domain in its passage through the phospholipid bilayer (see Figure 7.7).

Question 4 Fick's law (Section 7.2.1) shows that the flux of urea across the membrane i.e. from blood to dialysis fluid (DF) is directly proportional to *concentration difference* (i.e. urea concentration in blood minus urea concentration in DF) and to *area* (i.e. area of cellophane membrane to which the blood is exposed). It follows that dialysis can be speeded up (a) by increasing the concentration difference by replacing DF with fresh, urea-free solution and (b) by having as long and wide a dialysis membrane as possible—in practice, blood used to be pumped through a series of cellophane coils. Modern kidney machines use more complex membrane systems to maximize the exchange area that can be compressed into a small space.

Question 5 The amino acid transport protein can bind either Ser or Thr at its specific binding site, since they are structurally similar (Table 3.2 in Chapter 3). Because they *compete* for binding, uptake rate for Ser is reduced when both are present (as shown for sugars in Table 7.1). In effect, Thr becomes a *competitive inhibitor* for Ser transport. In contrast, molecule X is an *irreversible inhibitor* of both Thr and Ser transport, because they use the same transport protein and X permanently damages the structure of this protein—just as tetrodotoxin alters protein structure (as described at the end of Section 7.2.2).

Question 6 Both statements (a) and (b) are true. (See Section 7.2.2.)

Question 7 In order to get this calculation right, it is essential that you remember that Z is negative for an anion. So, substituting in Equation 7.7, we get

$$E_m = \frac{8.314 \times 291 \times 2.303}{-1 \times 96\,500} \times \log_{10} \frac{[\text{Cl}^-]_O}{[\text{Cl}^-]_I}$$

$$= -0.058 \times \log_{10} \frac{540}{60} \text{ volts}$$

$$= -58 \times \log_{10} 9 \text{ mV}$$

$$= -58 \times 0.95 \text{ mV}$$

$$= -55.1 \text{ mV}$$

Question 8 (a) criterion A; (b) both criteria A and B; (c) criterion C.

Question 9 (a) The intercept on the x-axis is $-1/K_t$ (see Chapter 4, Section 4.4.2 on the Lineweaver–Burk equation). Figure 7.18 shows that as $[\text{Na}^+]$ increases the value of $-1/K_t$ becomes more negative, i.e. $1/K_t$ becomes more positive (increases). This means that K_t must *decrease* as $[\text{Na}^+]$ is increased.

(b) The deoxyglucose molecule is bound more easily in the presence of Na^+. Recall that the *lower* K_t indicates that the affinity of the transport molecule for deoxyglucose has *increased* (Section 7.2.2).

(c) If deoxyglucose is bound more easily at high Na^+ concentrations, this supports the idea that the transport protein must have *both* binding sites occupied before it can undergo the necessary conformational change that moves deoxyglucose or glucose from one side of the membrane to the other. (This conformational change is similar to the one described in Figure 7.15 for the transport of Na^+ by the sodium pump.)

Question 10 (a) The La^{2+}-linked energy exchange system may be visualized as in Figure 7.36 (see also Figure 7.16). In the presence of La^{2+}, GTP hydrolysis is accompanied by the uptake of La^{2+} ions; this creates a transmembrane gradient whose energy can be dissipated (by La^{2+} ions flowing out), allowing the uptake of fucose against a concentration gradient.

Figure 7.36 Answer to Question 10a. Hypothetical La^{2+}-linked energy exchange system.

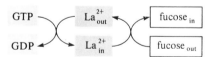

(b) If the bacterial cell membrane is made permeable to La^{2+} ions, GTP hydrolysis will not be accompanied by build-up of an ion gradient, and although the bacterium will still be able to hydrolyse GTP, it will no longer be able to link GTP hydrolysis to fucose transport.

Question 11 In normal circumstances, muscles are stimulated to contract when a wave of electrical activity passes to them from an adjoining nerve. At the molecular level, electrical activity in the nerve is sparked off by opening the sodium channel, which leads to an influx of Na^+ ions (Section 7.4.1). If the channel is permanently open, the nerve continually stimulates the muscle to contract. Thus the result is paralysis.

Question 12 Nerve excitability will be reduced if the mechanism for relaying electrical activity across the synapse (Figure 7.21) is impaired. If neurotransmitter (here acetylcholine) released from the pre-synaptic nerve membrane cannot readily bind to the post-synaptic membrane, there will be no influx of ions, and no action potential generated in the post-synaptic cell.

Question 13 In the absence of sorbitol, water would be drawn out of the diabetic lens cells to dilute the relatively high concentration of extracellular glucose. This efflux would continue until osmotic pressure across the membrane was zero i.e. intracellular and extracellular solute concentrations were equal (ΔC in Equation 7.8 is zero). (Decreased cell volume or cell shrinkage, in these dehydrated lens cells, would cause distortion of light rays passing through to the retina, and so impair vision.)

Question 14 Statements (a) and (b) are true (see Section 7.5.4).

Statement (c) is also true (see Figure 7.31).

Question 15 Statement (a) is true for receptor-mediated endocytosis and sometimes for protein sorting (see Section 7.5.5).

Statement (b) is true for all transport vesicles that are stabilized by scaffold protein, i.e. always true for receptor-mediated endocytosis and sometimes for protein sorting.

Question 16 (a) Docking protein is the specific receptor embedded in the ER membrane, that draws newly synthesized proteins into the ER and so initiates their journey along the secretory pathway. The docking protein binds only to the signal peptide of the secreted protein. So mutations here mean no secretion.

(b) Yes. The final transport step in the secretory pathway takes place in a transport vesicle budding from the Golgi (Figure 7.27). Targeting of the insulin precursor is thought to be more efficient if vesicle budding is triggered by ligand binding to its specific receptor in the Golgi membrane. Receptor then interacts with adaptor protein as postulated for the cell membrane in Figure 7.33a. Here the ligand is the insulin precursor protein, so mutations in amino acids that normally bind to receptor will greatly impair the formation of the transport vesicle, and hence less insulin will appear at the cell surface for secretion.

FURTHER READING

The following texts are recommended for background reading to accompany this book. None of these cover all the topics at the same level of detail as we do, and most go into much greater depth and have wider scope. Each title indicates the emphasis given, whether it be a cellular approach or a more molecular one. Several of these books are now standard texts and have run to more than one edition, so although this list will date, more current editions are likely to become available.

Alberts, B., Bray, D., Lewis, J., Raff, M., Roberts, K. and Watson, J.D. (1990) *Molecular Biology of the Cell*, 2nd edn, Garland Publishing, London. ISBN 0824036956.
This book is accompanied by:
Wilson, J. and Hunt, T. (1990) *Molecular Biology of the Cell—The Problems Book*, Garland Publishing, London. ISBN 0824036972.

Avers, C.J. (1986) *Molecular Cell Biology*, Addison-Wesley Publishing Co., Reading. ISBN 0201103079.

Darnell, J., Lodish, H. and Baltimore, D. (1990) *Molecular Cell Biology*, 2nd edn, Scientific American Books, New York. ISBN 0716720789.

DeDuve, C. (1985) *Guided Tour of the Living Cell*, W.H. Freeman and Co., Oxford. ISBN 0716760029.

DeRobertis, E.D.P. and DeRobertis, E.M.F. (1987) *Cell and Molecular Biology*, 8th edn, Lea and Febiger, Philadelphia. ISBN 0812110129.

Molecules of Life, Readings from Scientific American (1986) W.H. Freeman & Co., Oxford. ISBN 0716717832.

Stryer, L. (1988) *Biochemistry*, 3rd edn, W.H. Freeman and Co., New York. ISBN 0716719207.

Voet, D. and Voet, J.G. (1990) *Biochemistry*, John Wiley and Sons, Chichester. ISBN 0471512877.

Zubay, G. (1988) *Biochemistry*, 2nd edn, Collier Macmillan Publishers, London. ISBN 0024320803.

ACKNOWLEDGEMENTS

The series *Biology: Form and Function* (for Open University Course S203) is based on and updates the material in Course S202. The present Course Team gratefully acknowledges the work of those involved in the previous Course who are not also listed as authors in this book, in particular: Ian Calvert, Lindsay Haddon, Sean Murphy and Jeff Thomas.

Grateful acknowledgement is made to the following sources for permission to reproduce material in this book:

Cover: Eric Westhof, Centre Nationale de la Recherche Scientifique, IBMC Strasbourg.

FIGURES

Figure 2.2: from Caro, L. G. in the *Journal of Cell Biology*, **15**, 173–188, 1962, American Society for Cell Biology, Rockefeller University Press; *Figures 3.2, 3.11, 3.14, 3.15, 3.43, 3.54 and 3.57*: from Freifelder, D. (2nd edn) (1987) *Molecular Biology*, Jones and Bartlett Publishers; *Figure 3.3*: courtesy of Professor Arthur Landy, Browns University; *Figure 3.16*: adapted from Freifelder, D. (1985) *Essentials of Molecular Biology*, p. 54, Boston: Jones and Bartlett Publishers; *Figures 3.17, 3.18, 3.22, 3.24 and 4.9*: from *Biochemistry* (3rd edn) by Lubert Stryer. Copyright © 1975, 1981, 1988 by Lubert Stryer. Reprinted with permission from W. H. Freeman and Co. Ltd; *Figure 3.20(b)*: by Geis, I. from Kendrew, J. C. 'The three-dimensional structure of a protein molecule', copyright © 1961 by Scientific American, Inc. All rights reserved; *Figure 3.30*: reprinted by permission from *Nature*, Vol. 280, p. 561. Copyright © 1979, Macmillan Magazines Ltd; *Figure 3.32*: from Silverton, E. W. *et al*. 'Three-dimensional structure of an intact human immunoglobulin', *Proceedings of the National Academy of Sciences*, Vol. 74, No. 11, Nov. 1977 (this article is not subject to copyright); *Figures 3.27 and 3.33*: from Dickerson, R. E. and Geis, I. (1969) *The Structure and Action of Proteins*, Benjamin/Cummings Publisher, copyright © 1969 by Richard E. Dickerson and Irving Geis; *Figure 3.41*: from Gutell, R. R. *et al*. (1985) *Proceedings of the Nucleic Acids Res. Molec. Biol.*, Vol. 32, Academic Press, Inc; *Figure 3.55*: from Finch, J. T. *et al*. (1975) *Proceedings of the National Academy of Sciences*, 72:3321; *Figure 3.56*: from *Biochemistry* (3rd edn) by Lubert Stryer. Copyright © 1975, 1981, 1988 by Lubert Stryer. Reprinted by permission from W. H. Freeman and Co. Ltd (after A. Kornberg, *DNA Replication*, W. H. Freeman, 1980); *Figure 4.26*: from Steitz, T. A. *et al*. (1981) 'Structural dynamics of yeast hexokinase during catalysis', in *Philosophical Transactions of The Royal Society*, Vol. B293, The Royal Society; *Figure 4.28*: from Koshland, D. E. and Neet, K. E. (1968) 'The catalytic and regulatory properties of enzymes', reproduced, with permission, from the *Annual Review of Biochemistry*, Vol. 37, © 1968 by Annual Reviews Inc; *Figure 4.33*: from Latner, A. L. and Skillen, A. W. *Isoenzymes in Biology and Medicine*, Fig. 2, Academic Press, Inc.; *Figure*

4.34: from Markert, C. L. 'Lactate dehydrogenase isoenzymes: dissociation and recombination of subunits', *Science*, Vol. 140, American Association for the Advancement of Science; *Figure 4.35*: from Moss, D. W. (1982) *Isoenzymes*, Chapman & Hall Ltd; *Figure 5.1*: from Alberts, B. *et al.* (1983) *Molecular Biology of the Cell*, Garland Publishing Inc., copyright © 1983 by Bruce Alberts *et al*; *Figure 5.11*: reprinted with permission from an article by Gerhart, J. C. and Schachman, H. K. (1965) in *Biochemistry*, Vol. 4, p. 1059, copyright © 1965 American Chemical Society; *Figure 5.13*: from Krause, K. L. *et al.* (1985) 'Structure at 2.9-A resolution of aspartate carbamoyltransferase . . .', in *Proceedings of the National Academy of Sciences of the USA*, Vol. 82, March 1985, National Academy of Sciences, reproduced courtesy of Professor William N. Lipscomb; *Figure 5.14*: from Kantrowitz, S. T. *et al.* (1980) '*E. coli* aspartate transcarbamylase . . .', in *Trends in the Biochemical Sciences*, Vol. 5, Elsevier/North Holland Biomedical Press; *Figure 6.1*: adapted from *Biochemistry* (3rd edn) by Lubert Stryer. Copyright © 1975, 1981, 1988 by Lubert Stryer. Reprinted with with permission from W. H. Freeman and Co. Ltd; *Figure 7.2*: illustration by Dana Burns from Bretscher, M. S. (1985) 'The molecules of the cell membrane,' *Scientific American*, August 1985, copyright © 1985 by Scientific American, Inc. All rights reserved; *Figures 7.15, 7.20, 7.22, 7.27 and 7.30*: from Alberts, B. *et al.* (1983) *Molecular Biology of the Cell*, Garland Publishing Inc., *Figure 7.17*: illustration by Bunji Tagawa from Hinckle, C. and McCarty, R. E. (1978) 'How cells make ATP,' *Scientific American*, March 1978, copyright © 1978 by Scientific American, Inc. All rights reserved; *Figure 7.34(a)*: reprinted by permission from *Nature*, Vol. 289, p. 421, copyright © 1981 Macmillan Magazines Ltd; *Figure 7.34(c)*: from Perry, M. M. and Gilbert, A. B. (1979) 'Yolk transport in the ovarian follicle of the hen', in *Journal of Cell Science*, Vol. 39, The Company of Biologists Ltd.

PLATES

Plates 1.1 and *2.23*: Dr M. G. Stewart (Open University); *Plates 2.1, 2.3, 2.4, 2.7, 2.8, 2.10, 2.12* and *2.14*; Dr M. G. Stewart, Dr P. Mullins and J. Broadbent (Open University); *Plate 2.2*: Professor D. Bainton (University of California); *Plate 2.5*: Dr M. G. Stewart and J. Broadbent (Open University); *Plate 2.6*: courtesy of Biophotos Associates; *Plate 2.9*: Professor C. P. Leblond (McGill University); *Plate 2.11*: Professor L. Orci (University of Geneva); *Plate 2.13*: E. A. Anderson (Harvard Medical School); *Plate 2.15*: from Novikoff, A. B. (1970) *Cells and Organisms*, Holt, Rinehart & Winston, courtesy of Professor D. Branston; *Plate 2.16*; from Toner, P. G. and Carr, K. E. (1971) *Cell Structure* (2nd edn), Churchill Livingstone, courtesy of Dr P. Toner; *Plate 2.18*: from Novikoff, A. B. and Holtzmann, E. (1976) *Cells and Organelles*, Holt, Rinehart & Winston; *Plate 2.19*: from De Robertis, E. D. P. *et al.* (1976) *Cell Biology* (6th edn), W. B. Saunders, courtesy of H. Fernandez-Moran; *Plate 2.20*: from De Robertis, E. D. P. and De Robertis, E. M. F. (1987) *Cell and Molecular Biology* (8th edn), Lea & Febiger, courtesy of Professor Klaus Weber; *Plate 2.21*: from Novikoff, A. B. (1976) *Cells and Organisms*, Holt, Rinehart & Winston, courtesy of Professor B. R. Brinkley and E. Stubblefield; *Plate 2.22*: from Afzelius, B. A. in *Journal of Ultrastructure Research*, Vol. 37, 1971.

INDEX

Note Entries in **bold** are key terms. Indexed information on pages indicated by *italics* is carried mainly or wholly in a figure, table or plate.